A SYSTEMATIC APPROACH
TO DIGITAL LOGIC DESIGN

A SYSTEMATIC APPROACH
TO DIGITAL LOGIC DESIGN

Frederic J. Mowle

Purdue University

Addison-Wesley Publishing Company

Reading, Massachusetts

Menlo Park, California · London · Amsterdam · Don Mills, Ontario · Sydney

This book is in the
Addison-Wesley Series in Electrical Engineering

Third printing, July 1977

ISBN 0-201-04920-1
.BCDEFGHIJK-MA-79

To my wife, Mary Jane,
and to my Sons, Thomas
Michael, and William,
without whose understanding
this book might
still be in preparation.

Preface

The impact of digital logic on our way of life during recent years has been awesome; new digital devices are introduced daily. The most obvious product using digital logic is the digital computer, which has become a common tool of both industry and business. Digital logic systems are used daily in industry for such purposes as the control of the various steps in generating iron, steel, copper, and aluminum products. Municipalities employ digital logic systems to control and coordinate the flow of traffic through cities. Medical engineers are utilizing digital logic systems to monitor hospital patients. The telephone company possesses the largest distributed-logic system in the world. These advances would not have been possible without the invention of the transistor and the subsequent development of microelectronic devices.

The concepts of digital logic are by no means new. Even in the 1600's mechanical calculating machines were available. However, these early calculating units were slow and cumbersome to use. Today's hand-held calculators are direct consequences of the advances in microminiaturization made possible with current electronic technology.

The purpose of this book is not to dwell upon the uses and applications of computer systems, but rather to take a look inside the computer and examine the components which make up its various subsystems. For this reason, the author provides the mathematical background required by an engineer to successfully carry out a digital design project. Toward this goal the book provides the reader with a solid foundation in Boolean algebra and its direct application to digital-systems design. A secondary goal of this book is to provide the reader with the basic knowledge required to implement useful logic circuits. The four basic arithmetic functions—addition, subtraction, multiplication, and division—are discussed. Although binary arithmetic is stressed in the text for convenience, numerous examples using nonbinary arithmetic (for example decimal arithmetic) can be found throughout.

We can classify digital systems into two main categories: combinational logic systems and sequential logic systems. In this text we will classify as a combinational logic system any system whose outputs only respond (not necessarily instantaneously) to changes in its inputs. All other systems will be defined as sequential logic systems. We will see shortly that it is possible to design the four basic functions of addition, subtraction, multiplication, and division as either combinational or sequential circuits. In general, we will find the combinational logic circuits which realize these four functions require a large number of components and require a relatively short delay time from input to output. On the other hand, we will find in many cases that corresponding sequential circuits require fewer components but take longer to arrive at the desired result. The proper selection of design alternatives will be one of our objectives.

Throughout this book the emphasis will be on the implementation of digital systems using fundamental building blocks.

The material in this book begins with a discussion of Number Systems in Chapter 1 and of Base-R Arithmetic Systems in Chapter 2. Boolean Algebra is introduced in Chapter 3 using Huntington's postulates. The principle of duality, introduced in this chapter, will be used extensively in our discussions.

Chapter 4 introduces NAND and NOR logic as well as logic conventions to be used throughout the book. Digital multiplexers are introduced and their use as logical devices is emphasized. We will find the digital multiplexer to be one of our more versatile logic circuits.

Chapter 5 presents the classical minimization procedures for Boolean functions. Both map and tabular methods are included.

Although minimization techniques are presented in Chapter 5, the majority of the examples in this book are not minimal. This has been done so that the casual reader can easily follow the examples in later chapters. At the same time the complexity of the examples is reduced. It is the author's viewpoint that heavy emphasis on minimization distracts the student from the basic task of learning digital logic.

At Purdue, only the K-map minimization procedures are discussed in the introductory course, with the Quine-McCluskey procedure available for the students to pursue on their own should they so wish.

Binary arithmetic units are introduced in Chapter 6 as a means of teaching students combinational logic design. The material on carry-lookahead adders is important and should be covered in the first course; however, the material on ripple-lookahead adders and first-order carry-lookahead adders should be covered only as time permits. Array multiplication units are presented in Chapter 6 to demonstrate how full adders can be interconnected to form useful digital systems.

Chapter 7 is devoted to decimal arithmetic. This chapter serves as an application area for the material presented thus far in the book.

Chapters 8 and 9 provide the background necessary for the design of sequential logic circuits. The set-reset flip-flop is introduced in Chapter 8 along with a systematic procedure for using this element. Monostable multivibrators and digital clocks are also included in this chapter. The master-slave principle and edge-triggered flip-flops are introduced in Chapter 9 as a means of providing a practical use for the flip-flop. Extensive design procedures using JK flip-flops are included in this chapter.

Chapters 10 and 11 are primarily application chapters. Chapter 10 deals with the design of binary counters, and Chapter 11 deals with multifunction registers.

Chapter 12 makes use of the material in the first 11 chapters and deals with the design of serial arithmetic units. Detailed algorithms are presented for the design of serial addition, subtraction, multiplication, and division units.

The material in this book has been classroom-tested at Purdue University over the past four years with an enrollment of approximately 450 students a year. The course is primarily populated by sophomore students in Electrical Engineering and Computer Science. However, approximately 10% consists of freshmen interested in computer engineering. The course also has a goodly number of upper classmen not interested in computer engineering but rather in broadening their background. This course material was successfully presented at Indianapolis as part of the Continuing Education Program to approximately 60 engineers during the Spring semester, 1973.

At the present time, all the material in this book is presented to the students with the exception of the Quine-McCluskey procedure, the advanced carry-lookahead-adder material in Chapter 6, and the material contained in Chapter 12. A one-hour laboratory course using 7400-series logic elements as building blocks is taught in parallel with this course. Approximately 70% of the students taking the course elect to take the laboratory. An outline of the laboratory course can be obtained from the author.

We wish to express our thanks to the numerous students in EE 266 at Purdue who classroom-tested this material. I especially would like to gratefully acknowledge the various instructors, Owen Mitchell, Violet Haas, Kai Hwang, George Hughes, Phillip Swain, Anthony Akers, Robert Clark, David Meyer, Alan Desrochers, and S. Diane Smith for their helpful suggestions and critiques on this text. Special thanks is also given to Jean Weisbach and Virginia Jacko, who uncomplainingly typed from the handwritten manuscript.

West Lafayette, Indiana   F.J.M
April 1976

Contents

Chapter 1   Number Systems **1**

    1.1 Introduction 1
    1.2 Base conversion using decimal arithmetic 3
    1.3 Conversion between base A and base B using base-A arithmetic . 14
    1.4 Negative numbers 19
    1.5 Special problems with fixed n-digit systems 26
    1.6 Problems 27
    1.7 Bibliography 29

Chapter 2   Base-R Arithmetic **30**

    2.1 Introduction 30
    2.2 Fixed-point arithmetic in base-R 30
    2.3 Hardware complexity of base-R arithmetic 48
    2.4 Problems 57
    2.5 Bibliography 59

Chapter 3   Boolean Algebra **60**

    3.1 Introduction 60
    3.2 An introduction to Boolean algebra using Huntington's postulates . 60
    3.3 The Boolean operators—AND, OR, and NOT 70
    3.4 Enumeration of logic functions 73
    3.5 Characteristic numbers 77
    3.6 Truth table specification of Boolean functions 81
    3.7 Specification of Boolean functions using maps 82
    3.8 The principle of duality 90
    3.9 Function complementation 95
    3.10 Standard forms for Boolean functions 98
    3.11 Fundamental forms from truth tables or K-maps 101
    3.12 Conversion between Boolean forms 104

3.13 Problems   105
3.14 Bibliography   109

Chapter 4   Special Boolean Functions and Basic Logic Conventions **110**

4.1   Introduction   110
4.2   The NAND and NOR functions.   111
4.3   Logic design using NAND and NOR gates   113
4.4   Exclusive OR and exclusive NOR   119
4.5   Logical completeness.   124
4.6   Positive and negative logic conventions   125
4.7   The wired gates   126
4.8   Logic symbol conventions   129
4.9   Logic equivalent circuits   132
4.10 Digital multiplexers as logic devices   136
4.11 Binary demultiplexers or decoders   147
4.12 Problems   148
4.13 Bibliography   154

Chapter 5   Minimization Procedures for Boolean Functions **155**

5.1   Introduction   155
5.2   Minimization—when and why   155
5.3   Background material.   160
5.4   The Karnaugh-Map method for minimizing multivariable single-output
      Boolean functions   163
5.5   Quine-McCluskey method for minimizing multiple-input single-output
      functions   172
5.6   Incompletely specified functions   186
5.7   Problems   189
5.8   Bibliography   192

Chapter 6   Binary Arithmetic Units **193**

6.1   Introduction   193
6.2   Combinational logic circuits   193
6.3   Logical design of binary arithmetic circuits   197
6.4   Binary ripple arithmetic units   207
6.5 n-bit carry-lookahead adders, $\text{CLA}(n)$'s   208
6.6   Ripple-carry-lookahead adders, $\text{RLA}(n:a, b)$ 222
6.7   First-order carry-lookahead adders, $\text{CLA}(n:a, b)$ 226
6.8   Parallel multiplication techniques   233
6.9   Problems   235
6.10 Bibliography   240

Chapter 7   Decimal Arithmetic **241**

7.1   Introduction   241
7.2   Binary coding of decimal data   241

7.3 Binary coding of decimal data with error control 247
7.4 Code conversion and display circuits 254
7.5 Decimal adding circuits 261
7.6 BCD ripple adders 267
7.7 Problems 270
7.8 Bibliography 273

Chapter 8   Introduction to Sequential Circuit Design **274**

8.1 Introduction 274
8.2 Basic circuit characteristics 274
8.3 Hazards in combinational logic circuits 281
8.4 Introduction to sequential circuits 287
8.5 Binary-storage elements 296
8.6 NOR implemented set–reset (SR) flip-flops 298
8.7 NAND implemented set–reset ($\overline{\text{SR}}$) latch 303
8.8 Clocked-storage elements 305
8.9 Analysis of SR flip-flop circuits 308
8.10 Synthesis of sequential circuits using SR flip-flops 311
8.11 Excitation table for SR flip-flops 313
8.12 Specialized digital-logic circuits—monostable multivibrators . . 319
8.13 Digital clocks 324
8.14 Switch-interrupt circuits 328
8.15 Problems 330
8.16 Bibliography 335

Chapter 9   Practical Flip-Flop Circuits **336**

9.1 Introduction 336
9.2 The SR delay flip-flop 336
9.3 The toggle or T flip-flop 341
9.4 The JK flip-flop 342
9.5 Improved flip-flop design techniques 346
9.6 The master-slave principle 346
9.7 Edge-triggered SR flip-flops 356
9.8 Analysis and synthesis of sequential circuits using JK flip-flops . . 367
9.9 Excitation table for a JK flip-flop 370
9.10 Preset and preclear inputs 377
9.11 Problems 380
9.12 Bibliography 384

Chapter 10   Binary Counters **386**

10.1 Introduction 386
10.2 A basic binary counting cell 386
10.3 Synchronous binary counters 389
10.4 Arbitrary modulo-N counter 399
10.5 Direct resetting counter circuits 408

10.6	Binary down counters	414
10.7	Special binary counting units	419
10.8	Design of a digital stop watch	427
10.9	Problems	434
10.10	Bibliography.	438

Chapter 11 Register Design Techniques **439**

11.1	Introduction.	439
11.2	Data-storage registers	439
11.3	Special register units	451
11.4	Timing pattern generators	462
11.5	Problems	473
11.6	Bibliography.	478

Chapter 12 Advanced Arithmetic Units **479**

12.1	Introduction.	479
12.2	A binary serial adder	479
12.3	Serial-parallel multiplication techniques	487
12.4	Serial-parallel division of binary integers	506
12.5	Problems	527
12.6	Bibliography.	532

Index to Algorithm's and Rules	**533**
Index	**535**

Chapter 1

Number Systems

1.1 INTRODUCTION

There are many different ways of representing numbers, with positional notation the method used most frequently. In this chapter we will restrict our attention to fixed-point number systems using arbitrary number bases. In Section 1.2, we will develop several algorithms for converting numbers between an arbitrary number base R and base 10. In Section 1.3, these algorithms will be generalized to allow direct conversion between any two arbitrary number bases. The three most commonly used notations for representing negative numbers will be introduced in Section 1.4, namely; sign-and-magnitude, radix, and diminished-radix.

Finally, in Section 1.5, a few concluding remarks will be made to help the reader to better understand the problems which can arise when converting numbers between arbitrary number bases having different negative number notations.

Before establishing a set of rules for converting numbers, a few introductory remarks are in order. A fixed-point decimal number is a number written with the decimal point included in its proper place. For example, the number $+14\frac{3}{4}$ would be written as $(+14.75)_{10}$. As a floating-point number, using scientific notation, we could write this same number as $(+0.1475 \times 10^2)_{10}$ or as $(+1.475 \times 10^1)_{10}$ or even as $(+1475. \times 10^{-2})_{10}$; with the correct location of the decimal point determined by the location of the floating decimal point and the exponent of the multiplier. The actual multiplier would be suppressed, its value understood, and only the exponent written. Thus in a computer the same number might be represented as $(0.1475, 2)_{10}$.

Since we are all familiar with sign-and-magnitude number notation, let us use this notation to illustrate some of the peculiarities inherent in machine number

1

representations. In a fixed-point, base-R number system, a sign-and-magnitude number is written as

$$(N)_R = S(d_{n-1}d_{n-2} \cdots d_1 d_0.d_{-1}d_{-2} \cdots d_{-k})_R,$$

$$0 \le d_i \le (R - 1)_R.$$

The corresponding decimal number is

$$(N)_{10} = S(c_{n-1} \cdot R^{n-1} + c_{n-2} \cdot R^{n-2} + \cdots + c_1 \cdot R + c_0$$
$$+ c_{-1} \cdot R^{-1} + \cdots + c_{-k} \cdot R^{-k})_{10}.$$

$$(N)_{10} = S \sum_{i=-k}^{n-1} c_i \cdot R^i; \qquad 0 \le c_i \le (R - 1)_{10}.$$

Each d_i is a digit or symbol in the base-R number system, and each c_i is its equivalent base-10 number obtained by using a table of correspondence similar to that given in Table 1.1. R denotes the radix, or base, of the number system and indicates how many symbols are used in its number alphabet. For example, from Table 1.1 we see that the binary number system used only two symbols 0 and 1, while base 16 or the hexadecimal number system uses the number alphabet (0, 1, 2, 3, 4, 5, 6, 7, 8, 9, A, B, C, D, E, F). The number of *integer* positions in the number is indicated by n, and k indicates the number of *fractional* positions in the number. S will denote the sign of the number, $+$ or $-$.

The convention of enclosing the numbers in parenthesis and adding the subscript R will be used to indicate the base of the number. Thus $(+235)_7$ would indicate a positive base-7 number and $(-10110.)_2$ would indicate a negative base-2 or binary number.

Although we are in the habit of neglecting leading zeros in a number by placing the sign position to the left of the most significant position, this will not happen in a computer, in a mechanical adding machine, or in an electronic calculator. These devices have a fixed number of positions available for representing a number, including its sign indicator. A calculator, which has four digit positions and a sign indicator available, would represent the decimal number $+4$ as $+0004$.

The addition of $+0005$ to $+9997$ will yield $+0002$ instead of $+10002$, since a four digit calculator can only represent numbers in the range of $-9999 \le N \le +9999$. The high-order digit which results from the addition of two numbers whose sum lies outside the range of the representable numbers will be lost. This means that methods will have to be devised which will tell us when an incorrect answer occurs due to an arithmetic overflow condition. An arithmetic overflow occurs whenever the answer requires more digit positions than the computer or calculator has available. An N-digit calculator is an excellent example of a device which uses modular arithmetic. A modular number system divides the set of numbers into M classes, or subsets, using the rule of correspondence which states that a number X belongs to class A if X, when divided by M, has a remainder of

TABLE 1.1

Table of Correspondence Between Various Base Number Systems

N_{10}	N_2	N_3	N_4	N_5	N_6	N_8	N_{12}	N_{16}
0	0	0	0	0	0	0	0	0
1	1	1	1	1	1	1	1	1
2	10	2	2	2	2	2	2	2
3	11	10	3	3	3	3	3	3
4	100	11	10	4	4	4	4	4
5	101	12	11	10	5	5	5	5
6	110	20	12	11	10	6	6	6
7	111	21	13	12	11	7	7	7
8	1000	22	20	13	12	10	8	8
9	1001	100	21	14	13	11	9	9
10	1010	101	22	20	14	12	U	A
11	1011	102	23	21	15	13	V	B
12	1100	110	30	22	20	14	10	C
13	1101	111	31	23	21	15	11	D
14	1110	112	32	24	22	16	12	E
15	1111	120	33	30	23	17	13	F
16	10000	121	100	31	24	20	14	10

A for $0 \le A \le M - 1$. M is referred to as the modulus, and the basic operations are carried out modulo M.

1.2 BASE CONVERSION USING DECIMAL ARITHMETIC

All of the algorithms stated in this section will be valid only for positive numbers. The procedure for handling negative numbers will be first, to convert the negative numbers to positive numbers; then, to carry out the conversion to the new number base using positive numbers. Finally the resulting positive number should be reconverted to a negative number in the new base. Since most of us are familiar with the decimal number system and its arithmetic, and not too familiar with arithmetic in other bases, the algorithms presented in this section will use the decimal number system for arithmetic calculations.

There are four cases which must be considered:

Case 1—conversion of integers from base R to base 10.
Case 2—conversion of integers from base 10 to base R.
Case 3—conversion of fractions from base R to base 10.
Case 4—conversion of fractions from base 10 to base R.

If the reader has access to a digital computer, he is encouraged to implement each of the algorithms presented in this chapter on the computer using FORTRAN or another appropriate language.

CASE 1. Let us consider the problem of converting a base-R integer number to a decimal integer. In order to use decimal arithmetic in the conversion process, it will be necessary to use a Table of Correspondence. This Table of Correspondence gives us the equivalent base-10 value of each of the base-R symbols. Consider the four-digit base-R number $(d_3d_2d_1d_0)_R$ written using base-R symbols. We can rewrite this number using equivalent base-10 symbols as

$$(N)_{10} = c_3 \cdot R^3 + c_2 \cdot R^2 + c_1 \cdot R^1 + c_0 \cdot R^0 \qquad (1.1)$$

with c_i being the base-10 symbol equivalent of the base-R symbol d_i.

This particular form of $(N)_{10}$, Eq. (1.1), will require a large number of calculations. We can, however, rewrite this expression in nested form yielding

$$(N)_{10} = (((c_3 \cdot R + c_2) \cdot R + c_1) \cdot R + c_0). \qquad (1.2)$$

This latter form leads to Algorithm 1.1.

ALGORITHM 1.1

Conversion of a k-digit base-R integer to a base-10 integer.

$$(N)_R = (d_{k-1}d_{k-2} \cdots d_1d_0)_R.$$

STEP   OPERATION
1.   Using a table of correspondence, convert each of the base-R digits, d_i's to their base-10 equivalents, c_i's.

$$(N)_R = (c_{k-1}c_{k-2} \cdots c_1c_0)_R.$$

2.   Set $i = 0$ and $X_0 = 0$.

3.   Increment i by 1.

4.   Calculate $X_i = X_{i-1} \cdot R + c_{k-i}$

5.   If $i < k$ go back to step 3, otherwise proceed to step 6.

6.   STOP. The desired base-10 integer is $(N)_{10} = X_k$.

Example 1.1 Convert $(3AF2)_{16}$ to a decimal number.

For this problem, we have $k = 4$ and $R = 16$. Using the Table of Correspondence on page 3, we have $c_3 = 3$, $c_2 = 10$, $c_1 = 15$, and $c_0 = 2$.

Applying Algorithm 1.1 yields:

$$X_1 = 0 \cdot 16 + 3 = 3$$
$$X_2 = 3 \cdot 16 + 10 = 58$$
$$X_3 = 58 \cdot 16 + 15 = 943$$
$$X_4 = 943 \cdot 16 + 2 = 15090$$

Thus $(3AF2)_{16} = (15090)_{10}$.

Example 1.2 Convert $(1011011)_2$ to a decimal number.

For this problem, we have $k = 7$ and $R = 2$. Using the Table of Correspondence on page 3, we have $c_6 = 1$, $c_5 = 0$, $c_4 = 1$, $c_3 = 1$, $c_2 = 0$, $c_1 = 1$, and $c_0 = 1$. Applying Algorithm 1.1 yields:

$$X_1 = 0 \cdot 2 + 1 = 1$$
$$X_2 = 1 \cdot 2 + 0 = 2$$
$$X_3 = 2 \cdot 2 + 1 = 5$$
$$X_4 = 5 \cdot 2 + 1 = 11$$
$$X_5 = 11 \cdot 2 + 0 = 22$$
$$X_6 = 22 \cdot 2 + 1 = 45$$
$$X_7 = 45 \cdot 2 + 1 = 91$$

Thus $(1011011)_2 = (91)_{10}$.

CASE 2. Let us now consider the problem of converting an integer from base 10 to base R. The following brief discussion will serve as a means of motivation for the conversion procedure we will use. Let us examine the following equivalent statements.

$$(N)_R = (d_{k-1}d_{k-2} \cdots d_1 d_0)_R, \tag{1.3}$$

$$(N)_{10} = (c_{k-1}R^{k-1} + c_{k-2}R^{k-2} + \cdots + c_1 R + c_0)_{10}, \tag{1.4}$$

$$(N)_{10} = \sum_{J=0}^{k-1} c_J R^J. \tag{1.5}$$

In the latter expressions the c_i's are the base-10 equivalent values for the base-R symbols, d_i's.

We would like to be able to calculate the values of the c_i's, and hence the d_i's, by performing simple arithmetic operations on the decimal number $(N)_{10}$. If we divide both sides of Eq. (1.5) by R, we obtain an integer $(N_1)_{10}$ and a remainder $(P_1)_{10}$, with $(N_1)_{10} = \sum_{j=1}^{k-1} c_j R^{j-1}$ and $(P_1)_{10} = c_0$. Thus the remainder we have obtained in this manner is the lowest order coefficient of the desired base-R number. If we repeat the division process again, this time using $(N_1)_{10}$; we will obtain another integer $(N_2)_{10}$ and a second remainder $(P_2)_{10}$, with $(N_2)_{10} = \sum_{j=2}^{k-1} c_j R^{j-2}$ and $(P_2)_{10} = c_1$.

We have now obtained the next coefficient of the desired base-R number. In general, the ith coefficient will be found by dividing $(N_i)_{10}$ by R. The resulting remainder will be the desired coefficient c_i expressed as a base-10 number. To complete the conversion, we must convert the c_i's (equivalent base-10 coefficients) to their corresponding base-R symbols, d_i's. using a Table of Correspondence. The preceding discussion leads to the following algorithm.

ALGORITHM 1.2

Conversion of a base 10 integer to a base R integer.

STEP   OPERATION
1.   Set $i = 0$ and $x_0 = (N)_{10}$; the base-10 number to be converted.
2.   Divide x_i by R; the radix of the new number base.
3.   Set $x_{i+1} = \lfloor x_i/R \rfloor$; the integer part of the division in step 2.
4.   Set c_i equal to the remainder resulting from the division in step 2.
5.   If $x_{i+1} \neq 0$, then increment i by 1 and go back to step 2; otherwise proceed to step 6.
6.   Using a Table of Correspondence, convert the c_i's to their proper base-R symbols, d_i's, and STOP.

The base-R integer obtained is:

$$(N)_R = (d_i d_{i-1} \cdots d_1 d_0)_R.$$

Example 1.3 Convert $(727)_{10}$ to an octal number (base 8).
Applying Algorithm 1.2 yields:

i	R	x_i	c_{i-1}	d_{i-1}
0	8	727	base 10 remainders	base R remainders
1		90	$7 = c_0$	$d_0 = 7$
2		11	$2 = c_1$	$d_1 = 2$
3		1	$3 = c_2$	$d_2 = 3$
4		0	$1 = c_3$	$d_3 = 1$

Thus $(727)_{10} = (1327)_8$.

The reader should be cautioned that the coefficients produced by this algorithm are obtained with the least significant digit first, and the most significant digit last. A common error is to write the final answer with the coefficients reversed.

Example 1.4 Convert $(2803)_{10}$ to a hexadecimal number (base 16).
Applying Algorithm 1.2 yields:

i	R	x_i	c_{i-1}	d_{i-1}
0	16	2803	base 10 remainders	base 16 remainders
1		175	$3 = c_0$	$d_0 = 3$
2		10	$15 = c_1$	$d_1 = F$
3		0	$10 = c_2$	$d_2 = A$

Thus, $(2803)_{10} = (AF3)_{16}$.

1.2.1 Fractional Conversion Accuracy

Unlike integers, which convert exactly from one base to another, fractions do not convert exactly. For example, the decimal fraction 1/3 yields, as a decimal number, an infinite string of 3's. The decimal fraction 1/7 yields the repeating string of numbers $.\overline{142857}$. The bar underlines the group of repeating digits. Our problem is compounded by the fact that a fraction, which can be exactly represented in one number base, may require an infinite string of digits in another base. For example, the decimal fraction $(1/10)_{10} = (.1)_{10}$ cannot be expressed exactly by a finite string of binary digits. In fact $(.1)_{10} = (0.00\overline{011})_2$ with the string of digits 0011 infinitely repeating. We can easily verify this as follows:

$$\text{If } (x)_{10} = (.0001100110011\ldots)_2 \tag{1.6}$$
$$\text{Then } (2^4 x)_{10} = (0001.1001100110011\ldots)_2. \tag{1.7}$$

Subtracting Eq. (1.6) from Eq. (1.7) yields

$$(2^4 x - x)_{10} = (1.1)_2 = (1.5)_{10}$$
$$(15x)_{10} = (1.5)_{10}$$
$$x_{10} = (.1)_{10}.$$

In general, we will be required to convert decimal numbers which are accurate to ± 1 in their least significant position. Let us determine a relationship which will yield an acceptable limit on the number of places required in the new number base, while retaining this accuracy. In effect, we must solve the equation

$(.1)_B^j = (.1)_A^k$ for j, in terms of k, the limit on the accuracy of the number in base A, and the radixes of the present and new number bases A and B. Taking the log Base A, of both sides of the equation yields

$$\log_A (.1)_B^j = \log_A (.1)_A^k. \tag{1.8}$$

Since $\log (x)^m = m \log (x)$, we have

$$j \log_A (.1)_B = k \log_A (.1)_A. \tag{1.9}$$

Since $\log (x) = -\log (1/x)$, we have

$$-j \log_A (10)_B = -k \log_A (10)_A. \tag{1.10}$$

Since in any number base, $R = (10_R)$ we have

$$j \log_A (B) = k \log_A (A). \tag{1.11}$$

We also know that $\log_A (A) = 1$

$$\therefore \quad j = \frac{k}{\log_A (B)} \quad \text{or} \quad k = j \log_A (B).$$

Finally, using the identity

$$\log_A (B) = \frac{\log_{10} (B)}{\log_{10} (A)},$$

we have

$$j = k \frac{\log_{10} (A)}{\log_{10} (B)}. \tag{1.12}$$

However, since j in general will not be an integer, we will select j as the integer which satisfies the inequality

$$k \frac{\log_{10} (A)}{\log_{10} (B)} \le j < k \frac{\log_{10} (A)}{\log_{10}(B)} + 1 \tag{1.13}$$

or we will select k as the integer which satisfies the inequality

$$j \frac{\log_{10} (B)}{\log_{10} (A)} \le k < j \frac{\log_{10} (B)}{\log_{10} (A)} + 1. \tag{1.14}$$

In this section we are working with an arbitrary base, R, and base 10, therefore we can rewrite these inequalities as

$$\frac{k}{\log_{10} (R)} \le j < \frac{k}{\log_{10} (R)} + 1 \tag{1.15}$$

and

$$j \log_{10} (R) \le k < j \log_{10} (R) + 1. \tag{1.16}$$

Table 1.2 has been prepared as an aid to be used when working the examples and problems in this chapter.

TABLE 1.2
Table of Common Logarithms

R	$\log_{10}(R)$	$\dfrac{1}{\log_{10}(R)}$	R	$\log_{10}(R)$	$\dfrac{1}{\log_{10}(R)}$
1	0.000	∞	11	1.041	0.960
2	0.301	3.322	12	1.079	0.927
3	0.477	2.096	13	1.114	0.898
4	0.602	1.661	14	1.146	0.873
5	0.699	1.431	15	1.176	0.850
6	0.778	1.285	16	1.204	0.830
7	0.845	1.183	17	1.230	0.813
8	0.903	1.107	18	1.255	0.797
9	0.954	1.048	19	1.279	0.782
10	1.000	1.000	20	1.301	0.769

CASE 3. The procedure for converting the fractional part of a base-R number to a decimal number is almost the same as the procedure used for integer conversion. One major difference will be the accuracy requirements placed on the conversion. Thus if we have a j-digit base-R fractional number, which we must convert to a decimal number holding an accuracy of $\pm (.1)_R^j$, then we must select, k, the number of places to be retained in the decimal number, to be the integer which satisfies the inequality

$$j \log_{10}(R) \le k < j \log_{10}(R) + 1. \qquad (1.17)$$

We can establish a conversion procedure as follows. Let $(N)_R = (.d_{-1}d_{-2}\cdots d_{-j})_R$ where the d_i's are base-R symbols. Using a Table of Correspondence, we can convert the d_i's to their equivalent base-10 values, c_i's. After rewriting the number in series form, we obtain,

$$(N)_{10} = \sum_{i=-j}^{-1} c_i R^i = \sum_{i=1}^{j} \frac{c_{-i}}{R^i}. \qquad (1.18)$$

Factoring out the denominator yields

$$(N)_{10} = \frac{1}{R^j} \sum_{i=1}^{j} c_{-i} R^{j-i}. \qquad (1.19)$$

This latter form leads directly to Algorithm 1.3.

ALGORITHM 1.3

Conversion of a j-digit base-R fraction to a base-10 fraction holding an accuracy of $\pm(.1)_R^j$.

$$(N)_R = (.d_{-1}d_{-2} \cdots d_{-j+1}d_{-j})_R.$$

STEP   OPERATION

1.   Convert the base-R coefficients, d_i's, to their corresponding base-10 values, c_i's, using a Table of Correspondence.

2.   Calculate k, the number of decimal places to be retained after the conversion. Pick k as the integer satisfying the inequality,

$$j \log_{10}(R) \leq k < j \log_{10}(R) + 1.$$

3.   Set $i = 0$ and $x_0 = 0$.

4.   Increment i by 1.

5.   Calculate $x_i = x_{i-1} \cdot R + c_{-i}$.

6.   If $i < j$, go back to step 4; otherwise proceed to step 7.

7.   Divide x_j by R^j retaining only k places and STOP. $(N)_{10} \simeq x_j/R^j$

Example 1.5 Convert $(.10101)_2$ to base 10, holding an accuracy of $\pm(.1)_2^5$.
Applying Algorithm 1.3 we obtain

$$c_{-1} = 1, \quad c_{-2} = 0, \quad c_{-3} = 1, \quad c_{-4} = 0, \quad \text{and} \quad c_{-5} = 1.$$

Since $j = 5$, we must select k as the integer such that

$$5 \cdot \log_{10}(2) \leq k < 5 \cdot \log_{10}(2) + 1,$$

$$1.505 < k < 2.505.$$

Thus $k = 2$.

$$x_1 = 0 \cdot 2 + 1 = 1$$
$$x_2 = 1 \cdot 2 + 0 = 2$$
$$x_3 = 2 \cdot 2 + 1 = 5$$
$$x_4 = 5 \cdot 2 + 0 = 10$$
$$x_5 = 10 \cdot 2 + 1 = 21$$
$$(N)_{10} = \frac{21}{2^5} = \frac{21}{32} = .65625.$$

Since $k = 2$, we retain only 2 places.

$$(.10101)_2 \simeq (.65)_{10}.$$

Example 1.6 Convert $(.ABC)_{16}$ to base 10 holding an accuracy of $\pm(.1)_{16}^3$.
Applying Algorithm 1.3 we obtain

$$c_{-1} = 10, \quad c_{-2} = 11, \quad \text{and} \quad c_{-3} = 12.$$

Since $j = 3$, we must select k as the integer such that

$$3 \log_{10}(16) \leq k < 3 \log_{10}(16) + 1,$$

$$3.612 < k < 4.612.$$

Thus $k = 4$.

$$x_1 = 0 \cdot 16 + 10 = 10$$

$$x_2 = 10 \cdot 16 + 11 = 171$$

$$x_3 = 171 \cdot 16 + 12 = 2748$$

$$(N)_{10} = \frac{2748}{16^3} = \frac{2748}{4096} = .6708 \text{ to 4 places.}$$

$$\therefore \quad (.ABC)_{16} \simeq (.6708)_{10}.$$

CASE 4. In this section we will derive a procedure for converting decimal fractions to base-R fractions. Again we must be careful to retain as much accuracy as possible during the conversion process. Therefore in order to preserve an accuracy of ± 1 in the least significant position of a k-digit decimal number, we must select j, the number of base-R places to be retained, to be the integer which satisfies the inequality

$$\frac{k}{\log_{10}(R)} \leq j < \frac{k}{\log_{10}(R)} + 1. \tag{1.20}$$

We can determine the coefficients of the base-R number in the following manner. Let $(N)_{10} = (.a_{-1}a_{-2}\cdots a_{-k})_{10}$ and the desired base R number expressed to j places be $(N)_R = (.d_{-1}d_{-2}d_{-3}\cdots d_{-j})_R$. Expressing this latter equation as an infinite series with the base-R symbols replaced by their decimal equivalents we have

$$(N)_{10} = \sum_{i=1}^{\infty} \frac{c_{-i}}{R^i}. \tag{1.21}$$

multiplying both sides of this equation by R yields an integer part, c_{-1}, and a fractional part, $(N_1)_{10}$, with

$$(N_1)_{10} = \sum_{i=2}^{\infty} \frac{c_{-i}}{R^{i-1}}.$$

Thus we have been able to determine the decimal equivalent of the first coefficient of our base-R fraction. Continuing in this manner we can recursively obtain the desired k coefficients, with

$$(N_0)_{10} = (N)_{10} \tag{1.22}$$

and

$$(N_{i+1})_{10} = \text{fractional part of } R \cdot (N_i)_{10} \tag{1.23}$$

The desired coefficient, $c_{-(i+1)}$, is set equal to the integer part of $R \cdot (N_i)_{10}$. with

$$c_{-(i+1)} = \lfloor R \cdot (N_i)_{10} \rfloor. \qquad (1.24)$$

Once we have determined the c_i's, it is a simple matter to convert these symbols to their corresponding base-R values using a Table of Correspondence.

ALGORITHM 1.4

Conversion of a k-digit decimal fraction to a base-R fraction, holding an accuracy of $\pm(.1)_{10}^k$.

$$(N)_{10} = (.a_{-1}a_{-2}\cdots a_{-k})_{10}.$$

STEP   OPERATION
1.   Calculate j, the number of digits to be retained in the base-R number after conversion. Pick j as the integer satisfying the inequality

$$\frac{k}{\log_{10}(R)} \le j < \frac{k}{\log_{10}(R)} + 1.$$

2.   Set $i = 0$ and $x_0 = (N)_{10}$.

3.   Increment i by one.

4.   Calculate $Y = R \cdot x_{i-1}$.

5.   Set c_{-i} equal to the integer part of Y.

6.   Set x_i equal to the fractional part of Y.

7.   If $i < j$ go back to step 3; otherwise proceed to step 8.

8.   Using a Table of Correspondence convert each c_i to its proper base-R symbol d_i and STOP. $(N)_R = (.d_{-1}d_{-2}\cdots d_{-j})_R$.

Example 1.7 Convert $(.742)_{10}$ to an octal number (base 8) holding an accuracy of $\pm(.1)_{10}^3$.

Applying Algorithm 1.4 yields j to be the integer satisfying the inequality

$$\frac{3}{\log_{10}(8)} \le j < \frac{3}{\log_{10}(8)} + 1$$

$$3.321 \le j < 4.321, \quad \text{thus } j = 4.$$

Octal symbol d_{-i}	Decimal symbol c_{-i}	x_i	R	i
-	-	.742	8	0
5	5	.936		1
7	7	.488		2
3	3	.904		3
7	7	.232		4

$$\therefore \quad (.742)_{10} = (.5737)_8 \text{ to 4 places.}$$

Example 1.8 Convert $(.4321)_{10}$ to hexadecimal holding an accuracy of $\pm(.1)_{10}^4$.

Applying Algorithm 1.4 yields j to be the integer satisfying the inequality

$$\frac{4}{\log_{10}(16)} \le j < \frac{4}{\log_{10}(16)} + 1$$

$$3.322 \le j < 4.322; \quad \text{thus } j = 4.$$

Hexadecimal symbol d_{-i}	decimal symbol c_{-i}	x_i	R	i
-	-	.4321	16	0
6	6	.9136		1
E	14	.6176		2
9	9	.8816		3
E	14	.1056		4

$$\therefore \quad (.4321)_{10} = (.6E9E)_{16} \text{ to 4 places.}$$

1.2.2 Short Cut Techniques

Since base 8 and base 16 are actually powers of the base 2, we can reduce the amount of work required when converting between number bases which are powers of 2. This can be accomplished by examining each digit or group of digits as a unit. The conversion will also be exact in every case. Because each octal digit corresponds to a unique group of three binary digits, it is an easy matter to convert octal numbers to binary numbers and binary numbers to octal numbers. In the former case, each octal digit is replaced with its unique three-digit binary expansion. In the latter case, groups of three binary digits are marked off in both directions starting at the binary point. Leading or trailing zeroes are added as needed to make complete three-digit groups.

Example 1.9 Convert $(1254.173)_{10}$ to a binary number.

Let us first convert this number to base 8.

For the integer part use Algorithm 1.2	For the fractional part use Algorithm 1.4 $k = 3$ and $j = 4$

8	1254						
	156	$6 = d_0$				.173	8
	19	$4 = d_1$		d_{-1}	1	.384	
	2	$3 = d_2$		d_{-2}	3	.072	
	0	$2 = d_3$		d_{-3}	0	.576	
				d_{-4}	4	.608	

$$\therefore \quad (1254.173)_{10} = (2346.1304)_8$$

To finish the conversion, it is only necessary to convert the octal number into binary by inspection.

$$(\quad 2 \quad\quad 3 \quad\quad 4 \quad\quad 6 \,. \quad 1 \quad\quad 3 \quad\quad 0 \quad\quad 4)_8$$

$$(010 \quad 011 \quad 100 \quad 110 \,. 001 \quad 011 \quad 000 \quad 100)_2$$

This two step conversion process has actually resulted in our retaining too many places in the binary fraction, as only 10 binary digits are required. Therefore we can either truncate or round the previous result to 10 places and we can also eliminate the leading zero. Thus we obtain the result

$$(1254.173)_{10} = (10 \quad 011 \quad 100 \quad 110 \,. 001 \quad 011 \quad 000 \quad 1)_2$$

accurate to $\pm(.1)_{10}^3$.

The conversion to or from hexadecimal numbers is the same, except that groups of four digits are used instead of groups of three. *When converting numbers between hexadecimal and octal, it is recommended that the number first be converted into binary.*

Example 1.10   Convert $(A73.4F2)_{16}$ to octal.

Solution: Replace each hexadecimal digit with its unique 4-bit binary representation.

$$(\quad A \quad\quad 7 \quad\quad 3 \,. \quad 4 \quad\quad F \quad\quad 2)_{16}$$

$$(1010 \quad 0111 \quad 0011 \,. 0100 \quad 1111 \quad 0010)_2$$

Now mark off the resulting binary number in groups of three in both directions starting at the binary point. Then replace each group of three binary digits with its corresponding octal symbol.

$$(101 \quad 001 \quad 110 \quad 011 \,. 010 \quad 011 \quad 110 \quad 010)_2$$

$$(\quad 5 \quad\quad 1 \quad\quad 6 \quad\quad 3 \,. \quad 2 \quad\quad 3 \quad\quad 6 \quad\quad 2)_8$$

Thus $(A73.4F2)_{16} = (5163.2362)_8$.

These short cuts can also be advantageously used when converting between the decimal and binary number systems. This is done by using either base 8 or base 16 as an intermediate base. We can convert any number, by inspection, between base 8 or base 16 and base 2, using the short cut just discussed. Since there are fewer arithmetic calculations required when converting numbers between base 8 or base 16, and base 10, we are less likely to make errors.

1.3 CONVERSION BETWEEN BASE *A* AND BASE *B* USING BASE-A ARITHMETIC

In Section 1.2, we considered the conversion of numbers from any base to decimal, and from decimal to any base, with the arithmetic calculations being carried out in decimal arithmetic. Since the majority of modern computers use binary, octal, or hexadecimal arithmetic for calculations, let us generalize the algorithms in Section 1.2 so that we can better appreciate how the number conversions are done by a computer. In each of these algorithms, the arithmetic will be carried out using base-A arithmetic. The development of each of these algorithms will not be given, as they follow directly from the material in Section 1.2.

ALGORITHM 1.5

Conversion of an k-digit base-B integer to base-A numbers.

$$(N)_B = (d_{k-1}d_{k-2} \cdots d_1d_0)_B.$$

All arithmetic is to be done using base-A numbers.

STEP	OPERATION
1.	Using a Table of Correspondence, convert each of the base-B digits, d_i's, to their corresponding base-A values, c_i's.

$$(N)_B = (c_{k-1}c_{k-2} \cdots c_1c_0)_B.$$

2.	Set $i = 0$ and $x_0 = 0$.
3.	Increment i by one.
4.	Calculate $x_i = x_{i-1} \cdot B + c_{k-i}$.
5.	If $i < k$ go back to step 3; otherwise proceed to step 6.
6.	STOP. The desired base-A integer is:

$$(N)_A = x_k.$$

ALGORITHM 1.6

Conversion of a base-A integer to a base-B integer. All arithmetic is to be done using base-A numbers.

STEP	OPERATION
1.	Set $i = 0$ and $x_0 = (N)_A$.
2.	Calculate $Y = x_0/B$.
3.	Set x_{i+1} equal to the integer part of Y.
4.	Set c_i equal to the remainder resulting from the division in step 2.
5.	If $x_{i+1} \neq 0$, then increment i by one and go back to step 2; otherwise proceed to step 6.
6.	Using a Table of Correspondence, convert the c_i's to their proper base-B symbols, d_i's, and STOP.

The base-B integer obtained is:

$$(N)_B = (d_id_{i-1} \cdots d_1d_0)_B.$$

ALGORITHM 1.7

Conversion of a j-digit base-B fraction to a k-digit base-A fraction holding an accuracy of $\pm(.1)_B^j$.

$$(N)_B = (.d_{-1}d_{-2} \cdots d_{-j+1}d_{-j})_B.$$

All arithmetic is to be done using base-A numbers.

STEP	OPERATION
1.	Convert the base-B coefficients, d_i's, to their corresponding base-A values, c_i's, using a Table of Correspondence.
2.	Calculate k, the number of base-A digits to be retained after the conversion. Pick k as the integer satisfying the inequality

$$j\frac{\log_{10}(B)}{\log_{10}(A)} \le k < j\frac{\log_{10}(B)}{\log_{10}(A)} + 1.$$

3.	Set $i = 0$ and $x_0 = 0$.
4.	Increment i by one.
5.	Calculate $x_i = x_{i-1} \cdot B + c_{-i}$.
6.	If $i < j$ go back to step 4; otherwise proceed to step 7.
7.	Divide x_j by B^j, retaining only k places, and STOP. $(N)_A = x_j/B^j$.

ALGORITHM 1.8

Conversion of a k-digit base-A fraction to a j-digit base-B fraction, holding an accuracy of $\pm(.1)_A^k$.

$$(N)_A = (.a_{-1}a_{-2} \cdots a_{-k})_A.$$

All arithmetic is to be done using base-A numbers.

STEP	OPERATION
1.	Calculate j, the number of digits to be retained in the base-B number after conversion. Pick j as the integer satisfying the inequality:

$$k\frac{\log_{10}(A)}{\log_{10}(B)} \le j < k\frac{\log_{10}(A)}{\log_{10}(B)} + 1.$$

2.	Set $i = 0$ and $x_0 = (N)_A$.
3.	Increment i by one.
4.	Calculate $Y = B \cdot x_{i-1}$.
5.	Set c_{-i} equal to the integer part of Y.
6.	Set x_i equal to the fractional part of Y.
7.	If $i < j$ go back to step 3; otherwise proceed to step 8.

8.   Using a Table of Correspondence convert each c_i to its proper base-B symbol, d_i, and STOP.

$$(N)_B = (.d_{-1}d_{-2} \cdots d_{-j})_B.$$

It should be pointed out that the algorithms given in Section 1.2 can always be used to convert numbers between arbitrary bases, providing one uses base 10 as an intermediate base. However, for the readers who are able to master arithmetic in other number bases, the algorithms given in this section will yield much quicker results. Since most of our computers operate in either binary, octal, or hexadecimal, let us illustrate how the algorithms presented in this section can be used by a computer. For the convenience of the reader, the Table of Correspondence (Table 1.3) and tables of octal addition (Table 1.4), multiplication (Table 1.5), and division (Table 1.6) have been prepared.

TABLE 1.3
Table of Correspondence—Octal–Decimal.

Octal	0	1	2	3	4	5	6	7	10	11
Decimal	0	1	2	3	4	5	6	7	8	9

TABLE 1.4
Octal Addition Table

	0	1	2	3	4	5	6	7	10	11	12
0	0	1	2	3	4	5	6	7	10	11	12
1	1	2	3	4	5	6	7	10	11	12	13
2	2	3	4	5	6	7	10	11	12	13	14
3	3	4	5	6	7	10	11	12	13	14	15
4	4	5	6	7	10	11	12	13	14	15	16
5	5	6	7	10	11	12	13	14	15	16	17
6	6	7	10	11	12	13	14	15	16	17	20
7	7	10	11	12	13	14	15	16	17	20	21

Example 1.11 Convert $(742.711)_8$ to a decimal number using octal arithmetic, holding an accuracy of $\pm(.1)_8^3$.

Let us work on the fractional part first using Algorithm 1.8. In this problem $k = 3$. We must select j as the integer satisfying the inequality:

$$3\frac{\log_{10}(8)}{\log_{10}(10)} \le j < 3\frac{\log_{10}(8)}{\log_{10}(10)} + 1,$$

$$2.709 < j < 3.709; \quad \text{thus } j = 3.$$

TABLE 1.5
Octal Multiplication Table

	1	2	3	4	5	6	7	10	11	12
1	1	2	3	4	5	6	7	10	11	12
2	2	4	6	10	12	14	16	20	22	24
3	3	6	11	14	17	22	25	30	33	36
4	4	10	14	20	24	30	34	40	44	50
5	5	12	17	24	31	36	43	50	55	62
6	6	14	22	30	36	44	52	60	66	74
7	7	16	25	34	43	52	61	70	77	106
10	10	20	30	40	50	60	70	100	110	120
11	11	22	33	44	55	66	77	110	121	132
12	12	24	36	50	62	74	106	120	132	144

TABLE 1.6
Octal Division by 12; $N = 12Q + R$

N	Q	R	N	Q	R	N	Q	R	N	Q	R
0	0	0	24	2	0	50	4	0	74	6	0
1	0	1	25	2	1	51	4	1	75	6	1
2	0	2	26	2	2	52	4	2	76	6	2
3	0	3	27	2	3	53	4	3	77	6	3
4	0	4	30	2	4	54	4	6	100	6	4
5	0	5	31	2	5	55	4	5	101	6	5
6	0	6	32	2	6	56	4	6	102	6	6
7	0	7	33	2	7	57	4	7	103	6	7
10	0	10	34	2	10	60	4	10	104	6	10
11	0	11	35	2	11	61	4	11	105	6	11
12	1	0	36	3	0	62	5	0	106	7	0
13	1	1	37	3	1	63	5	1	107	7	1
14	1	2	40	3	2	64	5	2	110	7	2
15	1	3	41	3	3	65	5	3	111	7	3
16	1	4	42	3	4	66	5	4	112	7	4
17	1	5	43	3	5	67	5	5	113	7	5
20	1	6	44	3	6	70	5	6	114	7	6
21	1	7	45	3	7	71	5	7	115	7	7
22	1	10	46	3	10	72	5	10	116	7	10
23	1	11	47	3	11	73	5	11	117	7	11

Also $B = (10)_{10} = (12)_8$.

d_i	c_i	x_i	B	i
-	-	.711	12	0
8	10	.732		1
9	11	.204		2
2	2	.450		3

$$\therefore \quad (.711)_8 = (.892)_{10}.$$

We will use Algorithm 1.6 to convert the integer part of the number. The reader should also refer to the octal division table.

i	B	x_i	Base 8 Remainders	Base 10 Symbols
0	12	742	c_{i-1}	d_{i-1}
1		60	2	2
2		4	10	8
3		0	4	4

$$\therefore \quad (742)_8 = (482)_{10} \quad \text{and} \quad (742.711)_8 = (482.892)_{10}.$$

When converting integers between decimal and an arbitrary base R, the algorithms in Section 1.2 are easier to use. However, when working with fractions, Algorithm 1.8 has the advantage of not requiring the division by R^i to obtain the desired result.

1.4 NEGATIVE NUMBERS

Up until now we have only concerned ourselves with positive numbers. In this section we will examine three of the more common notations used to represent negative numbers. We are all familiar with the *sign-and-magnitude* notation since this is the notation we use daily when working with decimal numbers. However, as we will see in Section 1.5, sign-and-magnitude arithmetic is not the most efficient or economical for use by computing machines. Radix arithmetic and diminished-radix arithmetic are easier to implement.

Although we find it convenient to use the symbols "+" and "−" to indicate positive and negative numbers, they are a luxury and not really needed. Since the signs occupy just as much space as a digit, we can establish a simple convention. The first digit of a number will indicate its sign, with the number 0 being reserved for positive numbers, and the number $R - 1$, in base R, being reserved for negative numbers. This sign notation has the added advantage of reserving the symbols + and − unambiguously for the operations of addition and subtraction. This notation is not without its problems. For example, the number $(+72.3)_{10}$ should be written as $(072.3)_{10}$, and the number $(+9.25)_{10}$ should be written as $(09.25)_{10}$. However, we all have the habit of leaving off both leading and trailing zeroes and thus might write $(+72.3)_{10}$ as $(72.3)_{10}$ and $(+9.25)_{10}$ as $(9.25)_{10}$. In the former case we might correctly assume that the leading zero has been forgotten; however, in the latter case we will incorrectly interpret the result as the negative number $(-.25)_{10}$. *Careful adherence to this sign convention is the only way to avoid ambiguity and errors.*

The restriction that positive numbers start with a leading zero and negative numbers start with a leading $(R - 1)$ does limit the range of numbers which can be represented. Other conventions allow leading digits of between 0 and $\lfloor R/2 \rfloor$ for positive numbers, and leading digits between $\lfloor R/2 \rfloor + 1$ and $R - 1$ for negative numbers. The convention used in this book is not unreasonable, since most electronic calculators today use the sign-and-magnitude convention with plus and minus signs. Our sign convention does not restrict the range of numbers that can be represented using the binary number system used by the vast majority of digital computers.

1.4.1 Sign-And-Magnitude Notation

Sign-and-magnitude notation is the one we use in our decimal arithmetic system. In order to *negate*, or *complement*, a number using this notation, it is only necessary to change the sign indicator. The magnitude of the number remains unchanged. The negation, or complement, operation in effect multiplies a number by (-1). The term complement is the preferred name for this operation. The term negation will be used in Chapter 3 when we discuss Boolean Algebra.

ALGORITHM 1.9

Sign-and-magnitude complement (negation) of a base-R number. Let N be a base-R number with the sign indicator $S = 0$ if N is positive and $S = (R - 1)$ if N is negative.

$$(N)_R = (Sd_{n-2}d_{n-3} \cdots d_1 d_0.d_{-1} \cdots d_{-k})_R.$$

STEP   OPERATION
1.   Subtract S from $(R - 1)$ to yield the new sign.
2.   Append to the new sign indicator the magnitude of the original number and STOP.

$$(-N)_R = ((R - 1 - S)d_{n-2}d_{n-3} \cdots d_1 d_0.d_{-1} \cdots .d_{-k})_R.$$

Example 1.12 If $(N)_5 = (0124.341)_5$, then the sign-and-magnitude complement of N is

$$(-N)_5 = (4124.341)_5.$$

If $(N)_{10} = (9345)_{10}$, then the sign-and-magnitude complement of N is

$$(-N)_{10} = (0345)_{10}.$$

If $(N)_2 = (0.1011)_2$, then the sign-and-magnitude complement of N is

$$(-N)_2 = (1.1011)_2.$$

1.4.2 Diminished-Radix Notation

The next arithmetic notation we will consider is diminished-radix notation, also referred to as radix-minus-one notation. It is so named because to negate or complement a diminished-radix (radix-minus-one) number, it is only necessary to subtract separately each digit, including the sign digit, from $(R - 1)_R$, i.e., the radix diminished by one. Positive numbers are written in the same form as positive sign-and-magnitude numbers. In a formal manner, negative numbers are formed by subtracting the corresponding n-digit positive number from $(R^n)_R$, and then subtracting one from the least significant position. Thus if $(N)_R$ is an m-digit number with n-integer positions, including the sign and k-fractional positions, then

$$(-N)_R = (R^n)_R - (N)_R - (.1)_R^k. \tag{1.25}$$

For example, if $N = (0123.45)_{10}$, then $n = 4$, and $m = 6$, and $k = 2$;

$$(-N)_{10} = (10^4)_{10} - (N)_{10} - (.1)_{10}^2,$$
$$(-N)_{10} = (10000.00 - 0123.45 - 0000.01)_{10},$$
$$= (9999.99 - 0123.45)_{10} = (9876.54)_{10}.$$

If N has a total of m digits, then we only retain m digits when we negate it. Note that the sign convention is preserved using this notation and that there are two representations for the number zero. $(+0)_R$ is the number with all digits equal to zero, and $(-0)_R$ is the number with all digits equal to $(R - 1)_R$. One minor irritation which this double zero will cause is seen by adding $(N)_R$ to $(-N)_R$.

$$(N)_R + (-N)_R = (N)_R + (10)_R^n - (N)_R - (.1)_R^k$$
$$= (10)_R^n - (.1)_R^k = (-0)_R.$$

Thus the result is $(-0)_R$, the number with all digits equal to $(R - 1)_R$.

ALGORITHM 1.10

For taking the diminished-radix complement (negation) of a base-R number.

Let $(N)_R$ be an m-digit base-R number, n of which are integer positions including the sign and k fractional digits.

STEP   OPERATION
1.   Subtract each digit including the sign digit from $(R - 1)_R$.

2.   STOP

$$(-N)_R = (10^n)_R - (N) - (.1)_R^k.$$

Example 1.13 The following is a list of base-R numbers and their corresponding diminished-radix complements. The leftmost digit is the sign indicator.

$(N)_R$	$(-N)_R$
$(0143.21)_{10}$	$(9856.78)_{10}$
$(7421.37)_8$	$(0356.40)_8$
$(2121.11)_3$	$(0101.11)_3$
$(1010.11)_2$	$(0101.00)_2$
$(FA43.0B)_{16}$	$(05BC.F4)_{16}$

1.4.3 Radix Notation

Using radix notation, positive numbers are represented in the same form as positive sign-and-magnitude numbers, with the first digit being zero and the remaining digits corresponding to the magnitude of the number. Negative numbers are formed by subtracting the positive number from $(R^n)_R$, where n is the number of integer positions in the number including the sign position. Thus if N is an m-digit base-R number with n integer positions including the sign digit, then

$$(-N)_R = (R^n)_R - (N)_R. \tag{1.26}$$

For example, if $(N)_{10} = (0123.456)_{10}$, then $m = 7$, $n = 4$, and $(-N)_{10} = (10000.000 - 0123.456)_{10} = (9876.544)_{10}$. Note that we always truncate the result of this subtraction back to the same number of digits that were in the original number $(N)_R$.

It is easily seen that our sign convention is preserved by this notation, with positive numbers having a zero sign indicator and negative numbers having the number $(R - 1)_R$ as a sign indicator. There are two problem numbers using this notation. The number +0 is represented as the number with all zeroes. If we apply our formula for negation, we would obtain the number $(R^n)_R$ for the number $(-0)_R$. However, this number will have one more place than the original number. By convention, we will ignore the extra leading one, truncating the result back to the all-zero number. Doing this yields a unique representation of the number zero. In effect our addition operation is modulo $(R^n)_R$.

The second problem is the number with a negative sign and the remaining digits all zero. Applying our formula for negation to this negative number should yield a positive number. However, instead we obtain the number with a zero magnitude and a sign indicator of 1. That is, we have obtained an invalid number. For example, if $(N)_{10} = (9.00)_{10}$ then $(-N)_{10} = (1.00)_{10}$. By examining the following sequence of positive numbers and their corresponding negative values $[(+N)_{10}, (-N)_{10}]$: $[(0.97)_{10}, (9.03)_{10}]$; $[(0.98)_{10}, (9.02)_{10}]$; $[(0.99)_{10}, (9.01)_{10}]$;

$[(1.00)_{10}, (9.00)_{10}]$ we can determine that the number in question is the negative of $(R'')_R$. However, this number will require an extra place to be correctly written if we are to include its sign indicator. The problem with this number is easy to spot due to the invalid sign indicator as long as we are not working in base 2. In base 2, we find that the negation of $1.00 \cdots 0$ is itself.

These two problem numbers are interrelated. Since there is only a single representation for the number zero, assuming a fixed number of digits, there will exist a negative number which does not have a corresponding valid positive representation. This extra negative number will cause a great deal of trouble when it is necessary to convert numbers between machines using 1's complement (diminished-radix notation) and 2's complement (radix notation) binary numbers, since it cannot be represented as a valid 1's complement number.

ALGORITHM 1.11

For taking the radix complement of a base-R number. Let $(N)_R$ be a base-R number.

STEP   OPERATION
1.   Locate the least significant nonzero digit. If all digits are zero, STOP. $(-N)_R = 0$. If (N_R) is the most negative number, $R - 1$ followed by all zeros, STOP, as this number has no valid radix complement. Otherwise go to step 2.
2.   Subtract the least significant nonzero digit from R.
3.   Subtract each of the remaining digits to the left, including the sign digit, from $R - 1$.
4.   STOP. $(-N)_R = 10^n - (N)_R$, where n is the number of integer positions in the number including the sign position.

It should be noted that using radix notation

$$(N)_R + (-N)_R = (N)_R + (R'')_R - (N)_R = (R'')_R.$$

Which is equal to zero by our length convention.

Example 1.14 Take the 2's complement of $(0111.010)_2$.
The solution using Eq. 1.26 is:

$$(-N)_R = (10^4)_2 - N = \begin{array}{r} 10000.000 \\ -0111.010 \\ \hline 1000.110 \end{array}$$

This is the same result that we would obtain by applying Algorithm 1.11.

$$(-N)_R = \begin{array}{r} 1111.120 \\ -0111.010 \\ \hline 1000.110 \end{array}$$

Example 1.15 The following is a list of base-R numbers and their corresponding radix complement representation. The leftmost digit is the sign indicator.

$(N)_R$	$(-N)_R$
$(0432.12)_{10}$	$(9567.88)_{10}$
$(4312.41)_5$	$(0132.04)_5$
$(7142.11)_8$	$(0635.67)_8$
$(0A37.11)_{16}$	$(F5C8.EF)_{16}$

1.4.4 Comparison Among the Various Number Notations

In all three of the notations, positive numbers are written in the same manner. It is only the negative numbers which have different representations. Table 1.7 compares the three notations using binary fractions.

From the discussion in the previous two sections, it is clear that we can form the radix complement of a number by first forming the diminished-radix complement of the number, and then adding one to the least significant position. Thus if $(N)_R$ equals $(Sd_{n-2} \cdots d_0.d_{-1} \cdots d_{-k})_R$, then the diminished-radix complement, $(-N)_R$ equals $(10^n)_R - (N)_R - (.1)_R^k$. Now adding $(.1)_R^k$ gives $(10^n)_R - (N)_R$ which is the radix complement of the number.

A quick inspection of Table 1.7 reveals that, for the binary number system, we can use a set of simplified rules for forming the two's complement or the one's complement of a number $(N)_2$.

RULE 1. For forming the 1's complement of a binary number $(N)_2$, change each zero to a one and each one to a zero.

RULE 2. For forming the 2's complement of a binary number $(N)_2$, take the 1's complement using rule 1 and then add 1 to the least significant position. Ignore any carry from the sign position.

RULE 3. For forming the 2's complement of a binary number $(N)_2$, scan the number from right to left stopping if a one occurs. Do nothing to this digit and all digits to its right. Change each one, to the left of the first one, to a zero, and each zero, to the left of the first one, to a one.

You should satisfy yourself that these rules are correct.

In Table 1.8 several comparisons between the three notations are shown using binary fractions of the form $(S.d_{-1}d_{-2} \cdots d_{-k})_2$. From this table we again note that the 2's complement notation has a single representation for zero. The second problem, inherent to radix notation can be illustrated by the following example.

If we were to multiply the number (-1) by (-1) using two's complement binary numbers, we would not obtain the number $(+1)$ as an answer since this number cannot be represented in fractional notation. Since the number (-1) is such a popular number among engineers and scientists, we are going to have to be careful when designing arithmetic hardware to not overlook this problem. It is

TABLE 1.7

Base-2 Representations of the Decimal Fractions Using the Various Arithmetic Notations

$(N)_{10}$	Sign and magnitude	2's complement	1's complement
+7/8	0.111	0.111	0.111
+6/8	0.110	0.110	0.110
+5/8	0.101	0.101	0.101
+4/8	0.100	0.100	0.100
+3/8	0.011	0.011	0.011
+2/8	0.010	0.010	0.010
+1/8	0.001	0.001	0.001
+0/8	0.000	0.000	0.000
-0/8	1.000	-	1.111
-1/8	1.001	1.111	1.110
-2/8	1.010	1.110	1.101
-3/8	1.011	1.101	1.100
-4/8	1.100	1.100	1.011
-5/8	1.101	1.011	1.010
-6/8	1.110	1.010	1.001
-7/8	1.111	1.001	1.000
-8/8	-	1.000	-

TABLE 1.8

Comparison of Binary Number Representations

		Sign and Magnitude	2's Complement	1's Complement
Most Positive Number		0.111 $\cdots$ 11 $(1 - 2^{-k})$	0.111 $\cdots$ 11 $(1 - 2^{-k})$	0.111 $\cdots$ 11 $(1 - 2^{-k})$
Most negative Number		1.111 $\cdots$ 11 $- (1 - 2^{-k})$	1.000 $\cdots$ 00 -1	1.000 $\cdots$ 00 $- (1 - 2^{-k})$
Zero	+	0.000 $\cdots$ 00	0.000 $\cdots$ 00	0.000 $\cdots$ 00
	-	1.000 $\cdots$ 00	----	1.111 $\cdots$ 11

also clear from Table 1.7 that (-1) cannot be represented as an m-digit fractional sign-and-magnitude or diminished-radix number. This will cause some difficulty when passing m-digit numbers between machines using different negative number notations.

1.5 SPECIAL PROBLEMS WITH FIXED n-DIGIT SYSTEMS

Let us summarize the material in this chapter by considering some of the problems associated with converting an arbitrary decimal number into another base, using a different arithmetic notation. Assuming that only m, base-R, positions are available, and that we are free to place the radix point, we will be able to accurately convert all decimal numbers with less than or equal to $m \log_{10}(R)$ digits. If we are not free to select the radix point, but instead are only allowed n integer positions including the sign and k-fractional positions, then, in any of the three notations, we can only accurately represent and interchange decimal numbers in the range $|N| < (R^n + R^{-k} - 1)_{10}$ and only to an accuracy of $\pm(R^{-k})_{10}$. The extra radix negative number $(-R^n)_{10}$, should be outlawed to prevent conversion difficulties with diminished-radix and sign-and-magnitude numbers.

Example 1.16 Convert $(924.135)_{10}$ expressed in sign-and-magnitude notation to a two's complement binary number. Assume a 16 bit binary number is to be used, of which only 9 are fractional positions.

Solution. First we must convert the decimal number into a positive number so that we can apply the algorithms in Section 1.2.

$$(-N)_{10} = (024.135)_{10}.$$

To convert this number with an accuracy of one in the least significant position will require ten binary digits for the fractional part of the number. Since we are only allowed 9 fractional positions, our accuracy will be reduced to $\pm(2^{-9})_{10} = \pm 1/512$ or approximately $(\pm.002)_{10}$. In order to save time, we will use the short cut technique given in Section 1.2. Since 9 binary digits convert to 3 octal, we will only retain 3 octal fractional positions in our number conversion.

```
    Algorithm 1.2          Algorithm 1.4

    Integer part           Fractional part k = 3 due to
                           digit limitation.
  8 | 24 | Remainders      Integer part    .135|  8
  ---------------------    d_-1      1       .080
      3 | 0 = d_0          d_-2      0       .640
      0 | 3 = d_1          d_-3      5       .120
```

Thus we see that $(024.135)_{10}$ converts to $(030.105)_8$. We must now convert this octal number into a 16-bit binary number, filling in the missing leading-integer positions with zeroes.

$$(0 \quad 3 \quad 0 . \quad 1 \quad 0 \quad 5)_8$$

$$(0 \quad 011 \quad 000 . 001 \quad 000 \quad 101)_2$$

Note: If the integer part of our decimal number had been greater than $(63)_{10}$, we could not have represented the integer part of the number as a binary number with only six integer binary positions. Since our original number was negative, we must now take the 2's complement of the result. Thus $(924.135)_{10}$ sign-and-magnitude converts to $(1\ 100\ 111\ .\ 110\ 111\ 011)_2$ 2's complement, $(\pm.002)_{10}$.

A second problem occurs when we must add either leading, or trailing, digits to a signed base-R number. A simple set of rules for the three arithmetic conventions are summarized in Table 1.9.

TABLE 1.9
Rules for Adding Extra Digits to Signed Base-R Numbers

	ARITHMETIC CONVENTION		
	SIGN AND MAGNITUDE	RADIX	DIMINISHED RADIX
POSITIVE NUMBER	Add as many leading or trailing zeroes as required.		
Negative number leading digits	Insert as many zeroes as required between the sign position and the most significant digit.	Copy, or extend, the sign digit to the left as many places as required.	
Negative number trailing digits	Append as many zeroes as required to the least significant digit.	Copy, or append, the signed digit to the least significant digit as many times as required.	

Example 1.17 Increase the precision of the following numbers by four places, by adding two additional digits to the most significant and least significant positions.

N	Arithmetic notation	Result
$(0045.2)_{10}$	sign-and-magnitude radix diminished-radix	$(000045.200)_{10}$
$(7045.2)_8$	sign-and-magnitude	$(700045.200)_8$
$(1011.11)_2$	radix	$(111011.1100)_2$
$(210.11)_3$	diminished-radix	$(22210.1122)_3$

1.6 PROBLEMS

1.1 Suppose we allowed negative radix number systems. Then the number $(125)_{-10}$ would represent the number $1 \cdot (-10)^2 + 2 \cdot (-10)^1 + 5 \cdot (-10)^0 = (85)_{10}$.

    a) Express the number $(138)_{10}$ as a base (-10) number.

    b) What base-10 number does $(1245)_{-10}$ represent?

1.2 Using Algorithm 1.1 convert the following base-R numbers to base-10 integers.
   a) $(1234)_8$ b) $(7261)_8$
   c) $(AD2)_{16}$ d) $(1100111)_2$
   e) $(2367)_8$ f) $(12012)_3$

1.3 Using Algorithm 1.2, express the following base 10 integers in bases 2, 3, 5, 8, and 16.
   a) $(1264)_{10}$ b) $(321)_{10}$
   c) $(741)_{10}$ d) $(630)_{10}$

1.4 Using Algorithm 1.3, convert the following fractional base-R number to decimal fractions maintaining an accuracy of ±1 in the least-significant position.
   a) $(.110111)_2$ b) $(.437)_8$
   c) $(.AF2)_{16}$ d) $(.011011)_2$
   e) $(.12012)_3$ f) $(.5416)_8$

1.5 Using Algorithm 1.4, convert the following base-10 fractions to bases 2, 3, 5, 8, and 16 holding an accuracy of ±1 in the least significant position.
   a) $(.431)_{10}$ b) $(.1245)_{10}$
   c) $(.21)_{10}$ d) $(.742)_{10}$

1.6 Convert the following numbers holding an accuracy of ±1 in the least significant position.
   a) $(1011.110)_2$ to base 10
   b) $(243.167)_{10}$ to base 8
   c) $(374.126)_8$ to base 10
   d) $(AB1.EF)_{16}$ to base 8 and base 2
   e) $(102.11)_3$ to base 5

1.7 Write out a general rule for rounding a base-R number to a fixed number of places. What problems are encountered when R is an odd number?

1.8 How does the accuracy estimate given by Eq. 1.13 change if rounding is allowed during the conversion process?

1.9 Algorithms 1.3 and 1.4 truncate the answers to a fixed number of places. Rewrite these two algorithms to allow for a rounding operation.

1.10 Using the short cut techniques given in Section 1.2.2, convert the following base-R numbers holding an accuracy of ±1 in the least-significant position.
   a) $(AB7.4216)_{16}$ to base 8 and base 2
   b) $(742.113)_8$ to base 16 and base 2
   c) $(134.116)_{10}$ to base 16, base 8, and base 2

1.11 Form the sign-and-magnitude complement of:
   a) $(101101.11)_2$ b) $(03456)_{10}$
   c) $(F234.12)_{16}$ d) $(4121.11)_5$

1.12 Form the diminished-radix complement of:
   a) $(7314.27)_8$ b) $(101101.11)_2$
   c) $(0A21.6)_{16}$ d) $(03489)_{10}$

1.13 Form the radix complement of:

  a) $(32311)_4$ b) $(024.78)_{10}$
  c) $(010110.1)_2$ d) $(7345.72)_8$

1.14 Convert the following signed numbers:

  a) $(1011)_2$ sign-and-magnitude notation to base-10 diminished-radix notation;
  b) $(7426)_8$ radix notation to base-2 sign-and-magnitude notation;
  c) $(9124)_{10}$ diminished-radix notation to base-8 radix notation;
  d) $(0123)_4$ sign-and-magnitude to base-2 radix notation.

1.15 Increase the precision of the following numbers by four places, adding two
      additional positions to the most significant and least significant positions.

	N	Arithmetic notation
a)	$(012.114)_{10}$	sign-and-magnitude
b)	$(741.110)_8$	radix
c)	$(FA2.1B)_{16}$	diminished-radix
d)	$(1011.1101)_2$	sign-and-magnitude

1.7 BIBLIOGRAPHY

Booth, T. L., *Digital Networks and Computer Systems.* New York: Wiley, 1971.

Braun, E. L., *Digital Computer Design.* New York: Academic Press, 1963.

Cardenas, A. L., L. Presser, and M. A. Marin, *Computer Science.* New York: Wiley, 1972.

Chu, Y., *Digital Computer Design Fundamentals.* New York: McGraw-Hill, 1962.

Dietmeyer, D. L., *Logic Design of Digital Systems.* Boston, Mass.: Allyn and Bacon, 1971.

Flores, I., *The Logic of Computer Arithmetic.* Englewood Cliffs, N.J.: Prentice-Hall, 1963.

Knuth, D. C., *The Art of Computer Programming: Vol. 2, Seminumerical Algorithms.*
      Reading, Mass.: Addison-Wesley, 1969.

McCluskey, E. J., *Introduction to the Theory of Switching Circuits.* New York: McGraw-
      Hill, 1965.

Chapter 2

Base-R Arithmetic

2.1 INTRODUCTION

In Chapter 1, the three most common arithmetic notations, radix, diminished-radix, and sign-and-magnitude, were presented. In Section 2.2 of this chapter the rules for addition and subtraction using each of these notations will be derived. The material will be presented in general base-R notation. In Section 2.3, simple block diagrams are provided for the design of a combined adder-subtracter unit using the three arithmetic notations. A comparison of the three units is also made in Section 2.3. In Chapter 6 the actual logical design of these units using binary components will be presented.

2.2 FIXED-POINT ARITHMETIC IN BASE-R

Three different methods for representing negative numbers were presented in Chapter 1. In this section, we will derive a set of rules for addition and subtraction using each of these notations. The rules for multiplication and division will be included in a later chapter along with their hardware implementation. Our primary objective in this section will be to develop rules which are relatively easy to implement using digital hardware. The order of discussion for the various arithmetic notations will be in the order of increasing algorithmic complexity, radix, diminished-radix, and, lastly, sign-and-magnitude.

For notational convenience in this chapter, the following conventions will be used. Since we will be using arbitrary base-R numbers throughout this chapter, the radix subscript R will be deleted from our numbers, except in those cases where ambiguity might result.

Script letters, $\mathscr{A}$, $\mathscr{B}$, $\mathscr{C}$, will be used to indicate an arbitrary base-R number without regard to its sign. Upper case letters will be used to denote the absolute value of a number. Lower case letters will be used to denote the individual digits of a number. For convenience we will also assume that all base-R numbers have a fixed number of positions including the sign digit and are of the form

$$\mathscr{A} = a_s a_{n-2} \cdots a_1 a_0 . a_{-1} \cdots a_{-k}$$

with the leading digit a_s indicating the sign. The sign convention of a zero for plus and $R-1$ for minus will be adhered to in this chapter.

2.2.1 Rules for Unsigned Base-R Addition

One problem I have noted while teaching radix arithmetic to students is the large number of errors which result during the addition or subtraction of base-R numbers using paper and pencil techniques. These errors no doubt are due to our extensive training in base-10 arithmetic. I have found the following crutch extremely useful when adding pairs of base-R numbers, particularly hexadecimal numbers. The crutch method, by using a Table of Correspondence, allows us to carry out the actual addition using decimal arithmetic. The simple correction procedure needed to convert the resulting sum back to a base-R number can be performed during the actual addition process.

ALGORITHM 2.1

Crutch addition of two unsigned base R numbers.

$$\mathscr{A} = (a_{n-1} a_{n-2} \cdots a_1 a_0 . a_{-1} \cdots a_{-k})_R.$$
$$\mathscr{B} = (b_{n-1} b_{n-2} \cdots b_1 b_0 . b_{-1} \cdots b_{-k})_R.$$

STEP	OPERATION
1.	Using a Table of Correspondence, convert each a_i and b_i to its corresponding decimal value a_i^* and b_i^*.
2.	Set the carry in to the least significant position c_{-k-1} to 0. Set i to $-k$.
3.	Calculate $(x_i)_{10} = (a_i^* + b_i^* + c_{i-1}^*)_{10}$ using decimal arithmetic.
4.	Set the carry out of position i, c_i^*, equal to the integer part of the result obtained when x_i is divided by R using decimal arithmetic.
5.	Set the sum digit s_i^* equal to the remainder obtained when x_i is divided by R using decimal arithmetic.

6.   If $i < n - 1$, then increment i by one and go back to step 3; otherwise proceed to step 7.

7.   Using the Table of Correspondence convert c_{n-1}^*, the carry out of the most significant position, and each sum digit s_i^* into their proper base-R symbols c_{n-1} and s_i's and STOP.

$$\mathcal{S} = \mathcal{A} + \mathcal{B} = (c_{n-1}s_{n-1}s_{n-2} \cdots s_1s_0.s_{-1} \cdots s_{-k})_R.$$

It is left as an exercise for the reader to modify this algorithm so that negative as well as positive numbers can be added.

It should also be clear to the reader that as long as $R < 10$, steps 1 and 7 are not needed since in this case the two sets of symbols are the same.

Example 2.1 Add $(ABC7)_{16}$ to $(14DF)_{16}$.
Using the crutch method of Algorithm 2.1 we have

$$
\begin{array}{c}
\text{Base-16} \\
\text{symbols} \quad\quad\quad \text{Base-10 symbols}
\end{array}
$$

		Base-16 symbols				Base-10 symbols			
$\mathcal{A} =$	(A	B	C	$7)_{16}$ =	(10	11	12	$7)_{16}$
$\mathcal{B} =$	(1	4	D	$F)_{16}$ =	(1	4	13	$15)_{16}$

0	1	1	1	0	(carry)$_{16}$
	12	16	26	22	(sum)$_{10}$

$$\mathcal{S} = \mathcal{A} + \mathcal{B} = (\text{C} \quad \text{O} \quad \text{A} \quad 6)_{16} = (12 \quad 0 \quad 10 \quad 6) \ (\text{sum})_{16}$$

$$\text{note: carry}_{16} = \text{integer part of } \frac{(\text{sum})_{10}}{16}$$

$$\text{sum}_{16} = \text{remainder of } \frac{(\text{sum})_{10}}{16}$$

Example 2.2 Add $(134.57)_8$ to $(731.45)_8$.
Since R is less than 10 we can eliminate steps 1 and 7 of Algorithm 2.1. Doing so we obtain,

$\mathcal{A} =$ (	1	3	4	.	5	$7)_8$
$\mathcal{B} =$ (	7	3	1	.	4	$5)_8$

1	0	0	1	.	1	0	(carry)$_8$
	8	6	6	.	10	12	(sum)$_{10}$

$$\mathcal{S} = \mathcal{A} + \mathcal{B} = (1 \quad 0 \quad 6 \quad 6 \quad . \quad 2 \quad 4)_8 \ (\text{sum})_8$$

2.2.2 Radix Arithmetic

Using radix-arithmetic notation, positive and negative numbers can be translated into the familiar sign-and-magnitude notation as follows: if $\mathcal{A}$ is positive, then

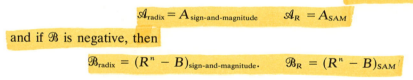

$$\mathcal{A}_{\text{radix}} = A_{\text{sign-and-magnitude}} \qquad \mathcal{A}_R = A_{\text{SAM}}$$

and if $\mathcal{B}$ is negative, then

$$\mathcal{B}_{\text{radix}} = (R^n - B)_{\text{sign-and-magnitude}}. \qquad \mathcal{B}_R = (R^n - B)_{\text{SAM}}$$

We will use the subscript R with script letters to denote that it is in radix notation. Assuming n integer and k fractional positions including the sign position, the radix numbers in this section will lie in the range $(-R^{n-1})_{10} \leq \mathcal{A}_R \leq (R^{n-1} - R^{-k})_{10}$. We must investigate the following three cases:

Case 1: Both $\mathcal{A}_R$ and $\mathcal{B}_R$ are positive.
Case 2: Both $\mathcal{A}_R$ and $\mathcal{B}_R$ are negative.
Case 3: $\mathcal{A}_R$ and $\mathcal{B}_R$ are of opposite signs.

PRELIMINARY ADDITION RULE.   Add directly all digits including the sign digits.

Let us determine if this rule is valid for each of the three cases. If necessary, this rule will be modified as each case is examined.

CASE 1. $\mathcal{A}_R$ and $\mathcal{B}_R$ are both positive.

	Radix notation	Sign-and-magnitude notation
$\mathcal{A}_R =$	$0\ a_{n-2}a_{n-3} \cdots a_0.a_{-1} \cdots a_{-k}$	$= A$
$\mathcal{B}_R =$	$0\ b_{n-2}b_{n-3} \cdots b_0.\ b_{-1} \cdots b_{-k}$	$= B$
$\mathcal{S}_R = \mathcal{A}_R + \mathcal{B}_R =$	$s_s\ s_{n-2}\ s_{n-3} \cdots s_0.\ s_{-1} \cdots s_{-k}$	$= (A + B)$

Since we only have a fixed number of positions available, the resulting sum $\mathcal{S}$ will only be correct as long as the resulting sign s_s equals zero. The sign digit s_s will be nonzero only if a carry out of the most significant position occurs when adding the two numbers. It is clear that as long as $A + B < R^{n-1}$, no carry into the sign position can result. If $A + B \geq R^{n-1}$, then an overflow will occur and an error signal should be given. The following test procedures for detecting this overflow are equivalent. The reader should verify this.

Tests for overflow when adding two positive radix numbers:

TEST 1.   If $\mathcal{A}_R$ and $\mathcal{B}_R$ are both positive and there is a carry out of the most significant position, then an overflow has occurred.

TEST 2. If $\mathscr{A}_R$ and $\mathscr{B}_R$ are both positive and the resulting sign digit s_s is nonzero, then an overflow has occurred.

TEST 3. If $\mathscr{A}_R$ and $\mathscr{B}_R$ are both positive and the carry out of the sign position differs from the carry into the sign position, then an overflow has occurred.

Example 2.3

$$\mathscr{A}_R = (034.25)_8$$
$$\mathscr{B}_R = (040.21)_8$$

$$\mathscr{A}_R + \mathscr{B}_R = (074.46)_8$$

No overflow has occurred and the answer is correct.

Example 2.4

$$\mathscr{A}_R = (0.1101)_2 \qquad A = (13/16)_{10}$$
$$\mathscr{B}_R = (0.1011)_2 \qquad B = (11/16)_{10}$$

$$\mathscr{A}_R + \mathscr{B}_R = (1.1000)_2 \qquad A + B = (24/16)_{10} > 1$$

Since $A + B > 1^0$, an overflow has occurred and is easily detected using any of the overflow tests.

CASE 2. Both $\mathscr{A}_R$ and $\mathscr{B}_R$ are negative. The addition of two negative numbers yields:

Sign-and-magnitude notation	Radix notation

$$\mathscr{A}_R = \quad R^n - A \qquad = (R-1)a_{n-2}\cdots a_0.a_{-1}\cdots a_{-k}$$
$$\mathscr{B}_R = \quad R^n - B \qquad = (R-1)b_{n-2}\cdots b_0.b_{-1}\cdots b_{-k}$$

$$\mathscr{S}_R = \mathscr{A}_R + \mathscr{B}_R = 2 \cdot R^n - (A + B) = \qquad c_s s_s s_{n-2}\cdots s_0.\ s_{-1}\cdots s_{-k} = S^*$$

The correct form for the answer is:

$$R^n - (A + B)$$

Let us try to modify the direct sum, S^*, in order to obtain the correct answer.

Ignoring the carry from the sign position is equivalent to dividing the resulting direct sum S^* by R^n and retaining only the remainder. Thus the result is

$$\frac{S^*}{R^n} = \frac{2 \cdot R^n - (A + B)}{R^n} = 1 + \frac{R^n - (A + B)}{R^n} \quad \text{for} \quad 2 \le A + B \le 2 \cdot R^{n-1}$$

with $1 \le A \le R^{n-1}$ and $1 \le B \le R^{n-1}$.

As long as $2 \le A + B \le 2 \cdot R^{n-1}$ the sum obtained by direct addition and ignoring the carry from the sign position yields $S = R^n - (A + B)$, the correct form for the answer desired. It is also clear that the direct sum S^* is of the form

$$S^* = (1 \overset{c_s}{R} - 1\overset{s_s}{s_{n-2}} s_{n-3} \cdots s_0 . s_{-1} \cdots s_{-k})_R \quad \text{if} \quad 2 \le A + B \le R^{n-1} \quad (2.1)$$

and that

$$S^* = (1 \overset{c_s}{R} - 2\overset{s_s}{s_{n-2}} s_{n-3} \cdots s_0 . s_{-1} \cdots s_{-k})_R \quad \text{if} \quad R^{n-1} < A + B \le 2 \cdot R^{n-1} \quad (2.2)$$

Thus as long as $A + B \le R^{n-1}$, the resulting sign indicator s_s will equal $R - 1$. Since a_s and b_s are equal to $R - 1$ for negative numbers, the sign of the sum can only equal $R - 1$ if a carry occurs out of the most significant position. The lack of a carry out of the most significant position when adding two negative numbers indicates that $A + B > R^{n-1}$, and that an overflow has occurred.

It should also be noted that the carry out of the sign position is always equal to one when adding two negative numbers. Therefore we again have three equivalent tests for determining if an overflow has occurred.

TEST 1. When adding two negative numbers, the lack of a carry out of the most significant position indicates that overflow has occurred.

TEST 2. When adding two negative numbers, if the resulting sign is not equal to $R - 1$, an overflow has occurred.

TEST 3. When adding two negative numbers, if the carry into the sign position is not equal to the carry out of the sign position, then an overflow has occurred.

Example 2.5

$$\begin{aligned}
\mathscr{A}_R &= (-4/16)_{10} = (1.1100)_2 \\
\mathscr{B}_R &= (-7/16)_{10} = (1.1001)_2 \\
\hline
\mathscr{S}_R &= (-11/16)_{10} = (1\,1.0101)_2
\end{aligned}$$

ignore this digit.

Since a carry does occur out of the most significant position, overflow has not occurred and the resulting answer, $= (1.0101)_2$, is correct.

Example 2.6

$$\begin{aligned}
\mathscr{A}_R &= (FA2.4)_{16} & (-93.75)_{10} \\
\mathscr{B}_R &= (F2B.C)_{16} & (-212.25)_{10} \\
\hline
\mathscr{S}_R &= \mathscr{A}_R + \mathscr{B}_R = (1ECE.0)_{16} = & (-306.00)_{10}
\end{aligned}$$

ignore this digit.

Since the carry into the sign position is not equal to the carry out of the sign position, overflow has occurred and the resulting sum $\mathcal{S} = (ECE.0)_{16}$ is incorrect. It is also clear in this case that $|\mathcal{A}_R| + |\mathcal{B}_R| > 16^2$; in fact, $A + B = (93.75 + 212.25)_{10} = (306)_{10}$.

CASE 3.   Addition of two numbers with opposite signs. Without loss of generality, let us assume $\mathcal{A}_R$ is positive and $\mathcal{B}_R$ is negative. Since $0 \leq A < R^{n-1}$ and $1 \leq B \leq R^{n-1}$, the magnitude of the resulting sum $S = |B - A|$ must be less than or equal to R^{n-1}, with equality only occurring if $A = 0$ and $B = R^{n-1}$. Thus with $\mathcal{A}$ positive and $\mathcal{B}$ negative, we obtain:

$$\mathcal{A}_R = A$$
$$\mathcal{B}_R = R^n - B$$

$$\overline{\mathcal{S}_R = \mathcal{A}_R + \mathcal{B}_R = R^n - (B - A) = S^*}$$

Clearly, if $B > A$, the resulting sum will be negative and in the correct format. However, if $B \leq A$, the result will be of the form $R^n + (A - B)$ with $0 \leq |A - B| < R^{n-1}$. In this situation there will be no carry into the sign position, since $|A - B| < R^{n-1}$, and the resulting carry out of the sign position and the sign indicator will be 1 and 0 respectively. By ignoring the carry out of the sign position, i.e., by dividing S^* by R^n and retaining only the remainder, we obtain the desired positive result. It should be obvious that an overflow cannot occur when we add two numbers with unlike signs.

Example 2.7

$$\mathcal{A}_R = (023.1)_5 = (+13.2)_{10}$$
$$\mathcal{B}_R = (413.3)_5 = (-16.4)_{10}$$

$$\overline{\mathcal{A}_R + \mathcal{B}_R = (441.4)_5 = \quad (-3.2)_{10}}$$

Since $B > A$ we will obtain a negative result. Thus $\mathcal{A}_R + \mathcal{B}_R = (441.4)_5$

Example 2.8

$$\mathcal{A}_R = (\ 072.3)_8 = (+58\tfrac{3}{8})_{10}$$
$$\mathcal{B}_R = (\ 721.4)_8 = (-46\tfrac{4}{8})_{10}$$

$$\overline{\mathcal{A}_R + \mathcal{B}_R = (1013.7)_8 = (+11\tfrac{7}{8})_{10}}$$

ignore this digit

Since $A > B$ we will obtain a positive result, with $\mathcal{A}_R + \mathcal{B}_R = (013.7)_8$

Summarizing the remarks presented in this section, the rules for addition of radix numbers can be stated as:

RULE 2.1. Addition of two base-R numbers using radix notation.

$$\mathcal{A}_R = (sa_{n-2}a_{n-3} \cdots a_1 a_0 . a_{-1} \cdots a_{-k})_R$$
$$\mathcal{B}_R = (sb_{n-2}b_{n-3} \cdots b_1 b_0 . b_{-1} \cdots b_{-k})_R$$

1. Add directly all digits including the sign digits ignoring any carry which might occur from the sign position.
2. The answer obtained in Step 1 will be correct as long as an overflow has not occurred. An overflow can be detected by any of the following three methods.

   a) If both numbers to be added are positive, then an overflow has occurred if there is a carry out of the most significant position. If both numbers to be added are negative, then an overflow has occurred if there is no carry out of the most significant position.

   b) If both numbers to be added are positive, then an overflow has occurred if the resulting sign indicator is nonzero. If both numbers to be added are negative, then an overflow has occurred if the resulting sign indicator is not equal to $(R-1)_R$.

   c) Regardless of the signs, an overflow has occurred if the carry out of the sign position differs from the carry into the sign position.

These three different but equivalent methods for detecting overflow conditions have been included to allow the hardware designer as much freedom as possible in the implementation of this arithmetic. Methods (a) and (c) are most frequently used in computer designs.

Rather than attempt to devise a set of special rules for subtraction, the familiar procedure of negating (complementing) the subtrahend and then adding the resulting number to the minuend will be used.

$$
\begin{array}{r}
\mathcal{A}_R \text{ minuend} \\
- \mathcal{B}_R \text{ subtrahend} \\
\hline
\mathcal{A}_R - \mathcal{B}_R \text{ difference}
\end{array}
$$

RULE 2.2 Subtraction of two base-R numbers using radix notation.

1. Form the radix complement of the subtrahend. Add the resulting number to the minuend using the rules for radix addition.
2. Form the diminished-radix complement of the subtrahend, add the resulting number to the minuend using the rules for radix addition, then add R^{-k} to the resulting sum.

It should be noted that using method 2, the addition of R^{-k} can be accomplished by initially setting the carry into the least significant position to a one instead of its normal value of zero. Method 2 is also easier to implement with hardware, since it is easier to design economical circuits for forming the diminished-radix complement.

Example 2.9 Subtract $(02122.11)_3$ from $(01212.12)_3$

$$
\begin{array}{cc}
\text{direct method} & \text{complement method} \\[4pt]
\mathscr{A}_R = (01212.12)_3 & \mathscr{A}_R = (01212.12)_3 \\
\mathscr{B}_R = (02122.11)_3 & +(-\mathscr{B}_R) = (20100.12)_3 \\[2pt]
\hline
\mathscr{A}_R - \mathscr{B}_R = (22020.01)_3 & \mathscr{A}_R + (-\mathscr{B}_R) = (22020.01)_3
\end{array}
$$

In Section 2.3 a block diagram for use in designing an arithmetic unit using radix addition and subtraction is shown. A comparison of the three-types of arithmetic in terms of hardware complexity and calculation time will also be made at that time.

2.2.3 Diminished-Radix Arithmetic

Using diminished-radix arithmetic, positive and negative numbers can be translated into our familiar sign-and-magnitude format as follows: if $\mathscr{A}$ is positive, then

$$\mathscr{A}_{\text{diminished-radix}} = A_{\text{sign-and-magnitude}};$$

and if $\mathscr{B}$ is negative, then:

$$\mathscr{B}_{\text{diminished-radix}} = (R^n - B - R^{-k})_{\text{sign-and-magnitude}}.$$

The subscript DR will be used to indicate diminished radix numbers. Assuming n integer positions, including the sign, and k fractional positions, the diminished-radix numbers in this section will lie in the range $0 \le |\mathscr{A}|_{DR} \le R^{n-1} - R^{-k}$. As was the situation with radix arithmetic there are three cases to be considered:

Case 1. Both $\mathscr{A}_{DR}$ and $\mathscr{B}_{DR}$ are positive.
Case 2. Both $\mathscr{A}_{DR}$ and $\mathscr{B}_{DR}$ are negative.
Case 3. $\mathscr{A}_{DR}$ and $\mathscr{B}_{DR}$ are of opposite sign.

PRELIMINARY ADDITION RULE. Add directly all digits including the sign digits.

Let us determine when this rule is valid for each of the three cases. If necessary, this preliminary rule will be modified as each case is examined.

CASE 1. $\mathscr{A}_{DR}$ and $\mathscr{B}_{DR}$ are both positive

Sign-and-
magnitude

Diminished-radix notation   notation

$$\mathscr{A}_{DR} = 0\,a_{n-2}a_{n-3} \cdots a_0.a_{-1} \cdots a_{-k} = A$$
$$\mathscr{B}_{DR} = 0\,b_{n-2}b_{n-3} \cdots b_0.b_{-1} \cdots b_{-k} = B$$
$$\mathscr{S}_{DR} = \mathscr{A}_{DR} + \mathscr{B}_{DR} = s_s s_{n-2} s_{n-3} \cdots s_0.s_{-1} \cdots s_{-k} \quad = (A + B)$$

The addition of two positive numbers will always result in a zero carry out of the sign position. As long as we do not get a carry into the sign position during the addition process, the sign indicator for the sum will also remain zero and the resulting sum will be correct.

As long as $A + B < R^{n-1}$, no carry into the sign position will occur, indicating that an overflow has occurred. It is obvious that the sign indicator for the sum of two positive numbers should always be zero. Thus a nonzero sign indicator can be used to detect an overflow condition. It should also be noted that the carry out of the sign position will differ from the carry into the sign position if an overflow occurs during the addition of two positive numbers.

Example 2.10

$$\mathscr{A}_{DR} = (042.37)_8$$
$$\mathscr{B}_{DR} = (037.42)_8$$

$$\mathscr{S}_{DR} = \mathscr{A}_{DR} + \mathscr{B}_{DR} = (102.01)_8$$

Since we have a nonzero sign indicator, overflow has occurred.

Example 2.11

$$\mathscr{A}_{DR} = (0.0100)_2 \qquad A = +\tfrac{4}{16}$$
$$\mathscr{B}_{DR} = (0.1011)_2 \qquad B = +\tfrac{11}{16}$$

$$\mathscr{S}_{DR} = \mathscr{A}_{DR} + \mathscr{B}_{DR} = (0.1111)_2 \qquad A + B = +\tfrac{15}{16}$$

No overflow conditions exist, hence the answer is correct.

CASE 2. Both $\mathscr{A}_{DR}$ and $\mathscr{B}_{DR}$ are negative. The direct addition of two negative diminished-radix numbers yields the result

$$\mathscr{A}_{DR} = \quad R^n - \quad A - \quad R^{-k}$$
$$\mathscr{B}_{DR} = \quad R^n - \quad B - \quad R^{-k}$$

$$\mathscr{S}_{DR} = \mathscr{A}_{DR} + \mathscr{B}_{DR} = 2 \cdot R^n - (A + B) - 2 \cdot R^{-k} = S^*$$

The correct answer is of the form $\mathcal{A}_{DR} + \mathcal{B}_{DR} = R^n - (A + B) - R^{-k}$. This is a difference of $R^n - R^{-k}$. We can eliminate the extra R^n from the direct sum S^* by truncating our result, i.e., by ignoring the carry out of the sign position. This is equivalent to dividing the direct sum by R^n and then retaining only the remainder. Dividing the direct sum S^* by R^n yields:

$$\frac{S^*}{R^n} = 1 + \frac{R^n - (A + B) - 2 \cdot R^{-k}}{R^n} \qquad \text{for} \qquad 0 \le A + B \le 2 \cdot [R^{n-1} - R^{-k}].$$

$$(2.3)$$

The extra $-R^{-k}$ can be eliminated by adding R^{-k} to the direct sum S^*. Note that the carry out of the sign position will always be one when adding two negative numbers, and that it will always be zero when adding two positive numbers. Suppose we add the carry out of the sign position to the least significant position of the direct sum, and then retain only the resulting sum digits for our answer. Doing this yields $S^{**} = S^* + c_s \cdot R^{-k}$, with $c_s = 0$ if $\mathcal{A}$ and $\mathcal{B}$ are both positive and $c_s = 1$ if $\mathcal{A}$ and $\mathcal{B}$ are both negative. Therefore $S^{**} = A + B$ if $\mathcal{A}$ and $\mathcal{B}$ are both positive, and $S^{**} = 2 \cdot R^n - (A + B) - R^{-k}$ if $\mathcal{A}$ and $\mathcal{B}$ are both negative. Finally dividing S^{**} by R^n and retaining only the remainder; $S = A + B$ if $\mathcal{A}$ and $\mathcal{B}$ are both positive, and $S = R^n - (A - B) - R^{-k}$ if $\mathcal{A}$ and $\mathcal{B}$ are both negative. These are the correct formats for both cases. The carry out of the sign position to be added to the least significant position will be called the *end-around-carry*. It should be noted that unless $A + B \ge R^{n-1}$ the resulting sign indicator will be equal to $(R-1)_R$. If $A + B \ge R^{n-1}$, then the carry into the sign position will always be zero and the resulting sign indicator will always be equal to $(R - 2)_R$. Therefore in order to detect an overflow when adding two negative numbers, an examination of the resulting sign indicator can be made. If it is not equal to $R - 1$ an overflow has occurred. Alternatively, one could check the carry into the sign position. If it is equal to zero, an overflow condition exists. Lastly, a check on both the carry into and carry out of the sign position could be made. If they are different an overflow condition exists. However, the overflow tests should not be applied until after the end-around-carry has been added to the direct sum.

Example 2.12

$$\mathcal{A}_{DR} = (11.1011)_2 = (-\tfrac{4}{16})_{10}$$
$$\mathcal{B}_{DR} = (10.1000)_2 = (-1\tfrac{7}{16})_{10}$$

Since the carry out is a 1, add in the end-around-carry

$\textcircled{1}10.0011 = (-1\tfrac{11}{16})_{10}$

$\longrightarrow 1$

$$\mathcal{S}_{DR} = \mathcal{A}_{DR} + \mathcal{B}_{DR} = (10.0100)_2$$

Example 2.13

$$\mathcal{A}_{DR} = (974.35)_{10} = (-25.64)_{10}$$
$$\mathcal{B}_{DR} = (912.47)_{10} = (-87.52)_{10}$$

$$\mathcal{S}^*_{DR} = 1886.82 \quad = -113.16_{10}$$

End-around-carry $\quad \longrightarrow 1 \quad$ Note: $A + B = 113.16 > 10^2$

$$\mathcal{S}_{DR} = (886.83)_{10} \Rightarrow \text{Overflow}$$

Sign digit
$\neq 9$ implies overflow.

Example 2.14

$$\mathcal{A}_{DR} = (7342.)_8 = (-285)_{10}$$
$$\mathcal{B}_{DR} = (7435.)_8 = (-226)_{10}$$

$$\mathcal{S}^*_{DR} = (16777.)_8 = (-511)_{10}$$

End-around-carry $\quad \longrightarrow 1.$

$$\mathcal{S}_{DR} = (7000.)_8$$

In this example a false overflow condition exists prior to adding in the end-around-carry. This false overflow is eliminated by the end-around-carry.

CASE 3. $\mathcal{A}$ and $\mathcal{B}$ are of opposite sign. Without loss of generality, let us assume $\mathcal{A}$ is positive and that $\mathcal{B}$ is negative. Since $0 \le A < R^{n-1}$ and $0 \le B < R^{n-1}$, the resulting sum $S = |B - A|$ lies in the range $0 \le S < R^{n-1}$ and overflow is impossible. The direct addition of a positive $\mathcal{A}$ and a negative $\mathcal{B}$ yields:

$$\mathcal{A}_{DR} = A$$
$$\mathcal{B}_{DR} = R^n - \quad B \quad - R^{-k}$$

$$\mathcal{S}_{DR} = \mathcal{A}_{DR} + \mathcal{B}_{DR} = R^n - (B - A) - R^{-k} = S^*$$

It is clear that if $B \ge A$, the resulting sum S^* will be correct since $|B - A| < R^{n-1}$. However, if $A > B$, then the correct answer should be $(A - B)$ instead of $R^n + (A - B) - R^{-k}$. Again we can eliminate the term R^n by dividing the direct sum by R^n and retaining only the remainder. Thus we have:

$$\frac{S^*}{R^n} = 1 + \frac{A - B - R^{-k}}{R^n} \qquad A > B$$

$$\frac{S^*}{R^n} = 0 + \frac{R^n - (B - A) - R^{-k}}{R^n} \qquad B \ge A$$

Since the carry out is equal to one when $A > B$, we can again use the carry out of the sign position as a signal to add 1 to the least significant position, thereby cancelling the extra $-R^{-k}$ term present in the direct sum. Overflow cannot occur when adding a positive and a negative number.

Example 2.15

$$\mathcal{A}_{DR} = (043.1)_5 \quad (+23.2)_{10}$$
$$\mathcal{B}_{DR} = (441.3)_5 \quad (-3.2)_{10}$$

$$\mathcal{S}^*_{DR} = (1034.4)_5 \quad (+20.0)_{10}$$
end-around-carry $\quad\longrightarrow 1$

$$\mathcal{S}_{DR} = (040.0)_5$$

Since $A > B$, we should get an end-around-carry.

Example 2.16

$$\mathcal{A}_{DR} = (0147)_8 \quad (+103)_{10}$$
$$\mathcal{B}_{DR} = (7421)_8 \quad (-238)_{10}$$

$$\mathcal{S}^*_{DR} = \mathcal{S}_{DR} = (7570)_8 \quad (-135)_{10}$$
Since there is
no carry out

Since $B > A$, we should
not get an
end-around-carry.

Summarizing the remarks made in this section, the rules for addition of diminished-radix numbers can be stated as:

RULE 2.3. Addition of two base-R numbers using diminished-radix notation.

$$\mathcal{A}_{DR} = (sa_{n-2}a_{n-3} \cdots a_1a_0.a_{-1} \cdots a_{-k})_R$$
$$\mathcal{B}_{DR} = (sb_{n-2}b_{n-3} \cdots b_1b_0.b_{-1} \cdots b_{-k})_R$$

1. Add directly all digits in the two numbers including the sign digits.
2. If a carry occurs from the sign position as a result of the addition in Step 1, then add $(R^{-k})_R$ to the result in Step 1; i.e., add 1 to the least significant position.
3. The answer obtained in Step 2, ignoring the carry from the sign position, will be correct as long as an overflow has not occurred. An overflow can be detected using any of the following three methods.
   a) If both numbers to be added are positive, then an overflow has occurred if there is a carry out of the most significant position. If both numbers to be added are negative, then an overflow occurred if there is no carry out of the most significant position.
   b) If both numbers to be added are positive, then an overflow has occurred if the resulting sign indicator is nonzero. If both numbers to be added are negative, then an overflow has occurred if the resulting sign indicator is not equal to $(R-1)_R$.
   c) Regardless of the signs, an overflow has occurred if the carry out of the sign position differs from the carry into sign position.

These three different but equivalent methods for detecting overflow conditions have been included to provide the circuit designer with as much design freedom as possible.

We will not attempt to derive a special set of rules for subtracting diminished-radix numbers. Instead, we will again use the concept of complement arithmetic. Thus to subtract $\mathscr{B}_{DR}$ from $\mathscr{A}_{DR}$ we will negate or take the diminished-radix complement of the subtrahend $\mathscr{B}_{DR}$ and then add it to the minuend $\mathscr{A}_{DR}$

$$
\begin{array}{rl}
\mathscr{A}_{DR} & \text{Minuend} \\
-\mathscr{B}_{DR} & \text{Subtrahend} \\
\hline
\mathscr{A}_{DR} - \mathscr{B}_{DR} & \text{Difference}
\end{array}
$$

RULE 2.4. Subtraction of two base-R numbers using diminished-radix notation.
1. Form the diminished-radix complement of the subtrahend.
2. Add the complemented subtrahend to the minuend using the rules for diminished-radix addition.

Example 2.17. Subtract $(10011)_2$ from $(11101)_2$

Direct method

$$
\begin{array}{l}
\mathscr{A}_{DR} = (11101)_2 \\
\mathscr{B}_{DR} = (10011)_2 \\
\hline
\mathscr{A}_{DR} - \mathscr{B}_{DR} = (01010)_2
\end{array}
$$

Complement method

$$
\begin{array}{rl}
\mathscr{A}_{DR} = & (11101)_2 \\
+ (-\mathscr{B}_{DR}) = & (01100)_2 \\
\hline
\mathscr{S}^*_{DR} = & (101001)_2 \\
\end{array}
$$

End-around-carry $\quad \longrightarrow 1$

$$
\mathscr{A}_{DR} + (-\mathscr{B}_{DR}) = (01010)_2
$$

In Section 2.3, a block diagram for use in designing an arithmetic unit using diminished-radix addition and subtraction is shown. At that time, a comparison among the three types of arithmetic units will be made in terms of hardware complexity and calculation time.

2.2.4 Sign-and-Magnitude Arithmetic

Using sign-and-magnitude notation, positive and negative numbers can be easily translated into our familiar sign-and-magnitude format. Thus if $\mathscr{A}$ is positive, then

$$
\mathscr{A}_{\text{sign-and-magnitude}} = A_{\text{sign-and-magnitude}};
$$

and if $\mathscr{B}$ is negative then

$$
\mathscr{B}_{\text{sign-and-magnitude}} = B_{\text{sign-and-magnitude}}.
$$

The subscript SAM will be used on script letters to indicate that the numbers are written using sign-and-magnitude notations. If $\mathscr{A}$ is a sign-and-magnitude number written in the form $\mathscr{A}_{SAM} = (s a_{n-2} a_{n-3} \cdots a_1 a_0 . a_{-1} \cdots a_{-k})_R$, then the magnitude of $\mathscr{A}_{SAM}$ will lie in the range $0 \leq A \leq R^{n-1} - R^{-k}$.

Since the sign indicator is not really a part of a sign-and-magnitude number, only two cases need to be considered when establishing the rules for addition:

Case 1: $\mathcal{A}_{SAM}$ and $\mathcal{B}_{SAM}$ are of the same sign.

Case 2: $\mathcal{A}_{SAM}$ and $\mathcal{B}_{SAM}$ are of opposite sign.

As was done for the two previous arithmetic notations let us derive a set of rules for addition which will be valid for all possible combinations of signed numbers.

CASE 1. $\mathcal{A}_{SAM}$ and $\mathcal{B}_{SAM}$ both have the same sign. This case is relatively simple. All that is necessary is that the sign digit be stripped off the two numbers; the magnitudes of the two numbers are then added directly. The sign indicator of the result is then set equal to the common sign of the two numbers being added. It is also clear that an overflow will occur only if $A + B \geq R^{n-1}$ which will always cause a carry out of the most significant position. Therefore the following preliminary rule is obtained.

PRELIMINARY ADDITION RULE. When adding two sign-and-magnitude numbers with like signs, add directly all digits except the sign indicators. Give the resulting sum the common sign of the numbers being added.

If a carry should occur out of the most significant position during the direct addition of the magnitudes, then an overflow has occurred.

Example 2.18   Add $\mathcal{A}_{SAM} = (3132.2)_4$ to $\mathcal{B}_{SAM} = (3101.2)_4$

Removing the signs and adding the magnitudes yields:

$$
\begin{aligned}
A &= (132.2)_4 \\
B &= (101.2)_4 \\
\hline
S &= (300.0)_4
\end{aligned}
$$

Since there is no carry from the sign position, overflow did not occur. Appending the common negative sign yields

$$\mathcal{S}_{SAM} = (3300.0)_4$$

Example 2.19   Add $\mathcal{A}_{SAM} = (011011.11)_2$ to $\mathcal{B}_{SAM} = (001100.11)_2$

Removing the signs and adding the magnitudes yields:

$$
\begin{aligned}
A &= (11011.11)_2 = (27.75)_{10} \\
B &= (01100.11)_2 = (12.75)_{10} \\
\hline
S &= (101000.10)_2 = (40.50)_{10}
\end{aligned}
$$

In this example the carry from the most significant sign position indicates overflow has occurred. In this case, $(A + B)_{10} = (40.50)_{10}$ exceeds the limit of 2^5.

CASE 2. $\mathcal{A}_{SAM}$ and $\mathcal{B}_{SAM}$ are of opposite signs. When we add two numbers with unlike signs we normally determine which is larger by inspection and then subtract the smaller number from the larger, appending to it the sign of the larger number. Unfortunately, this type of an approach requires a considerable amount of special hardware to implement the comparison. What is needed is a method which includes the comparison operation at the same time that the subtraction is being done.

This can be accomplished by taking the diminished-radix complement of the magnitude of the addend and adding it directly to the magnitude of the augend.

$$
\begin{array}{lll}
\text{Augend} & |\mathscr{A}_{\text{SAM}}| = |\mathscr{A}_{\text{SAM}}|_{\text{DR}} = A \\
\text{Addend} & -|\mathscr{B}_{\text{SAM}}|_{\text{DR}} & = R^{n-1} - B - R^{-k} \\
\hline
\mathscr{S}_{\text{DR}} = |\mathscr{A}_{\text{SAM}}|_{\text{DR}} - |\mathscr{B}_{\text{SAM}}|_{\text{DR}} & = R^{n-1} + (A - B) - R^{-k} = S^*
\end{array}
$$

The resulting sum will then be $S^* = R^{n-1} + (A - B) - R^{-k}$. It should be noted that the term R^{n-1} appears in the complement of $|\mathscr{B}_{\text{SAM}}|$ instead of R^n since we ignore the sign bit when adding magnitudes. Thus we are really adding unsigned positive numbers with $n - 1$ integer positions. Dividing S^* by R^{n-1} yields

$$
\frac{S^*}{R^{n-1}} = 1 + \frac{A - B - R^{-k}}{R^{n-1}} \qquad \text{if} \quad A > B \tag{2.4}
$$

$$
= 0 + \frac{R^{n-1} + (A - B) - R^{-k}}{R^{n-1}} \qquad \text{if} \quad B > A \tag{2.5}
$$

From these expressions, it is clear that if we obtain a carry out of the most significant position, we know that $A > B$ and that we actually subtracted the smaller number from the larger. The correct answer is $A - B$; however, the remainder is incorrect by the amount R^{-k}. This discrepancy can be remedied by again adding the end-around-carry to the direct result as was done with diminished-radix arithmetic. Since the carry out informed us that $A > B$ we now must attach the sign of the augend to the result.

It is also clear from expressions 2.4 and 2.5 that if $B \geq A$, no carry will occur out of the sign position. In this case we have in fact subtracted the larger number from the smaller. To remedy the error it is only necessary to recomplement the resulting answer. Thus if $B > A$, we obtain $S^* = R^{n-1} + A - B - R^{-k}$. Taking the diminished-radix complement yields

$$
S = R^{n-1} - S^* - R^{-k} = R^{n-1} - (R^{n-1} + A - B - R^{-k}) - R^{-k}.
$$

$S = B - A$. Since the lack of a carry out of the most significant position indicates that $B > A$, the sign of the addend must be attached to the recomplemented result. A few examples will help to illustrate this procedure.

Example 2.20

$$
\text{Add} \qquad \mathscr{A}_{\text{SAM}} = (0543)_8 \qquad \text{to} \qquad \mathscr{B}_{\text{SAM}} = (7546)_8
$$

Let us first add $|\mathscr{A}_{\text{SAM}}|$ to $-|\mathscr{B}_{\text{SAM}}|$

$$
\begin{aligned}
|\mathscr{A}_{\text{SAM}}|_{\text{DR}} &= (\ 543)_8 \\
-|\mathscr{B}_{\text{SAM}}|_{\text{DR}} &= (\ 231)_8 \\
\hline
\mathscr{S}^{*}_{\text{DR}} &= (0774)_8
\end{aligned}
$$

Since the carry out of the most significant position is zero, we know that $B > A$; and we must recomplement the result $(774)_8$, appending the negative sign of the addend.

$$\therefore \quad \mathscr{S}_{\text{SAM}} = (7003)_8$$

Example 2.21
Let us now switch these numbers and add them again.

$$\text{Add} \qquad \mathscr{A}_{\text{SAM}} = (7546)_8 \qquad \text{to} \qquad \mathscr{B}_{\text{SAM}} = (0543)_8$$

$$
\begin{aligned}
|\mathscr{A}_{\text{SAM}}|_{\text{DR}} &= (\ 546)_8 \\
-|\mathscr{B}_{\text{SAM}}|_{\text{DR}} &= (\ 234)_8 \\
\hline
\mathscr{S}^{*}_{\text{DR}} &= (1002)_8
\end{aligned}
$$

Since the carry out of the most significant position is a one and we know that $A > B$, we must now add back the end-around-carry and append to the result the negative sign of the augend.

$$
\begin{aligned}
\mathscr{S}^{*}_{\text{DR}} \quad & (002)_8 \\
\text{End-around-carry} \quad & (001)_8 \\
\hline
\mathscr{S}_{\text{DR}} = \ & 7(003)_8
\end{aligned}
$$

Sign of addend

$$\therefore \quad \mathscr{S}_{\text{SAM}} = \mathscr{A}_{\text{SAM}} + \mathscr{B}_{\text{SAM}} = (7003)_8$$

Before stating a formal set of rules for the addition of sign-and-magnitude numbers a few additional comments are needed. We are working with unsigned numbers in all of the arithmetic operations. The sign of the answer is separately determined. The rules for unsigned diminished-radix arithmetic can be extended to apply to cases of both like and unlike signed numbers. This will cause the addition of a one to the least significant position of like signed numbers should an overflow occur. However, if we take the attitude that one wrong answer is as bad as any other wrong answer no harm will result. The rule for determining the sign of the result for opposite signed numbers can also be used with like signed numbers. Lastly, the only reason diminished-radix arithmetic was selected over radix arithmetic was because a radix complement generally requires more complex hardware than does the diminished-radix complement.

RULE 2.5. Addition of two base-R numbers using sign-and-magnitude notation.

$$\mathcal{A}_{SAM} = (sa_{n-2}a_{n-3}\cdots a_1 a_0.a_{-1}\cdots a_{-k})_R$$
$$\mathcal{B}_{SAM} = (sb_{n-2}b_{n-3}\cdots b_1 b_0.b_{-1}\cdots b_{-k})_R$$

1. *Like Signed Numbers.* Add directly the magnitudes of the two numbers only. Do not add the sign indicators. Give the result the common sign. A carry out of the most significant position will indicate that overflow has occurred.

2. *Unlike Signed Numbers.* Add the diminished-radix complement of the addend (magnitude only) to the augend, adding all positions except the sign bit. If a carry is produced out of the most significant position, add one to the least significant position and give the result the sign of the augend. If no carry is produced out of the most significant position, take the diminished-radix complement of the result (magnitude only) and give the result the sign of the addend.

As was done with the other two arithmetic notations, complement arithmetic will be used for subtraction. Hence in order to subtract $\mathcal{B}_{SAM}$ from $\mathcal{A}_{SAM}$ we will first take the sign-and-magnitude complement of the subtrahend and then add the resulting number to the minuend.

RULE 2.6. Subtraction of two base-R numbers using sign-and-magnitude notation.

1. Form the sign-and-magnitude complement of the subtrahend.

2. Add the complemented subtrahend to the minuend using the rules for sign-and-magnitude addition.

Example 2.22 Subtract $(30112)_4$ from $(31212)_4$.

Direct method

$$\mathcal{A}_{SAM} = (31212)_4$$
$$\mathcal{B}_{SAM} = (30112)_4$$

$$\mathcal{A}_{SAM} - \mathcal{B}_{SAM} = (31100)_4$$

Complement method

$$\mathcal{A}_{SAM} = (31212)_4$$
$$+(-\mathcal{B}_{SAM}) = (00112)_4$$

Unlike signs

$$|\mathcal{A}_{SAM}|_{DR} = (1212)_4$$
$$-|-\mathcal{B}_{SAM}|_{DR} = (3221)_4$$

Since carry out = 1, add in end-around-carry and give the result the sign of the augend.

$$(11033)_4$$
$$\llcorner\!\!\rightarrow 1$$
$$(31100)_4$$
$$\therefore \quad \mathcal{A}_{SAM} - \mathcal{B}_{SAM} = (31100)_4$$

2.3 HARDWARE COMPLEXITY OF BASE-*R* ARITHMETIC

In this section a comparison of the three types of arithmetic will be presented. The comparison will be based on the hardware complexity of the electronic circuits required to implement a combined adder and subtracter using the rules given in the preceding section.

Since we have not as yet studied logical design techniques, a simplified building-block approach will be used to describe these arithmetic units. In Chapter 6, we will show how to logically design specific base-*R* building blocks using binary logic elements.

Before examining the overall structure of each of the three different arithmetic units, let us first define the operations which each of the subunits or building blocks will perform. A functional symbol will be given for each unit along with a word description of its operation. In subsequent chapters these functional units will be designed in considerable detail using binary logic units. In order to control the operation of many of these units, control input signals will be required. These control signals will be binary in nature with the signal state being described interchangeably by the pairs of words: (off, on), (inactive, active), (low, high), (false, true), (zero, one), (0, 1). The pair (0, 1) will be used whenever the control signals have been so designed as to allow their use as either a control input or as a numerical input having a value of 0 or 1.

BLOCK 1. Base-*R* Full Adder. The base-*R* full adder's functional symbol is shown in Fig. 2.1. Its purpose is to form the arithmetic sum of the augend input x, the addend input y, and the carry input z. The output of the full adder is a two-digit sum, cs, with carry c and sum s. The addend and augend inputs, x and y, and the sum output s can be any base-*R* number $0 \leq x, y, s, \leq R - 1$. The carry in, z, and the carry out, c, will be restricted to the values of 0 and 1, the only carry values possible. Due to the binary nature of the carry signals, the carry in input z

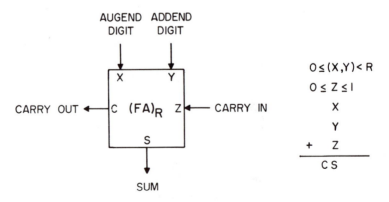

Fig. 2.1 Base-*R* Full-Adder Symbol

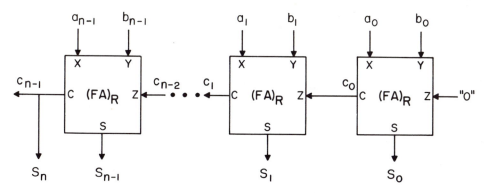

Fig. 2.2 Block Diagram of an *N*-Digit Unsigned Base-*R* Adder Using Radix Arithmetic

will accept control signal inputs. The carry out c will also be a binary signal and therefore it can also be used as a control signal.

By cascading together n-radix full adders, as shown in Fig. 2.2, an unsigned n-digit radix adder will be obtained. Since the carry into the ith stage is equal to the carry out of the preceding stage, the carries tend to ripple across the adder from right to left as they are determined. Because all of the addend and augend digits are also added simultaneously, this type of adder organization is commonly called a parallel-ripple-adder.

$$\mathcal{A}_R = (a_{n-1}a_{n-2} \cdots a_1 a_0)_R$$
$$\mathcal{B}_R = (b_{n-1}b_{n-2} \cdots b_1 b_0)_R$$
$$\overline{\phantom{\mathcal{S}_R = \mathcal{A}_R + \mathcal{B}_R = (s_n s_{n-1} s_{n-2} \cdots s_1 s_0)_R}}$$
$$\mathcal{S}_R = \mathcal{A}_R + \mathcal{B}_R = (s_n s_{n-1} s_{n-2} \cdots s_1 s_0)_R$$

BLOCK 2. Diminished Radix Complement Unit. The unit whose functional symbol is shown in Fig. 2.3 accepts any base-R digit as an input, x, and yields the

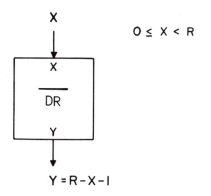

Fig. 2.3 Functional Symbol for a Diminished-Radix-Complement Unit

single output, y, the diminished-radix complement of input x. Thus

$$y = (R - x - 1)_R$$

with $0 \leq x < R$.

BLOCK 3. Controlled Diminished-Radix Complement Unit. This unit whose symbolic representation is shown in Fig. 2.4 differs only slightly from Block 2. A control input has been added to this unit. The diminished-radix complement of input x is performed only if the control input is activated, i.e., set equal to a one.

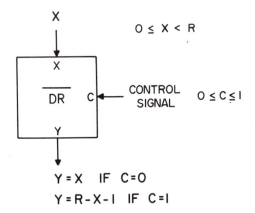

Fig. 2.4 Functional Symbol for a Controlled Diminished-Radix-Complement Unit

If the control input is not activated (set equal to a zero) then the input digit x is transmitted directly to the output y. Functionally this unit yields as an output y:

$$y = x \qquad\qquad \text{if} \quad c = 0\text{—Control line not active}$$
$$y = R - x - 1 \quad\text{if} \quad c = 1\text{—Control line active}$$

with $0 \leq x < R$.

The controlled diminished-radix complement unit is easily assembled using a diminished-radix complement unit and the electronic equivalent of a single-pole double-throw switch.

BLOCK 4. Controlled-Radix Complement Unit. Figure 2.5 shows the symbolic representation for a controlled radix complement unit. This unit is designed to operate as a typical stage of an N-unit cascade required to take the radix complement of an N-digit base-R number. If the complement is to be taken, the unit scans the number from right to left until a nonzero input digit (if one exists) is found. All the digits to its right are transferred unchanged to the outputs of their separate units. The radix complement of this rightmost nonzero digit is

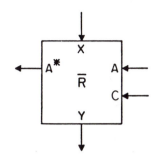

Fig. 2.5   Symbol for a Controlled-Radix-Complement Unit

passed to the output. The diminished-radix complement of all digits to the left of the first nonzero digit are passed directly to the outputs of their separate units. The control input signal c is used to determine if the complement is to be performed. Input A is used to signal a unit that at least one of the preceding digits was nonzero. The output signal A^* is set to a one if either the input signal A is nonzero or if the input x to the current stage is nonzero. The outputs y and A^* are determined as follows:

$y = x$	if	$c = 0$				
$y = 0$	if	$c = 1,$		$A = 0,$	and	$x = 0$
$y = R - x$	if	$c = 1,$		$A = 0,$	and	$x \neq 0$
$y = R - x - 1$	if	$c = 1,$	and	$A = 1$		
$A^* = 1$	if	$x \neq 0$	or	$A = 1$		
$A^* = 0$	if	$x = 0$	and	$A = 0$		

When N of these units are connected in a cascade to form the radix complement of an N-digit number, all of the c inputs are connected together. The A input of the rightmost unit is set equal to zero and the A^* output of each unit is connected to the A input of the unit on its left.

It should be pointed out that an alternate method for taking the radix complement of an N-digit number would be to take the diminished-radix complement of the number and then add 1 to the result. Thus one could also construct a radix complement unit by cascading together N controlled diminished-radix units with a common control line and adding $(1)_R$ to the result whenever the control line is equal to a one.

Let us now see how we can interconnect the preceding units to form useful arithmetic devices. Figures 2.6, 2.7, and 2.8 show basic block diagrams of radix, diminished-radix, and sign-and-magnitude addition and subtraction units. The control mode line M is set to a one for subtraction and zero for addition.

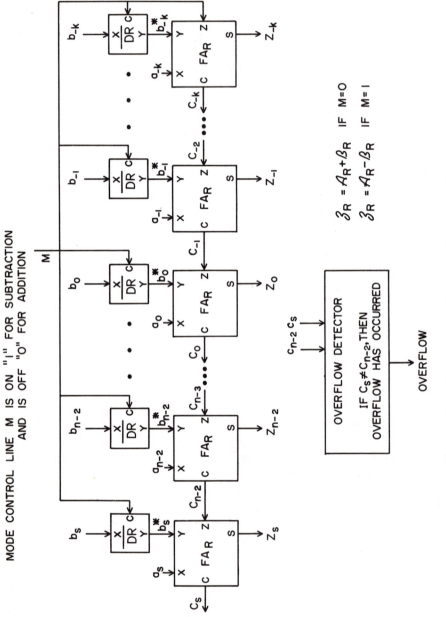

Fig. 2.6 Block Diagram of a Radix-Arithmetic Unit

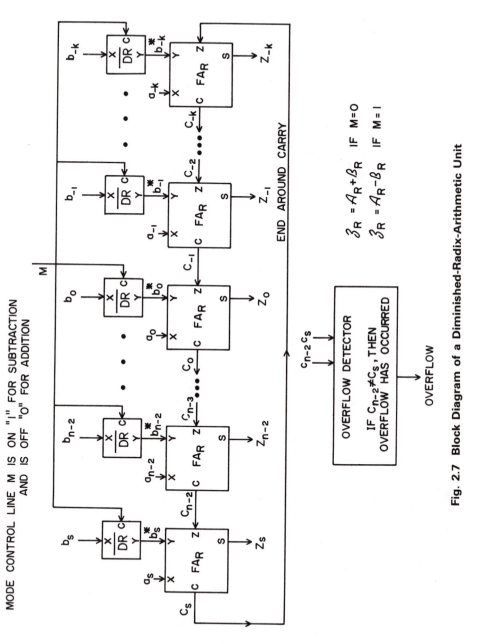

Fig. 2.7 Block Diagram of a Diminished-Radix-Arithmetic Unit

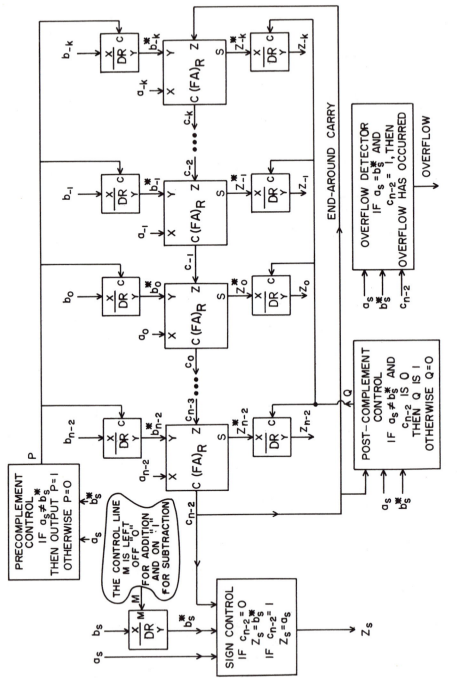

Fig. 2.8 Block Diagram of a Sign-and-Magnitude Adder-Subtracter

Examination of Fig. 2.6 reveals that with M equal to zero, the diminished-radix complement of $\mathcal{B}_R$ is not taken and the carry into the least significant position is also set to a zero. The resulting summation of $\mathcal{A}_R$ and $\mathcal{B}_R$ is then carried out using a cascaded set of full adders. An overflow logic block has been included to prevent errors from being passed undetected. When the mode control line M is set to a one for subtraction, the radix complement $\mathcal{B}_R$ is carried out by first taking the diminished-radix complement of $\mathcal{B}_R$ and then adding one to the result. The addition of one is accomplished by forcing the initial carry into the least significant position to be a one. The overflow detector can also be designed to use either of the alternate schemes mentioned in Section 2.2.1.

Examination of Fig. 2.7, the block diagram of the diminished-radix arithmetic unit, shows its resemblance to that of the radix adder-subtracter in Fig. 2.6. The only difference between the two block diagrams is the replacement of the mode control line M feeding the carry into the least significant position with the end-around-carry line, i.e., the carry out of the sign position.

The permanent attachment of this end-around-carry line on the diminished-radix arithmetic unit raises questions concerning the stability of the adder due to the feedback path. Assuming the carry out changes value at most once, this feedback line will not cause any oscillations to occur. The worst thing that can happen is that sometimes the wrong value of zero will result. This is only possible when adding two numbers $\mathcal{A}$ and $\mathcal{B}$ with $\mathcal{A} = -\mathcal{B}$, The expected answer is $(-0) \cdot \mathcal{A} + -\mathcal{B} = (A + R^n - A - R^{-k})_R = (R^n - R^{-k})_R$. However if c_s should be equal to one at the start of the addition process, the answer obtained will be $(R^n - R^{-k} + R^{-k})_R = R^n$, which is the number $+0$. The carry out c_s will also be held active, 1, preventing the answer from oscillating.

The diminished-radix adder-subtracter operates in much the same manner as the radix adder-subtracter. With the mode control line M equal to zero, the diminished-radix complement of $\mathcal{B}_R$ is not taken and the sum of $\mathcal{A}_R$ and $\mathcal{B}_R$ is calculated. With the mode control line M equal to one, the diminished-radix complement of $\mathcal{B}_R$ is taken and the result $\mathcal{B}_R^*$ added to $\mathcal{A}_R$ yielding the difference. The same overflow detector can be used with either radix or diminished-radix units. The astute reader will note that by providing another control line T, a dual arithmetic unit will result. If T is set to zero, radix mode, then the mode control line M is switched into the carry input of the least significant positions. If T is set to a one, diminished-radix mode, then the end around carry line c_s is switched into the carry input of the least significant position.

Finally, in Fig. 2.8, a block diagram for a sign-and-magnitude adder-subtracter is shown. Careful inspection of this figure reveals that a diminished-radix adder-subtracter is imbedded in this unit. The precomplement control unit determines whether addition or subtraction is to be performed. This adder-subtracter requires one less-full adder, since the sign of the result is separately determined. The unit, however, does require an additional set of diminished-radix complement units for post complementing the resulting sum whenever $|\mathcal{B}^*| > |\mathcal{A}|$

and $\mathcal{B}^*$ and $\mathcal{A}$ have opposite signs. Note that if an overflow should occur, the end-around-carry will be automatically added to the result. However, as was previously stated, one wrong answer is as bad as another wrong answer so that no harm is done.

Of the three units, the sign-and-magnitude adder unit requires the least amount of additional hardware to convert it into a adder-subtracter. Only a single controlled diminished-radix complement unit on the sign of $\mathcal{B}$ is required.

The reason for using a diminished-radix addition unit instead of a radix-complementing unit in the sign-and-magnitude unit can be seen by examining Fig. 2.8. Although the end-around-carry line could be replaced with a one for subtraction thus allowing for a radix precomplement; the post-complement operation will either require an additional row of full adders to add the extra one to the diminished radix complement or it will require the use of more complicated controlled radix complement units as shown in Fig. 2.5.

In terms of overall logic complexity, the radix full adder will be the most complex with the various complement and control units requiring relatively little hardware in comparison.

The time delay required to carry out the control functions and the complementing operations will at most take as long as the time delay required for one full adder unit to calculate its sum and carry. We will see shortly (in Chapter 7) that for $R > 2$ the time required will in fact be much less. An examination of the three block diagrams reveals that subtraction will take longer than addition. Using this crude time estimate, with all logic blocks and control units having equal time delays, upper bounds on the time required to calculate a sum or difference can be obtained. A summary of these estimates is presented in Table 2.1 along with a logic block count. In these calculations it was assumed that two m-digit signed numbers were being added or subtracted with $m = n + k$. The calculation time for the radix and diminished-radix units is based on the amount of time required

TABLE 2.1

Unit	Number of Full Adders	Number of Precomplement Boxes	Number of Post-complement Boxes	Number of Special Control Boxes	Total Unit Calculation Time
Radix	m	m	0	1	m + 2
Diminished Radix	m	m	0	1	m + 3
Sign and magnitude	m-1	m-1	m-1	5	m + 5

to pass the data through the precomplement boxes, 1 unit of delay, m full adder stages, and one additional unit of delay for the time required to verify the answer with the overflow box. The total calculation time is bounded by $m + 2$ units of delay for radix and $m + 3$ units for diminished radix due to the end around carry delay.

The worst case estimate for the sign-and-magnitude adder-subtracter is based on one unit of delay for complementing the sign of $\mathscr{B}$, one unit of delay for the precomplement control unit, one unit of delay for the precomplement boxes, m units of full adder delay including the end around carry delay, one unit of delay for the post-complement control box and one unit of delay for the post-complementing. The total calculation time is $m + 5$ units. The overflow delay is masked by the post-complement units delay time.

For a nontrivial value of m, it is clear that the total calculation time for these three units is roughly the same. In a later chapter we will discuss methods of speeding up the addition time by using parallel logic techniques as opposed to the serial ripple adders shown in Figs. 2.6 to 2.8.

Based on the small amount of hardware and calculation time differences between these three units, it appears that the user can take his choice without much penalty. However, there are two other factors to be considered before a final choice is made. The first is the use of these adder-subtracters as a basic building block in a multiplication and division unit. As we will see in Chapter 12 there will be no real design advantage to using any of these units provided the problem of multiplying $(-R^n)$ by itself using radix arithmetic is avoided. The last factor to be considered is the use of one of these arithmetic units as part of a sub-unit doing multiple-precision arithmetic calculations as is often done in computers. For example, we could add two $2m$ digit numbers by first adding the rightmost m digits saving any resulting carry and partial sum and then adding the leftmost m digits using the saved carry out as the carry in to the second addition pair.

Only slight modification would be necessary in order for the three adder-subtracters to provide this useful capability. However, in the latter two units, diminished-radix and sign-and-magnitude, the end-around-carry addition would have to be delayed until after the addition of the two left half parts of the number have been added. The end-around-carry would then have to be added requiring two more additions. Thus if multiple-precision arithmetic capability is required, the radix arithmetic unit will be twice as fast. Lastly, it should be clear that the location of the radix point has no effect on the final hardware design of any of these fixed-point arithmetic units.

2.4 PROBLEMS

2.1 Modify Algorithm 2.1 so that crutch addition can be used with signed as well as unsigned radix numbers.

   a) Assume radix notation.
   b) Assume diminished-radix notation.
   c) Assume sign-and-magnitude notation.

2.2 Using crutch addition, add the following unsigned positive numbers:

   a) $(ABC.42)_{16}$ b) $(721.65)_8$ c) $(110110.111)_2$
       $(F12.10)_{16}$ $(312.01)_8$ $(101101.100)_2$
       $\overline{\phantom{(ABC.42)_{16}}}$ $\overline{}$ $\overline{}$

2.3 Show that the three overflow tests listed on page 33 are equivalent.

2.4 Add the following numbers assuming Radix notation. Indicate the validity of the final
     result.

   a) $(11011)_2$ b) $(742.16)_8$ c) $(0AB.74)_{16}$
       $(01011)_2$ $(731.14)_8$ $(026.71)_{16}$
       $\overline{}$ $\overline{}$ $\overline{\phantom{(0AB.74)_{16}}}$

   d) $(0124.5)_8$ e) $(FA3.17)_{16}$ f) $(0101101)_2$
       $(0721.6)_8$ $(F72.16)_{16}$ $(0010011)_2$
       $\overline{}$ $\overline{\phantom{(FA3.17)_{16}}}$ $\overline{}$

   g) $(91245)_{10}$ h) $(4234.1)_5$ i) $(0123.4)_8$
       $(08617)_{10}$ $(0124.0)_5$ $(7654.4)_8$
       $\overline{\phantom{(91245)_{10}}}$ $\overline{}$ $\overline{}$

2.5 Repeat problem 2.4 assuming diminished-radix notation.

2.6 Repeat problem 2.4 assuming sign-and-magnitude notation.

2.7 Subtract the following numbers assuming radix notation. Indicate the validity of the
     final results.

   a) $(1011011)_2$ b) $(754321)_8$ c) $(951234)_{10}$
       $(0110111)_2$ $(712467)_8$ $(046178)_{10}$
       $\overline{}$ $\overline{}$ $\overline{\phantom{(951234)_{10}}}$

   d) $(0A21.6)_{16}$ e) $(2011.211)_3$ f) $(10111011)_2$
       $(FF12.3)_{16}$ $(2211.120)_3$ $(01010011)_2$
       $\overline{\phantom{(0A21.6)_{16}}}$ $\overline{}$ $\overline{}$

2.8 Repeat problem 2.7 assuming diminished-radix notation.

2.9 Repeat problem 2.7 assuming sign-and-magnitude notation.

2.10 Convert $A = (-15.75)_{10}$ and $B = (-21.125)_{10}$ into 12-bit signed binary numbers
     with 5 fractional positions. Determine the binary sum $A + B$ and the binary
     difference $A - B$.
     a) Assuming sign-and-magnitude notation base 2
     b) Assuming 2's complement notation
     c) Assuming 1's complement notation
     Hint: Reread Section 1.5.

2.11 Show that the feedback line in Fig. 2.7 will not cause an oscillation to occur.
     Hint: Consider $A = R^n - 1$, $B = 0$, $M = 0$ and c_{n-1} equal to either 0 or 1.

2.12 Modify the radix addition subtraction unit shown in Fig. 2.6 so that the unit will
     operate using $(T = 0)$ radix arithmetic or using $(T = 1)$ diminished-radix arithmetic.

2.5 BIBLIOGRAPHY

Cardenas, A. F., L. Presser, M. A. Marin, *Computer Science.* New York: Wiley, 1972.

Chu, Y., *Digital Computer Design Fundamentals.* New York: McGraw-Hill, 1962.

Flores, I., *The Logic of Computer Arithmetic.* Englewood Cliffs, N.J.: Prentice-Hall, 1963.

Knuth, D. E., *The Art of Computer Programming: Vol. 2, Seminumerical Algorithms.* Reading, Mass.: Addison-Wesley, 1969.

Richards, R. K., *Arithmetic Operations in Digital Computers.* New York: Van Nostrand Reinhold, 1956.

Chapter 3

Boolean Algebra

3.1 INTRODUCTION

Before we can begin to functionally design a digital subsystem, a short review of Boolean algebra and its applicability to logical design seems in order. In this chapter, the definitions and properties used to define a Boolean algebra will be presented in Sections 3.2 and 3.3. In Section 3.4 we will enumerate the number of Boolean functions of n variables and introduce a few shorthand notations for cataloging them. Sections 3.6 and 3.7 will introduce truth tables and K-maps as means of specifying Boolean functions. The concepts of functional duality and complementation will be introduced in Sections 3.8 and 3.9 along with a few comments on how these relationships might prove helpful in logic design.

Standard forms of representing Boolean functions will be introduced in Section 3.10. Section 3.11 will be devoted to techniques for quickly determining, by inspection, Boolean functions from truth tables or K-maps. Finally in Section 3.12, algorithms will be presented for transforming Boolean expressions from one standard form to another.

3.2 AN INTRODUCTION TO BOOLEAN ALGEBRA USING HUNTINGTON'S POSTULATES

In this section we will define a Boolean algebra based on Huntington's postulates, first proposed in 1904. The presentation will be intuitive rather than strictly rigorous. For more formal mathematical treatments, the reader is referred to the references listed at the end of this chapter. To understand the material in this section the reader needs to be familiar with the concept of a set and have some idea of the meanings of an equivalence relation and of the principle of substitution.

60

In simple terms an equivalence relation is some relation R defined on a set K which satisfies the following three basic properties:

a) Reflexive: For every x in the set K, the relationship xRx holds.

b) Symmetric: For every x and y in the set K, the relationship yRx holds whenever the relationship xRy holds.

c) Transitive: For every x, y, and z in the set K, if the relationships xRy and yRz hold, then the relationship xRz also holds.

It is easily seen that our intuitive idea of equality "=" satisfies these three properties, while the concept "<" (less than) does not, since it is neither reflexive nor symmetric. The relationship "is the parent of" fails to satisfy any of the three basic properties.

Whenever two expressions A and B are equivalent the law of substitution allows us to substitute the value of B for A in any expression involving A without affecting the validity of the resulting expression.

Definition 3.1 A *Boolean algebra* is a triplet $[k, +, \cdot]$ consisting of a finite set of elements K, subject to an equivalence relation "=" and two binary operators denoted "+" and "·", such that for every element x and $y \in K$, the operations $x + y$ and $x \cdot y$ are uniquely defined and postulates 1 through 6 are satisfied.

HUNTINGTON'S POSTULATES

P1. *The operations are closed.*
For all x and $y \in K$,
a) $x + y \in K$;
b) $x \cdot y \in K$.

P2. *For each operation there exists an identity element.*
a) There exists an element $0 \in K$ such that for all $x \in K$,

$$x + 0 = x;$$

b)   There exists an element $1 \in K$ such that for all $x \in K$,

$$x \cdot 1 = x.$$

P3. *The operations are commutative.*
For all x and $y \in K$,
a) $x + y = y + x$;
b) $x \cdot y = y \cdot x$.

P4. *The operations are distributive.*
For all x, y, and $z \in K$,
a) $x + (y \cdot z) = (x + y) \cdot (x + z)$;
b) $x \cdot (y + z) = (x \cdot y) + (x \cdot z)$.

P5. *For every element $x \in K$ there exists an element $\bar{x} \in K$ (called the complement of x) such that*

a) $x + \bar{x} = 1$

and

b) $x \cdot \bar{x} = 0$.

P6. *There exist at least two elements x and $y \in K$ such that $x \neq y$.*

Before we restrict our attention to the special case when the set K contains only two elements, let us establish a few additional properties characteristic of Boolean algebras. In terms of the basic postulates, we can recursively define a Boolean expression.

Definition 3.2 Letting x denote an arbitrary Boolean variable, then

$0, 1, x,$ and $\bar{x}$ are Boolean expressions.

If A and B are Boolean expressions, then

$\bar{A}, \bar{B}, A + B,$ and $A \cdot B$ are also Boolean expressions.

An examination of the basic postulates reveals that besides being given as pairs, the statements differ only by the simultaneous interchange of the operators "·" and "+" and the elements "0" and "1". *This special property is called duality.*

To help us establish the validity of a few of the more useful properties of a Boolean algebra, we will utilize the *principle of duality*. This principle states that if two expressions can be proven equivalent using a sequence of postulates, then the dual expressions can be proven to be equivalent by simply applying the sequence of dual postulates. In essence this means that for each Boolean property which we establish, the dual property will also hold without additional proof.

The commutative law, Postulate 3, is of fundamental importance in that it allows us to ignore the order in which the pair of arguments are presented. The reader is reminded that the commutative law does not hold for ordinary division, $a/b \neq b/a$ unless $a = b$. In general, matrix multiplication is also a noncommutative operation. We shall see shortly that a Boolean algebra also satisfies the associative law which will allow us to combine several variables using the same operators without regard to the arrangement of the variables.

In order to reduce the complexity of our proofs and also the statements of the various properties, it will be understood that the commutative law holds for all of the properties presented in the book without additional proof. Thus from Postulate 3,

$$x \cdot 0 = 0 \cdot x = 0$$

and

$$x + 1 = 1 + x = 1.$$

Property 3.1 The special law of zero and one. For all $x \in K$,
a) $x \cdot 0 = 0$;
b) $x + 1 = 1$.

Proof. Part a)

$$
\begin{aligned}
x \cdot 0 &= (x \cdot 0) + 0 && \text{Postulate 2a (identity)} \\
&= (x \cdot 0) + (x \cdot \bar{x}) && \text{Postulate 5b (law of complements)} \\
&= x \cdot (0 + \bar{x}) && \text{Postulate 4b (distributive law)} \\
&= x \cdot \bar{x} && \text{Postulate 2a (identity)} \\
&= 0. && \text{Postulate 5b (law of complements)}
\end{aligned}
$$

Part b) By simultaneously interchanging the operators $\cdot$ and $+$, and the elements 0 and 1, we can prove part b) by applying the principle of duality.

Original statement	Dual statement	Dual postulate
$x \cdot 0 = (x \cdot 0) + 0$	$x + 1 = (x + 1) \cdot 1$	2b
$= (x \cdot 0) + (x \cdot \bar{x})$	$= (x + 1) \cdot (x + \bar{x})$	5a
$= x \cdot (0 + \bar{x})$	$= x + (1 \cdot \bar{x})$	4a
$= x \cdot \bar{x}$	$= x + \bar{x}$	2b
$= 0.$	$= 1$	5a

The next three properties are included for the sake of completeness and also as illustrative examples of the type of reasoning required to verify the validity of Boolean expressions.

Property 3.2
a) The element 0 is unique;
b) The element 1 is unique.

Proof. Part a) Proof by contradiction. Let us assume there are two zero elements denoted 0_1 and 0_2. Postulate 2a states that

$$x + 0_1 = x$$

and

$$y + 0_2 = y.$$

Applying commutativity, postulate 3a, to the latter equation yields

$$x + 0_1 = x$$

and

$$0_2 + y = y.$$

Letting $x = 0_2$ and $y = 0_1$, we obtain

$$0_2 + 0_1 = 0_2$$

and

$$0_2 + 0_1 = 0_1.$$

Using the transistivity property of our equivalence relation we have $0_2 = 0_1$ contradicting our initial assumption. Therefore we must conclude that our assumption is incorrect and hence the element 0 is unique.

Part b) Apply the principle of duality.

Property 3.3

a) The complement of 0 is $\bar{0} = 1$;

b) The complement of 1 is $\bar{1} = 0$.

Proof. Part a)

$$x + 0 = x \qquad \text{Postulate 2a}$$

Thus,

$$\bar{0} + 0 = \bar{0}.$$

However,

$$\bar{0} + 0 = 1. \qquad \text{Postulate 5a}$$

Therefore

$$\bar{0} = 1.$$

Part b) Apply the principle of duality.

Property 3.4   Each element $x \in K$ has a unique complement denoted by $\bar{x}$.

Proof. Let us assume some element y has two complements $\bar{y}_1$ and $\bar{y}_2$. Then

$$
\begin{aligned}
\bar{y}_1 &= \bar{y}_1 + 0 && \text{Postulate 2a} \\
&= \bar{y}_1 + (y \cdot \bar{y}_2) && \text{Postulate 5b} \\
&= (\bar{y}_1 + y) \cdot (\bar{y}_1 + \bar{y}_2) && \text{Postulate 4a} \\
&= 1 \cdot (\bar{y}_1 + \bar{y}_2) && \text{Postulate 5a} \\
&= (\bar{y}_2 + y) \cdot (\bar{y}_2 + \bar{y}_1) && \text{Postulate 5a} \\
&= \bar{y}_2 + (y \cdot \bar{y}_1) && \text{Postulate 4a} \\
&= \bar{y}_2 + 0 && \text{Postulate 5b} \\
&= \bar{y}_2. && \text{Postulate 2a}
\end{aligned}
$$

Thus we have shown that $\bar{y}_1 = \bar{y}_2$, thereby contradicting our initial assumption that y has two complements. Therefore we must conclude that each element $x \in K$ has a unique complement $\bar{x}$.

The next three properties are particularly important in the area of logic design. Combined with Postulates 2 and 5, we will use these properties to remove redundancy from or add redundancy to our logic circuits.

Property 3.5 The idempotency law. For all $x \in K$,
a) $x + x = x$;
b) $x \cdot x = x$.

Proof. Part a)

$$
\begin{array}{ll}
x + x = (x + x) \cdot 1 & \text{Postulate 2b} \\
= (x + x) \cdot (x + \bar{x}) & \text{Postulate 5a} \\
= x + (x \cdot \bar{x}) & \text{Postulate 4a} \\
= x + 0 & \text{Postulate 5b} \\
= x & \text{Postulate 2a}
\end{array}
$$

Part b) Apply the principle of duality.

Property 3.6 First law of absorption. For all $x, y \in K$,
a) $x + (x \cdot y) = x$;
b) $x \cdot (x + y) = x$.

Proof. Part a)

$$
\begin{array}{ll}
x + (x \cdot y) = (x \cdot 1) + (x \cdot y) & \text{Postulate 2b} \\
= x \cdot (1 + y) & \text{Postulate 4b} \\
= x \cdot 1 & \text{Property 3.1b} \\
= x & \text{Postulate 2b}
\end{array}
$$

Part b) Apply the principle of duality.

Property 3.7 Second law of absorption. For all $x, y \in K$,
a) $x + (\bar{x} \cdot y) = x + y$
b) $x \cdot (\bar{x} + y) = x \cdot y$

Proof. Part a)

$$
\begin{array}{ll}
x + (\bar{x} \cdot y) = (x + \bar{x}) \cdot (x + y) & \text{Postulate 4a} \\
= 1 \cdot (x + y) & \text{Postulate 5a} \\
= x + y & \text{Postulate 2b}
\end{array}
$$

Part b) Apply the principle of duality.

The next two properties are included to provide a definite and simple test procedure for determining if two Boolean expressions are equivalents or complements of each other.

Property 3.8   The law of identity. For all x and $y \in K$, if

a) $x + y = y$

and also

b) $x \cdot y = y$,

then

$x = y$.

Proof. Substituting (b) into the left hand side of (a), we have

$$x + (x \cdot y) = y.$$

However, Property 3.6a states that

$$x + (x \cdot y) = x.$$

Therefore by transitivity, $x = y$.

Property 3.9   The law of complements. For all $x, y \in K$, if

a) $x + y = 1$

and

b) $x \cdot y = 0$,

then

$y = \bar{x}$

i.e., y is the complement of x.

Proof. This is nothing more than a restatement of Postulate 5 in a more convenient form. Postulate 5 states that the complement of x must satisfy the equations

$$x + \bar{x} = 1$$

and

$$x \cdot \bar{x} = 0.$$

The law of involution, Property 3.10, is included both for completeness and because it illustrates how we can make use of Property 3.8 when trying to determine if two expressions are equivalent.

Property 3.10 The law of involution. For all $x \in K$, $\bar{\bar{x}} = x$.

Proof. Let us show that property 3.8 holds, i.e.

$$\bar{\bar{x}} + x = \bar{x} \qquad \text{and that} \qquad \bar{\bar{x}} \cdot x = \bar{\bar{x}}.$$

$$\bar{x} = \bar{x} + 0 \qquad\qquad\qquad \text{Postulate 2a}$$
$$= \bar{x} + (x \cdot \bar{x}) \qquad\qquad \text{Postulate 5b}$$
$$= (\bar{x} + x) \cdot (\bar{x} + \bar{x}) \qquad \text{Postulate 4a}$$
$$= (\bar{x} + x) \cdot 1 \qquad\qquad \text{Postulate 5a}$$

Thus
$$\bar{x} = (\bar{x} + x). \qquad\qquad\qquad \text{Postulate 2b}$$

Also
$$\bar{x} = \bar{x} \cdot 1 \qquad\qquad\qquad\qquad \text{Postulate 2b}$$
$$= \bar{x} \cdot (x + \bar{x}) \qquad\qquad\; \text{Postulate 5a}$$
$$= (\bar{x} \cdot x) + (\bar{x} \cdot \bar{x}) \qquad \text{Postulate 4b}$$
$$= (\bar{x} \cdot x) + 0 \qquad\qquad\; \text{Postulate 5b}$$
$$= (\bar{x} \cdot x) \qquad\qquad\qquad \text{Postulate 2a}$$

Thus we have been able to show that

$$\bar{\bar{x}} + x = \bar{\bar{x}}$$

and that

$$\bar{\bar{x}} \cdot x = \bar{\bar{x}}.$$

Hence by Property 3.8

$$\bar{\bar{x}} = x.$$

Let us now show that the associative law holds for the operators defined in our Boolean algebra. Anticipating its need in the proof to follow, let us extend the law of absorption to handle a special three-variable expression.

Property 3.11 Special case of the law of absorption. For all x, y, $z \in K$,

a) $x \cdot ((x + y) + z) = x$;
b) $x + ((x \cdot y) \cdot z) = x$.

Proof. Part a)

$$x \cdot ((x + y) + z) = (x \cdot (x + y)) + (x \cdot z) \qquad \text{Postulate 4b}$$
$$= x + (x \cdot z) \qquad\qquad\qquad\qquad \text{Property 3.6b}$$
$$= x \qquad\qquad\qquad\qquad\qquad\quad\; \text{Property 3.6a}$$

Part b) Apply the principle of duality.

By evoking the commutative law it is also true that:

c) $((x + y) + z) \cdot x = x$

and

d) $((x \cdot y) \cdot z) + x = x$.

Property 3.12 The operators $+$ and $\cdot$ are associative. For all x, y, and $z \in K$,

a) $(x + y) + z = x + (y + z)$;
b) $(x \cdot y) \cdot z = x \cdot (y \cdot z)$.

Proof. Part a) Let $U = (x + y) + z$ and $V = x + (y + z)$. We will attempt to show that:

$$U \cdot V = V,$$

and also that:

$$U \cdot V = U.$$

This will allow us to evoke transitivity to show that $U = V$. Let us first show that $U \cdot V = V$.

$$((x + y) + z) \cdot (x + (y + z)) = (((x + y) + z) \cdot x) + (((x + y) + z) \cdot (y + z))$$

Postulate 4b

$$= x + (((x + y) + z) \cdot (y + z))$$

Property 3.11c

$$= x + ((((x + y) + z) \cdot y) + (((x + y) + z) \cdot z))$$

Postulate 4b

$$= x + (y + ((x + y) + z) \cdot z)$$

Property 3.11c

$$= x + (y + z) \qquad \text{Property 3.6b}$$

$$\therefore \quad U \cdot V = V.$$

To complete our proof we must also show that

$$U \cdot V = U.$$

$$((x + y) + z) \cdot (x + (y + z)) = ((x + y) \cdot (x + (y + z))) + (z \cdot ((x + (y + z)))$$

Postulate 4b

$$= ((x + y) \cdot (x + (y + z))) + z$$

Property 3.11a

$$= [(x \cdot (x + (y + z))) + (y \cdot (x + (y + z)))] + z$$

$$\begin{aligned}
&\text{Postulate 4b}\\
&= [((x \cdot (x + (y + z))) + y)] + z \quad\quad\\
&\hspace{6.5cm}\text{Property 3.11a}\\
&= (x + y) + z \quad\quad\quad\quad\quad\text{Property 3.6b}\\
\therefore\quad U \cdot V = U.&
\end{aligned}$$

The associative law simply states that when combining several variables using the *same* operator, the order in which the variables are combined is not important; i.e., the expression $x + y + z$ and $x \cdot y \cdot z$ are not ambiguous. However, the reader is cautioned not to be careless in the use of parenthesis. The expression $x \cdot y + z$ is ambiguous as written yielding two possible interpretations: $(x \cdot y) + z$ or $x \cdot (y + z)$ which are clearly not equivalent. If $x = y = 0$ and $z = 1$; then $(x \cdot y) + z = 1$, while $x \cdot (y + z) = 0$. It is common practice to give the operator $(\cdot)$ precedence over the operator $(+)$ and to omit the $(\cdot)$ altogether when combining variables. Using this convention the expression $xy + z$ must be interpreted as $((x \cdot y) + z)$.

Property 3.13 DeMorgan's law. For all $x, y \in K$,

a) $\overline{x + y} = \bar{x} \cdot \bar{y}$;

b) $\overline{x \cdot y} = \bar{x} + \bar{y}$.

Proof. Part a) The proof will utilize Property 3.9 and will be in two subparts. First we will show that $(x + y) \cdot (\bar{x} \cdot \bar{y}) = 0$ and then we will show that $(\bar{x} + \bar{y}) + (x \cdot y) = 1$; thereby showing that $x + y$ is the complement of $\bar{x} \cdot \bar{y}$; and hence that $\overline{x + y} = \bar{x} \cdot \bar{y}$.

     Subpart 1)

$$\begin{aligned}
(x + y) \cdot (\bar{x} \cdot \bar{y}) &= (x \cdot \bar{x} \cdot \bar{y}) + (y \cdot \bar{x} \cdot \bar{y}) \quad\quad &&\text{Postulate 4b,}\\
&&&\text{Property 3.12b}\\
&= 0 + 0 &&\text{Postulate 5b}\\
&= 0 &&\text{Postulate 2a}
\end{aligned}$$

     Subpart 2)

$$\begin{aligned}
(x + y) + \bar{x} \cdot \bar{y} &= (x + \bar{x} \cdot \bar{y}) + y \quad\quad &&\text{Property 3.12a}\\
&= x + \bar{y} + y &&\text{Property 3.7a}\\
&= x + 1 &&\text{Postulate 5a}\\
&= 1 &&\text{Property 3.1b}
\end{aligned}$$

     Part b) Apply the principle of duality.

     DeMorgan's Law is particularly important in that it shows us how to form the complement of expressions involving more than one variable.

3.3 THE BOOLEAN OPERATORS—AND, OR, AND NOT

In the preceding section no special restrictions were placed on the number of elements in the set K. However, when the set K is restricted to 2 elements, i.e., $K = \{0, 1\}$, then the resulting Boolean algebra is particularly suited for use as a switching algebra. Let us now formally define our two-element Boolean algebra by defining the basic operators "+" and "·".

Definition 3.3 A binary *variable* x is a two-valued quantity such that,

a) if $x \neq 1$, then $x = 0$;
b) if $x \neq 0$, then $x = 1$.

The analogy in switching terminology would be to use a variable to denote a switch such that,

a) if switch A is not open, then it is closed;
b) if switch A is not closed, then it is open.

Definition 3.4 The Boolean complement operator "‾" (overbar), also called NOT, is defined by the table below:

x	$\bar{x}$
0	1
1	0

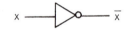

The complement operator logic symbol

Definition 3.5 The Boolean operators + (logical addition) and · (logical product) are defined by the tables below:

x	y	$x+y$
0	0	0
0	1	1
1	0	1
1	1	1

x	y	$x \cdot y$
0	0	0
0	1	0
1	0	0
1	1	1

+   ·

(OR)   (AND)

The logical OR symbol   The logical AND symbol

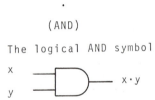

The logical addition operator (OR, +) yields a 1 any time either or both of its arguments are equal to one, while the logical product operator (AND, ·) yields a 1 only when both of its arguments are equal to one. The logical OR operator is also referred to as inclusive OR since it includes the case where both inputs are equal to one. We shall see in a later section that there is another logical OR operator, the eXclusive-OR operator (XOR, ⊕) which yields a one when either but not both arguments are equal to one.

A circuit realization from switching theory for these two operators is a pair of parallel single-pole single-throw switches for the OR operator and a pair of series single-pole single-throw switches for the AND operator.

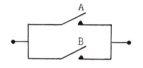

A	B	Resulting Circuit
open	open	open
open	closed	closed
closed	open	closed
closed	closed	closed

Switching circuit realization
for OR

A	B	Resulting Circuit
open	open	open
open	closed	open
closed	open	open
closed	closed	closed

Switching circuit realization
for AND

The NOT operator requires a double-pole single-throw switch.

A	$\bar{A}$
open	closed
closed	open

We must verify that the AND and OR operators satisfy the six basic postulates in order to be sure that these operators qualify as Boolean operators.

TABLE 3.1

x	y	x · y	y · x	x+y	y+x
0	0	0	0	0	0
0	1	0	0	1	1
1	0	0	0	1	1
1	1	1	1	1	1

Postulate 1 is satisfied since the operators are closed by definition. The identity elements 0 and 1 satisfy the conditions of Postulate 2 with

$$x + 0 = x \qquad x \cdot 1 = x$$
$$0 + 0 = 0 \quad \text{and} \quad 1 \cdot 1 = 1$$
$$1 + 0 = 1 \qquad 0 \cdot 1 = 0.$$

We can test the validity of Postulates 3 and 4 by means of *perfect induction*, i.e., we will evaluate the expressions for all possible inputs. From Table 3.1 it is clear that the commutative law holds.

The second distributive law is also valid for ordinary algebra, while the first distributive law is obviously invalid for ordinary algebra. Let us verify that the first distributive law is true by using the method of perfect induction; i.e., we will evaluate each expression separately using all possible combinations of input

TABLE 3.2

x	y	z	y·z	x+y·z	x+y	x+z	(x+y)(x+z)
				left side			right side
0	0	0	0	0	0	0	0
0	0	1	0	0	0	1	0
0	1	0	0	0	1	0	0
0	1	1	1	1	1	1	1
1	0	0	0	1	1	1	1
1	0	1	0	1	1	1	1
1	1	0	0	1	1	1	1
1	1	1	1	1	1	1	1

arguments and see if the two expressions agree in all cases. We will leave the verification of the second law as an exercise for the reader. (See Table 3.2.)

Since the two expressions agree for all possible arguments they are equal. Finally, Postulates 5 and 6 are satisfied by definition.

3.4 ENUMERATION OF LOGIC FUNCTIONS

In terms of the two-element Boolean algebra defined in the preceding section, we can define a Boolean function as follows.

Definition 3.6 A Boolean function of n variables, $x_1 \cdots x_n$, defined on the set $K = \{0, 1\}$, is an assignment of the value 0 or 1 to each of the possible 2^n combinations of the n variables.

The number 2^n arises because each of the n variables can only assume the values 0 or 1. A logical question at this point would be: How many distinct Boolean functions of n variables are there? Fortunately the number is quite easy to determine. Each of the n variables can be either a 0 or 1 yielding a total of 2^n distinct combinations. Each of these 2^n distinct combinations of n variables can be assigned the value 0 or 1 thereby yielding a total of $(2)^{2^n}$ distinct assignments.

Property 3.14 There are a total of $(2)^{2^n}$ distinct Boolean functions of n variables defined on the set $K = \{0, 1\}$. (See Table 3.3.)

A single variable Boolean function $f(x)$ can always be expressed in the form:

$$f(x) = f(0) \cdot \bar{x} + f(1) \cdot x \tag{3.1}$$

or in the form:

$$f(x) = (f(1) + \bar{x}) \cdot (f(0) + x). \tag{3.2}$$

TABLE 3.3
The Number of Functions of N Variables

n	2^n	2^{2^n}
1	2	4
2	4	16
3	8	256
4	16	65,536
5	32	$\approx 4.295 \times 10^9$
10	1024	$\approx 1.787 \times 10^{308}$

These expressions can be easily verified by means of perfect induction. Thus we have:

$$f(x) = f(0) \cdot \bar{x} + f(1) \cdot x \qquad f(x) = (f(1) + \bar{x}) \cdot (f(0) + x)$$
$$f(0) = f(0) \cdot 1 + f(1) \cdot 0 \qquad f(0) = (f(1) + 1) \cdot (f(0) + 0)$$
$$= f(0) \qquad\qquad\qquad\qquad = f(0)$$
$$f(1) = f(0) \cdot 0 + f(1) \cdot 1 \qquad f(1) = (f(1) + 0) \cdot (f(0) + 1)$$
$$= f(1) \qquad\qquad\qquad\qquad = f(1)$$

In a similar manner it can be shown that any two variable functions can be expressed in the forms:

$$f(x_1, x_0) = f(0,0)\bar{x}_1\bar{x}_0 + f(0,1)\bar{x}_1 x_0 + f(1,0)x_1\bar{x}_0 + f(1,1)x_1 x_0 \qquad (3.3)$$

or

$$f(x_1, x_0) = (f(1,1) + \bar{x}_1 + \bar{x}_0)(f(1,0) + \bar{x}_1 + x_0)$$
$$\cdot (f(0,1) + x_1 + \bar{x}_0)(f(0,0) + x_1 + x_0). \qquad (3.4)$$

Before we generalize these expressions, let us introduce a simplifying notation.

Definition 3.7 A single nonzero Boolean term which consists of the logical AND of some or all of the binary variables or their complements from an n-variable Boolean algebra will be called *a product term*. Product terms which contain all of the variables are also referred to as *canonic product terms, fundamental product terms,* or *minterms.*

The latter is most frequently used in connection with minimization methods which will be discussed in Chapter 5. For the three-variable case, expressions such as $x_2 x_1$, $\bar{x}_1 x_0$, and $\bar{x}_2\bar{x}_0$ are simple product terms; while product terms such as $x_2 x_1 \bar{x}_0$ and $\bar{x}_2\bar{x}_1 x_0$ are canonic product terms, fundamental product terms, or simply minterms. I prefer the name fundamental product terms.

There are n variables to be considered, each of which can appear in a product in two ways, complemented or uncomplemented.

There are two ways of picking the first variable, two ways of picking the next variable, ..., and two ways of picking the nth variable for a total of $(2)^n$ different ways. Thus we have shown that *there are 2^n distinct fundamental product terms in an n-variable Boolean algebra.*

For the three-variable case with variables x_2, x_1, x_0, the eight fundamental product terms or minterms are $\bar{x}_2\bar{x}_1\bar{x}_0$, $\bar{x}_2\bar{x}_1 x_0$, $\bar{x}_2 x_1\bar{x}_0$, $\bar{x}_2 x_1 x_0$, $x_2\bar{x}_1\bar{x}_0$, $x_2\bar{x}_1 x_0$, $x_2 x_1\bar{x}_0$, and $x_2 x_1 x_0$.

In order to be able to specify an arbitrary fundamental product term in our equations, we find the following notation convenient. Every fundamental product can be written as:

$$P_I^n = x_{n-1}^{i_{n-1}} \cdots x_2^{i_2} x_1^{i_1} x_0^{i_0}$$

where

$$x_j^{i_j} = \begin{cases} x_j & \text{if} & i_j = 1 \\ \bar{x}_j & \text{if} & i_j = 0 \end{cases} \tag{3.5}$$

n indicates the number of Boolean variables and

$$(I)_{10} = (i_{n-1} \cdots i_2 i_1 i_0)_2.$$

Using this notation we can express an arbitrary two-variable Boolean function in the Sum-of-Product form.

$$f(x_1, x_0) = a_0 \bar{x}_1 \bar{x}_0 + a_1 \bar{x}_1 x_0 + a_2 x_1 \bar{x}_0 + a_3 x_1 x_0 \tag{3.6}$$

or in the shortened form

$$f(x_1, x_0) = \bigvee_{I=0}^{3} a_I P_I^2. \tag{3.7}$$

The Boolean constants a_I are either 0 or 1 depending upon whether the corresponding fundamental product term P_I^2 is to be excluded from or included in the functions. The symbol $\bigvee$ will denote logical summation as opposed to $\sum$ which has the connotation of real summation.

Comparing Eqs. 3.3 and 3.6 we see that if

$$(I)_{10} = (i_1 i_0)_2, \quad \text{then} \quad f(i_1, i_0) = a_I.$$

The preceding result for 2 variables is easily generalized to functions of n variables. In the general case, there will be 2^n fundamental product terms and an arbitrary function of n variables can be written as

$$f(x_{n-1}, x_{n-2}, \ldots, x_1, x_0) = \bigvee_{J=0}^{2^n-1} a_J \cdot P_J^n. \tag{3.8}$$

Using Eq. 3.8, arbitrary Boolean functions can be specified as a set, A_n, consisting of the nonzero coefficients a_I's which determine the fundamental product terms, P_I^n's, of the function. The subscript n is used to indicate the number of independent binary variables in the function.

Example 3.1 Let $A_3 = [a_1, a_3, a_4, a_7]$. Write out the function in fundamental sum-of-product form.

The subscript 3 indicates the function has three binary variables. Therefore

$$\begin{aligned}
f(x_2, x_1, x_0) &= \bigvee_{J=0}^{7} a_J P_J^3 = a_1 \cdot P_1^3 + a_3 \cdot P_3^3 + a_4 \cdot P_4^3 + a_7 \cdot P_7^3 \\
&= P_1^3 + P_3^3 + P_4^3 + P_7^3 \\
&= x_2^0 x_1^0 x_0^1 + x_2^0 x_1^1 x_0^1 + x_2^1 x_1^0 x_0^0 + x_2^1 x_1^1 x_0^1 \\
&= \bar{x}_2 \bar{x}_1 x_0 + \bar{x}_2 x_1 x_0 + x_2 \bar{x}_1 \bar{x}_0 + x_2 x_1 x_0.
\end{aligned}$$

The function in Example 3.1 could also have been specified as $f(x_2, x_1, x_0) = \vee (a_1, a_4, a_3, a_7)$, the symbol $\vee$ being used to denote the logical OR operation of the fundamental product terms whose coefficients appear in the parenthesis.

Definition 3.8 A single nonunit Boolean term which consists of the logical OR of some or all of the binary variables or their complements from an n variable Boolean algebra will be called a *sum term*. Sum terms which contain all of the variables are called canonic sum terms, fundamental sum terms, or maxterms.

The latter is used frequently in connection with Boolean minimization procedures. Three-variable expressions such as $(x_1 + x_0)$, $(x_1 + x_2)$, and x_1 would be considered simple sum terms. Terms like $(x_2 + x_1 + \bar{x}_0)$ and $(\bar{x}_2 + x_1 + \bar{x}_0)$ would be considered as fundamental sum terms, or maxterms. As was the case for fundamental product terms there are a total of 2^n fundamental sum terms.

Using the notation developed for product terms, every fundamental sum can be written as

$$S_I^n = x_{n-1}^{i_{n-1}} + \cdots + x_1^{i_1} + x_0^{i_0} \tag{3.9}$$

where

$$x_j^{i_j} = \begin{cases} x_j & \text{if} & i_j = 1 \\ \bar{x}_j & \text{if} & i_j = 0 \end{cases}$$

n indicates the number of Boolean variables and

$$(I)_{10} = (i_{n-1} \cdots i_2 i_1 i_0)_2.$$

Using this notation we can express an arbitrary Boolean function in product-of-sum form as

$$f(x_1, x_0) = (a_3 + \bar{x}_1 + \bar{x}_0)(a_2 + \bar{x}_1 + x_0)(a_1 + x_1 + \bar{x}_0)(a_0 + x_1 + x_0)$$

or in the shortened form

$$f(x_1, x_0) = \bigwedge_{I=0}^{3} (a_{3-I} + S_I^2) = \bigwedge_{I=0}^{3} (a_I + S_{3-I}^2). \tag{3.10}$$

The symbol $\wedge$ will be used to denote the logical AND operation. The coefficients a_I are either zero or one. A zero coefficient indicates that the sum term will be included, while a coefficient of 1 will yield a unit sum term.

Comparing Eqs. 3.4 and 3.10 it is clear that if

$$(I)_{10} = (i_1 i_0)_2,$$

then

$$f(i_1, i_0) = a_I.$$

The preceding result for 2 variables can easily be generalized to functions of n

variables. In the general case there will be a total of 2^n fundamental sum terms. Therefore an arbitrary function of n variables can be written as

$$f(x_{n-1}, x_{n-2}, \ldots, x_1, x_0) = \bigwedge_{I=0}^{2^n-1} (a_I + S_{2^n-1-I}^n). \tag{3.11}$$

Using Eq. 3.11, as a guide, an arbitrary Boolean function can be specified as a set, $\bar{A}_n$, consisting of the zero coefficients, a_I's, which yield the nonunit fundamental sum terms, $S_{2^n-1-I}^n$, of the function.

Example 3.2 Let $\bar{A}_3 = [a_0, a_1, a_5, a_7]$. Write out the function in fundamental product-of-sum form.

The subscript 3 indicates the function has three binary variables. Therefore:

$$f(x_2, x_1, x_0) = \bigwedge_{I=0}^{7} (a_I + S_{7-I}^3)$$

Since the set $\bar{A}_3$ indicates the zero coefficients

$$
\begin{aligned}
f(x_2, x_1, x_0) &= (a_0 + S_7^3)(a_1 + S_6^3)(a_5 + S_2^3)(a_7 + S_0^3) \\
&= (S_7^3)(S_6^3)(S_2^3)(S_0^3) \\
&= (x_2^1 + x_1^1 + x_0^1)(x_2^1 + x_1^1 + x_0^0)(x_2^0 + x_1^1 + x_0^0)(x_2^0 + x_1^0 + x_0^0) \\
&= (x_2 + x_1 + x_0)(x_2 + x_1 + \bar{x}_0)(\bar{x}_2 + x_1 + \bar{x}_0)(\bar{x}_2 + \bar{x}_1 + \bar{x}_0).
\end{aligned}
$$

The reader should expand this last expression and verify that it yields the fundamental product form:

$$
\begin{aligned}
f(x_2, x_1, x_0) &= \bar{x}_2 x_1 \bar{x}_0 + \bar{x}_2 x_1 x_0 + x_2 \bar{x}_1 \bar{x}_0 + x_2 x_1 \bar{x}_0 \\
&= \bigvee [a_2, a_3, a_4, a_6].
\end{aligned}
$$

Thus

$$A_3 = [a_2, a_3, a_4, a_6]$$

is the set of nonzero coefficients used to specify the Boolean function.

3.5 CHARACTERISTIC NUMBERS

The shorthand notation presented in the preceding section can be carried one more step by noting that the coefficients, the a_i's, can only be zero or one. Hence, these coefficients can be written as a binary number,

$$(CN_n)_2 = (a_{2^n-1} a_{2^n-2} \cdots a_2 a_1 a_0)_2. \tag{3.12}$$

This binary number can in turn be converted into a decimal or even a hexadecimal number to reduce its length. As a decimal number

$$(CN_n)_{10} = \sum_{k=0}^{2^n-1} a_k \cdot 2^k. \tag{3.13}$$

This number is referred to as the *characteristic number* of the Boolean function. It is much easier to specify a Boolean function by its characteristic number than to write it out in expanded form. In order to avoid the use of double subscript notation, we will use B_n, O_n, D_n, and H_n to denote the characteristic number as being expressed in binary, octal, decimal, or hexadecimal form, respectively.

Example 3.3 The characteristic number of a Boolean function is $D_3 = 42$. Write out the corresponding function in sum-of-product form.

We must first convert 42 to binary adding two leading zeroes to yield $2^3 = 8$ binary digits.

$$(42)_{10} = (0 \quad 0 \quad 1 \quad 0 \quad 1 \quad 0 \quad 1 \quad 0)_2$$
$$= (a_7 \quad a_6 \quad a_5 \quad a_4 \quad a_3 \quad a_2 \quad a_1 \quad a_0)_2$$

Thus

$$A_3 = [a_5, a_3, a_1]$$

and therefore

$$f(x_2, x_1, x_0) = \bigvee (a_5, a_3, a_1)$$
$$= x_2^1 x_1^0 x_0^1 + x_2^0 x_1^1 x_0^1 + x_2^0 x_1^0 x_0^1$$
$$= x_2 \bar{x}_1 x_0 + \bar{x}_2 x_1 x_0 + \bar{x}_2 \bar{x}_1 x_0.$$

This fundamental sum-of-product expression can be reduced to

$$f(x_2, x_1, x_0) = x_2 \bar{x}_1 x_0 + \bar{x}_2 x_1 x_0 + \bar{x}_2 \bar{x}_1 x_0.$$

with $\bar{x}_1 x_0$ connecting the 1st and 3rd terms and $\bar{x}_2 x_0$ connecting the 2nd and 3rd terms.

The idempotency law allows us to use the term $\bar{x}_2 \bar{x}_1 x_0$ twice. Using the law of absorption $xy + xy = y$ on the 1st and 3rd terms and on the 2nd and 3rd terms yields

$$f(x_2, x_1, x_0) = \bar{x}_1 x_0 + \bar{x}_2 x_0.$$

Example 3.4 If the characteristic number of a Boolean function is expressed as $H_2 = E$, then write out the corresponding function in product-of-sum form.

Converting the hexadecimal number into binary, we obtain

$$H_2 = (E)_{16} = (1110)_2 = (a_3 a_2 a_1 a_0)_2.$$

Thus

$$A_2 = (a_3, a_2, a_1) \quad \text{and} \quad \bar{A}_2 = (a_0).$$

Therefore using the set $\bar{A}_2$ we have

$$f(x_1, x_0) = \bigwedge_{I=0}^{3} (a_I + S_{3-I}^2) = S_3^2,$$

$$f(x_1, x_0) = x_1^1 + x_0^1 = x_1 + x_0.$$

In terms of the characteristic numbers it is easy to tabulate the Boolean functions of n variables. Tables 3.4 and 3.5 list the one- and two-variable Boolean functions.

Given a Boolean function of n variables, there are several methods which we can use to obtain a characteristic number. The coefficients a_I's can always be obtained by evaluating the function with

$$a_I = f(i_{n-1}, i_{n-2}, \ldots, i_1, i_0),$$

$$(I)_{10} = (i_{n-1} i_{n-2} \cdots i_1 i_0)_2.$$

Once the 2^n coefficients have been obtained, the characteristic number can be calculated using the number conversion algorithms described in Chapter 1.

Example 3.5 Find the characteristic number for the Boolean function

$$f(x_2, x_1, x_0) = x_2 x_1 + \bar{x}_2 x_0.$$

Since there are 3 variables, we will have to determine $2^3 = 8$ coefficients. These can be determined by evaluating the function

$$a_I = f(i_2, i_1, i_0) \qquad (I)_{10} = (i_2 i_1 i_0)_2$$
$$a_0 = f(0, 0, 0) = 0 \qquad a_4 = f(1, 0, 0) = 0$$
$$a_1 = f(0, 0, 1) = 1 \qquad a_5 = f(1, 0, 1) = 0$$
$$a_2 = f(0, 1, 0) = 0 \qquad a_6 = f(1, 1, 0) = 1$$
$$a_3 = f(0, 1, 1) = 1 \qquad a_7 = f(1, 1, 1) = 1$$

These coefficients, when arranged in ascending order, yield the binary value of the characteristic number

$$B_3 = (a_7 a_6 a_5 a_4 a_3 a_2 a_1 a_0)_2$$

$$= (1\ 1\ 0\ 0\ 1\ 0\ 1\ 0)_2.$$

The decimal characteristic number for the function is $D_3 = 202$ while the hexidecimal characteristic number is $H_3 = CA$.

An alternate method for calculating the characteristic number is to expand the function until it is in the form of logical sums of fundamental product terms. A formal procedure for accomplishing this expansion will be given in Section 3.10.

TABLE 3.4
Single-Variable Boolean Functions

k	a_1	a_0	$D_1 = k = f(x_0)$
0	0	0	0
1	0	1	$\overline{x}_0$
2	1	0	x_0
3	1	1	1

TABLE 3.5
Two-Variable Boolean Functions

k	a_3	a_2	a_1	a_0	$f(x_1, x_0) = D_2 = k$
0	0	0	0	0	0
1	0	0	0	1	$\overline{x}_1 \overline{x}_0$
2	0	0	1	0	$\overline{x}_1 x_0$
3	0	0	1	1	$\overline{x}_1$
4	0	1	0	0	$x_1 \overline{x}_0$
5	0	1	0	1	$\overline{x}_0$
6	0	1	1	0	$x_1 \overline{x}_0 + \overline{x}_1 x_0$
7	0	1	1	1	$\overline{x}_1 + \overline{x}_0$
8	1	0	0	0	$x_1 x_0$
9	1	0	0	1	$x_1 x_0 + \overline{x}_1 \overline{x}_0$
10	1	0	1	0	x_0
11	1	0	1	1	$\overline{x}_1 + x_0$
12	1	1	0	0	x_1
13	1	1	0	1	$x_1 + \overline{x}_0$
14	1	1	1	0	$x_1 + x_0$
15	1	1	1	1	1

The procedure is based on the fact that a variable x_i which is missing from a product term can be inserted by multiplying the product term by $(x_i + \bar{x}_i)$. The laws of idempotency are then utilized to reduce the duplicate terms which might result from the multiplication process.

Example 3.6 Find the characteristic number for

$$f(x_2, x_1, x_0) = x_2 + x_1 x_0.$$

Using the alternate procedure, we can expand $f(x_2, x_1, x_0)$ into the desired form by multiplying the first product term by $(x_1 + \bar{x}_1)$ and by $(x_0 + \bar{x}_0)$ thereby yielding four fundamental product terms. The last term needs only to be multiplied by $(x_2 + \bar{x}_2)$ to obtain an additional pair of fundamental product terms. Thus we have

$$f(x_2, x_1, x_0) = x_2(x_1 + \bar{x}_1)(x_0 + \bar{x}_0) + (x_2 + \bar{x}_2)x_1 x_0$$

$$= x_2 x_1 x_0 + x_2 x_1 \bar{x}_0 + x_2 \bar{x}_1 x_0 + x_2 \bar{x}_1 \bar{x}_0$$

$$+ \; x_2 x_1 x_0 + \bar{x}_2 x_1 x_0.$$

The law of idempotency can be used to eliminate the second product term $x_2 x_1 x_0$.

$$f(x_2, x_1, x_0) = x_2 x_1 x_0 + x_2 x_1 \bar{x}_0 + x_2 \bar{x}_1 x_0 + x_2 \bar{x}_1 \bar{x}_0 + \bar{x}_2 x_1 x_0$$

$$= x_2^1 x_1^1 x_0^1 + x_2^1 x_1^1 x_0^0 + x_2^1 x_1^0 x_0^1 + x_2^1 x_1^0 x_0^0 + x_2^0 x_1^1 x_0^1$$

$$= a_7 \cdot P_7^3 + a_6 \cdot P_6^3 + a_5 \cdot P_5^3 + a_4 \cdot P_4^3 + a_3 \cdot P_3^3.$$

Thus we have ascertained which coefficients are equal to one and the characteristic number is therefore:

$$B_3 = (a_7 a_6 a_5 a_4 a_3 a_2 a_1 a_0)_2$$

$$= (1\ 1\ 1\ 1\ 1\ 0\ 0\ 0)_2$$

$$D_3 = (248)_{10} \quad \text{and} \quad H_3 = F8.$$

3.6 TRUTH TABLE SPECIFICATION OF BOOLEAN FUNCTIONS

A truth table is commonly used for specifying a Boolean function. All of the input variables are listed at the top of the table along with the function output. For each possible set of input values, the function output value for these values is specified as being either a zero or a one. It should be noted that if one associates each input row with a binary number and converts this number to its decimal equivalent i, then the output function's value is a_i, the coefficient for the ith fundamental product term. The complete output function is obtained by ORing all of the fundamental product terms together which yield an output of one and hence is written in fundamental sum-of-product form. Table 3.6 is a three-variable Truth Table.

TABLE 3.6
Three-Variable Truth Table

x_2	x_1	x_0	$f(x_2,x_1,x_0)$	fundamental product term	coefficient a_i	i
0	0	0	0	$\bar{x}_2\bar{x}_1\bar{x}_0$	0	0
0	0	1	1	$\bar{x}_2\bar{x}_1 x_0$	1	1
0	1	0	1	$\bar{x}_2 x_1\bar{x}_0$	1	2
0	1	1	0	$\bar{x}_2 x_1 x_0$	0	3
1	0	0	0	$x_2\bar{x}_1\bar{x}_0$	0	4
1	0	1	1	$x_2\bar{x}_1 x_0$	1	5
1	1	0	1	$x_2 x_1\bar{x}_0$	1	6
1	1	1	0	$x_2 x_1 x_0$	0	7

$$f(x_2,x_1,x_0) = \bar{x}_2\bar{x}_1 x_0 + \bar{x}_2 x_1\bar{x}_0 + x_2\bar{x}_1 x_0 + x_2 x_1\bar{x}_0$$

$$= V[a_1,a_2,a_5,a_6]$$

$$B_3 = (01100110)_2 \quad \text{and} \quad D_3 = (102)_{10}$$

A major disadvantage to using truth tables for specifying Boolean functions is the size of the table required. For two, three, and perhaps even four variables the table is easy to construct. However, because the number of lines in the table is growing at the rate of 2^n, for more than about four variables, truth table specification of Boolean functions is time consuming, uneconomical in terms of space, and also subject to error due to the large number of entries required. A major advantage of the truth table method for specifying Boolean functions is the relative ease in which the function can be obtained from the table. For, unlike the characteristic-number specification method, no intermediate calculations are required in order to determine the function.

3.7 SPECIFICATION OF BOOLEAN FUNCTIONS USING MAPS

There is still another method for specifying Boolean functions. This method, the map method, requires much less space than the truth table and requires only slightly more effort to determine the Boolean function represented. In this section

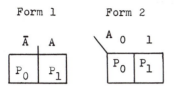

(a) 1 variable maps f(A)

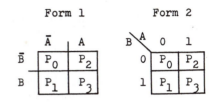

(b) 2 variable maps f(A,B)

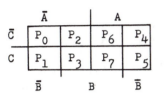

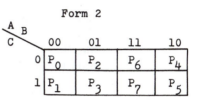

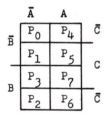

(c) 3 variable maps f(A,B,C)

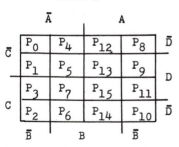

(d) 4 variable maps f(A,B,C,D)

Fig. 3.1 Karnaugh Maps (*K*-Maps) for Several Variables. (a) 1-Variable Maps—*f(A)*; (b) 2-Variable Maps—*f(A, B)*; (c) 3-Variable Maps—*f(A, B, C)*; (d) 4-Variable Maps—*f(A, B, C, D)*

we will show how maps are used to specify Boolean functions. In Chapter 5, we will show how these maps can also be used to simplify or minimize the amount of hardware required to implement Boolean functions.

Basically a map is a set of 2^n squares arranged in an ordered two-dimensional array.

Each square on the map corresponds to one of the 2^n possible fundamental product terms P_i^n. The most common maps are the Karnaugh maps which we will examine in this section. The Karnaugh or K-Maps for functions of 1, 2, 3, and 4 variables are shown in Fig. 3.1. The two forms for each map are equivalent, only the manner in which the labels are placed along the side of the maps differs. We will use both forms interchangeably in this book. You should adopt the form you find the most comfortable for your own use.

In order to determine the fundamental product term P_i specified by a given square, it is only necessary to examine the labels along the sides of the grid. Thus the square in the lower right hand corner of the 4-variable map, (Fig. 3.1), lies at the intersection of the labels A, $\bar{B}$, C, and $\bar{D}$ and thus corresponds to the fundamental product term $A\bar{B}C\bar{D} = P_{10}$. Because a single square represents the minimum amount of area which a single fundamental product term can specify; fundamental product terms are often referred to as *minterms*.

A closer examination of the structure of a K-map reveals that the labels were not just randomly placed along the sides. Instead they were placed in such a way that the fundamental product terms on any pair of adjacent squares differ by only one literal.* For the 3- and 4-variable maps, the top and bottom and right and left sides are also considered adjacent to one another.

When writing the output function for a Boolean function in sum of product form, any time a pair of adjacent squares are both equal to 1, only a single product term having $n - 1$ variables needs to be included. The variable which appears barred in one square and unbarred in the other is omitted. The reverse also holds, i.e., an $n - 1$ variable product term will cover $2^1 = 2$ squares on an n variable K-map.

Example 3.7 Write the output function for the K-map shown below.

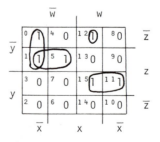

* A literal is any binary variable or the complement of a binary variable.

Term 12 cannot be combined. However terms (0 and 1), (1 and 5), and (11 and 15) can be combined with the variable z, x and x being eliminated from each pair of products respectively. Thus

$$f(w, x, y, z) = wx\bar{y}\bar{z} + \bar{w}\bar{x}\bar{y} + \bar{w}\bar{y}z + wyz.$$

Further inspection of the K-map in Fig. 3.2 reveals that it is possible to combine four overlapping pairs of squares into a group of four squares and eliminate two variables from each of the four product terms. Not all possible groups of four squares can be combined. In fact for a 3-variable map only six groupings are legal. An examination of Fig. 3.3 reveals that only product terms listed below can be

	$\bar{A}$		A		
$\bar{C}$	$\bar{A}\bar{B}\bar{C}\bar{D}$ 0000 0	$\bar{A}B\bar{C}\bar{D}$ 0100 4	$AB\bar{C}\bar{D}$ 1100 12	$A\bar{B}\bar{C}\bar{D}$ 1000 8	$\bar{D}$
	$\bar{A}\bar{B}\bar{C}D$ 0001 1	$\bar{A}B\bar{C}D$ 0101 5	$AB\bar{C}D$ 1101 13	$A\bar{B}\bar{C}D$ 1001 9	D
C	$\bar{A}\bar{B}CD$ 0011 3	$\bar{A}BCD$ 0111 7	$ABCD$ 1111 15	$A\bar{B}CD$ 1011 11	
	$\bar{A}\bar{B}C\bar{D}$ 0010 2	$\bar{A}BC\bar{D}$ 0110 6	$ABC\bar{D}$ 1110 14	$A\bar{B}C\bar{D}$ 1010 10	$\bar{D}$

       $\bar{B}$ B $\bar{B}$

Fig. 3.2 4-Variable K-Map

	$\bar{A}$		A	
$\bar{C}$	$\bar{A}\bar{B}\bar{C}$ 000 0	$\bar{A}B\bar{C}$ 010 2	$AB\bar{C}$ 110 6	$A\bar{B}\bar{C}$ 100 4
C	$\bar{A}\bar{B}C$ 001 1	$\bar{A}BC$ 011 3	ABC 111 7	$A\bar{B}C$ 101 5

       $\bar{B}$ B $\bar{B}$

Fig. 3.3 3-Variable K-Map

combined. The reader should locate these terms on the K-map in Fig. 3.3.

$$[P_0, P_1, P_2, P_3] \quad \bar{A} \qquad [P_2, P_3, P_6, P_7] \quad B$$
$$[P_4, P_5, P_6, P_7] \quad A \qquad [P_0, P_2, P_6, P_4] \quad \bar{C}$$
$$[P_0, P_1, P_4, P_5] \quad \bar{B} \qquad [P_1, P_3, P_7, P_5] \quad C$$

On a 4-variable map, 24 groups of four squares can be found corresponding to the 24 product terms.

$\bar{w}\bar{x}$	$\bar{w}\bar{y}$	$\bar{w}\bar{z}$	$\bar{x}\bar{y}$	$\bar{x}\bar{z}$	$\bar{y}\bar{z}$
$\bar{w}x$	$\bar{w}y$	$\bar{w}z$	$\bar{x}y$	$\bar{x}z$	$\bar{y}z$
$w\bar{x}$	$w\bar{y}$	$w\bar{z}$	$x\bar{y}$	$x\bar{z}$	$y\bar{z}$
wx	wy	wz	xy	xz	yz

The reader should locate these terms on a 4-variable K-map. This idea of grouping adjacent squares in order to eliminate the number of product terms in a function can be extended to include groups of 2^i squares with $i \le n$. Each such grouping eliminates i variables from each of the fundamental product terms to yield a single $(n - i)$-variable product term consisting of only the product of those literals common to each of the fundamental product terms. The number of product terms with i variables, $i \le n$ is

$$2^i \binom{n}{i} = 2^i \frac{n!}{(n-i)!\, i!} \qquad 0 \le i \le n. \tag{3.14}$$

This number can be reasoned out as follows. Consider that we can pick the first literal in $2n$ ways, the second in $2(n - 1)$ ways, and the ith in $2(n - i + 1)$ ways. However, in doing so, each product will be duplicated many times with only the literals appearing in a different order. It is well known that there are $i!$ different ways of arranging i objects. Thus we have the number product term with i variables equal to $[2(n)(2)(n-1)\cdots 2(n-i+1)]/i!$. Equation 3.14 is just a simple way to express this number.

Example 3.8 Write out the output function for the K-map below.

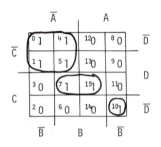

For this K-map the squares [0-1-4-5] can be combined to yield the term $\bar{A}\bar{C}$; the squares [7, 15] yield the term BCD and the single square 10 the term $A\bar{B}C\bar{D}$. Therefore

$$f(A, B, C, D) = \bar{A}\bar{C} + BCD + A\bar{B}C\bar{D}.$$

Each product term is obtained by examining the labels on the side of the map and only including those common to each square in the group. To prevent errors caused by using an invalid grouping, it is also suggested that the reader remember that a grouping of 2^i squares can only eliminate i variables. For example, if one were to attempt to group the four terms along the diagonal in Example 3.8, there would be no common terms. This is obviously an illegal combination since a group of four should only have eliminated two variables. Since normally the product terms are not numbered on the maps, form 1 (Fig. 3.1) of the K-maps is somewhat easier to use when a Boolean function expressed as a sum of products is to be placed on a map.

Form two is easier to use if the function is specified as a set of a_i numbers.

Example 3.9 Map $f(A, B, C) = \bar{A} + \bar{B}C + B\bar{C}$ on a 3 variable map.

Solution. Each term is mapped separately on the K-map with 1's placed in all squares which satisfy the logical expression and 0's in the remaining squares. In this example, separate maps are used for each term and then the three maps are logically ORed together to yield the final map. As your skill develops with practice, only the final map will need to be drawn.

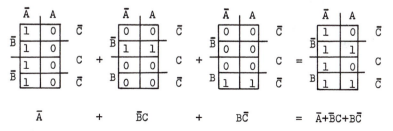

Let us consider what happens if we map a single fundamental sum term on a K-map.

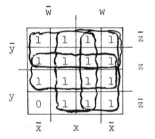

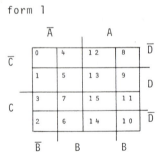

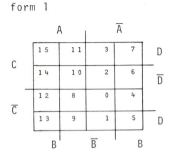

(a)

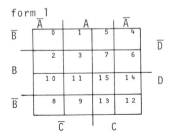

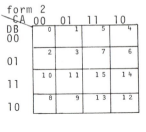

(b)

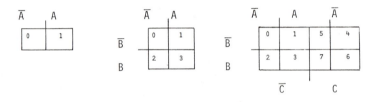

(c)

Fig. 3.4 Alternate K-Maps. (a) Standard Map—f(A, B, C, D); (b) Variation 1—f(A, B, C, D); (c) Variation 2—f(D, C, B, A).

Fig. 3.5 Mahoney Maps of Several Variables

A mapping of $f(w, x, y, z) = (w + x + \bar{y} + z) = S_{13}^4$ reveals that only the square associated with the product term P_2^4 is zero. In general, a single fundamental sum term $S_{2^n-1-i}^n$ will fill all squares on a K-map except the square associated with P_i^n. For this reason, fundamental sum terms are also referred to as *maxterms* since they occupy the maximum amount of space on a K-map.

Example 3.10 Is $AB + \bar{A}C + BC = AB + \bar{A}C$?

After mapping both functions we can tell by inspection if the two expressions are equal. Since the two maps are identical, the expressions are equal.

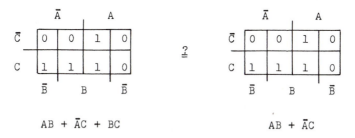

AB + $\bar{A}$C + BC   AB + $\bar{A}$C

The reader should be aware of the fact that the Karnaugh Map can be labeled in a variety of alternative ways. Two such alternative forms are shown in Fig. 3.4 for the 4-variable case along with the standard 4-variable map for comparison purposes.

Variation 1 is nothing more than a complementing of the labels on the maps. Variation 2, sometimes referred to as a Mahoney Map, has the advantage of imbedding the lower order K-maps without changing their position. This can be seen by examining the form-1 maps of 1-, 2-, and 3-variables shown in Fig. 3.5. The map is unfolded first to the right and then down to form each map of higher order. Form 1 of this map is easy to use, however form 2 leaves much to be desired. Its strong point is the ease with which an additional variable can be added, or deleted without necessitating the redrawing of the map.

To save time when using K-maps, make a master sheet containing several copies of the K-maps. These sheets can then be reproduced either by xerox or ditto to obtain work sheets.

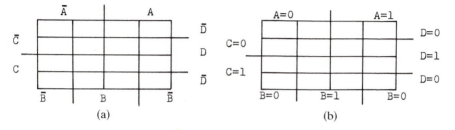

Fig. 3.6 (a) Shorthand Map; (b) Actual Map Specified

Do not misinterpret the shorthand notation used to label form 1 of the K-maps. The labels A and $\bar{A}$ mean $A = 1$ and $A = 0$ respectively and are not to be mistaken for variables.

3.8 THE PRINCIPLE OF DUALITY

In this section we will expand upon the principle of duality which holds for Boolean algebra. Using this principle we will show how DeMorgan's laws can be generalized to cover expressions of more than two variables.

Definition 3.9 Two Boolean $\begin{Bmatrix} \text{functions} \\ \text{expressions} \\ \text{tables} \\ \text{maps} \end{Bmatrix}$ will be called *duals* if they differ only by the simultaneous interchange of AND for OR, and "0" for "1." The following examples illustrate this principle.

Example 3.11 Law of Absorption.

$$x + xy = x. \qquad \text{Dual is} \qquad x(x + y) = x.$$

Example 3.12 Dual logic tables.

Table for "OR"

x	y	f(x,y)
0	0	0
0	1	1
1	0	1
1	1	1

$$f(x,y) = x + y$$

Dual table "AND"

x	y	f(x,y)
1	1	1
1	0	0
0	1	0
0	0	0

$$f(x,y) = x \cdot y$$

Example 3.13 Dual symbols.

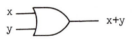

OR gate

dual symbol AND gate

Example 3.14 Dual K-Maps.

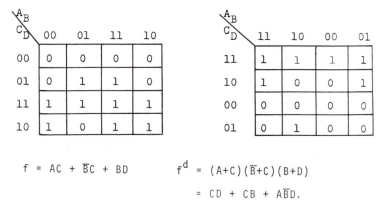

$$f = AC + \bar{B}C + BD$$

$$f^d = (A{+}C)(\bar{B}{+}C)(B{+}D)$$

$$= CD + CB + A\bar{B}D.$$

Definition 3.10 Axioms of Duality.

$$A(1) \quad 1^d = 0 \qquad 0^d = 1;$$
$$A(2) \quad x^d = x \qquad \bar{x}^d = \bar{x};$$

$A(3)$ if A, B, and C are Boolean expressions where

     a) $C = A + B$ then $C^d = A^d \cdot B^d$;
     b) $C = A \cdot B$ then $C^d = A^d + B^d$;
     c) $C = \bar{B}$ then $C^d = \overline{B^d} = (\bar{B}^d)$.

Let us examine axiom 2 since this axiom seems to give students the most trouble. The function $f(x) = x$ has the following truth table:

x	f(x)
1	1
0	0

If we now apply the principle of duality to this truth table, interchanging 0's for 1's and 1's for 0's, we obtain the table below which yields

$$f^d(x) = \bar{x}f^d(0) + xf^d(1) = x.$$

x	$f^d(x)$
0	0
1	1

Let us determine the dual of the function:

$$f(x_2, x_1, x_0) = \bar{x}_2 x_0 + \bar{x}_1 x_0$$

Using the axioms of duality, we have

$$f^d(x_2, x_1, x_0) = (\bar{x}_2 x_0)^d \cdot (\bar{x}_1 x_0)^d$$
$$= (\bar{x}_2 + x_0)(\bar{x}_1 + x_0)$$
$$= \bar{x}_2\bar{x}_1 + \bar{x}_2 x_0 + \bar{x}_1 x_0 + x_0$$
$$= \bar{x}_2\bar{x}_1 + x_0.$$

Noting that $f(x_2, x_1, x_0) = \bar{x}_2 x_0 + \bar{x}_1 x_0 = (\bar{x}_2 + \bar{x}_1)x_0$, by applying the axioms of duality to the second form, we obtain directly

$$f(x_2, x_1, x_0) = (\bar{x}_2 + \bar{x}_1)x_0$$
$$f^d(x_2, x_1, x_0) = (\bar{x}_2 + \bar{x}_1)^d + (x_0)^d$$
$$= \bar{x}_2 \cdot \bar{x}_1 + x_0.$$

The truth table for $f(x_2, x_1, x_0) = \bar{x}_2 x_0 + \bar{x}_1 x_0$ is shown in Fig. 3.7. The truth table for the dual function is shown in Fig. 3.8. Note that by using the dual symbol in a logic diagram or dual element in a circuit realization (Fig. 3.9), the dual function is realized directly.

We shall see in a later section how the principle of duality will aid us in our logic designs. Using the series form for representing a Boolean function:

$$f(x_{n-1}, \ldots, x_1, x_0) = \bigvee_{j=0}^{2^n-1} a_j P_j^n$$

x_2	x_1	x_0	$f(x_2, x_1, x_0)$
0	0	0	0
0	0	1	1
0	1	0	0
0	1	1	1
1	0	0	0
1	0	1	1
1	1	0	0
1	1	1	0

K-Map

$x_1 x_0$ \ x_2	0	1
00	0	0
01	1	1
11	1	0
10	0	0

Figure 3.7

x_2	x_1	x_0	$f^d(x_2, x_1, x_0)$
1	1	1	1
1	1	0	0
1	0	1	1
1	0	0	0
0	1	1	1
0	1	0	0
0	0	1	1
0	0	0	1

Dual K-Map

$x_1 x_0$ \ x_2	1	0
11	1	1
10	0	0
00	0	1
01	1	1

$$f^d(x_2, x_1, x_0) = \bar{x}_2 \bar{x}_1 + x_0$$

Figure 3.8

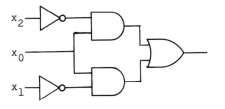

$$\bar{x}_2 x_0 + \bar{x}_1 x_0$$

(a)

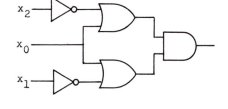

$$(\bar{x}_2 + x_0)(\bar{x}_1 + x_0)$$

(b)

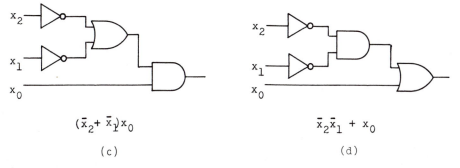

$$(\bar{x}_2 + \bar{x}_1) x_0$$

(c)

$$\bar{x}_2 \bar{x}_1 + x_0$$

(d)

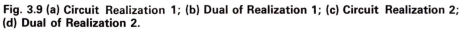

Fig. 3.9 (a) Circuit Realization 1; (b) Dual of Realization 1; (c) Circuit Realization 2; (d) Dual of Realization 2.

the dual function becomes:

$$f^d(x_{n-1}, \ldots, x_1, x_0) = \bigwedge_{j=0}^{2^n-1} (\bar{a}_j + [P_j^n]^d) = \bigwedge_{j=0}^{2^n-1} (\bar{a}_j + S_j^n) \qquad (3.15)$$

where the symbol $\bigwedge$ is used to denote a logical product and S_j^n is a fundamental sum term.

$$S_I^n = (x_{n-1}^{i_{n-1}} + x_{n-2}^{i_{n-2}} + \cdots + x_1^{i_1} + x_0^{i_0})$$

$$= \bigvee_{j=0}^{n-1} x_j^{i_j} \qquad (3.16)$$

with

$$I = (i_{n-1}i_{n-2} \cdots i_1 i_0)_2. \qquad (3.17)$$

The latter form of equation 3.15 for f^d is expressed as the product of sums rather than as the sum of products. Let us see if we can convert the dual function into a sum-of-product form. One way of doing this would be to expand the series; however, this will be messy and unnecessary. We can reason out the final result as follows: if $f(i_{n-1}, i_{n-2}, \ldots, i_1, i_0) = a_I$, then taking the dual of this expression yields:

$$f^d(i_{n-1}, i_{n-2}, \cdots, i_1, i_0) = f(\bar{i}_{n-1}, \bar{i}_{n-2}, \ldots, \bar{i}_1, \bar{i}_0) = \bar{a}_I$$

Remember i_j is a constant not a variable. Remember also that $(\bar{i}_{n-1}\bar{i}_{n-2} \cdots \bar{i}_1 \bar{i}_0)_2 = (2^n - 1 - I)_{10}$. Thus we see that in the dual function, $\bar{a}_I$ is the coefficient of the fundamental product term $P_{2^n-1-I}^n$. This simple argument yields the conclusion that

$$f^d(x_{n-1}, x_{n-2}, \cdots, x_1, x_0) = \bigvee_{I=0}^{2^n-1} \bar{a}_I \, P_{2^n-1-I}^n \qquad (3.18)$$

$$= \bigvee_{I=0}^{2^n-1} \bar{a}_{2^n-1-I} P_I^n. \qquad (3.19)$$

This also leads to the result that

$$\bigvee_{I=0}^{2^n-1} \bar{a}_{2^n-1-I} P_I^n = \bigwedge_{I=0}^{2^n-1} (\bar{a}_I + S_I^n). \qquad (3.20)$$

Equation 3.18 in essence says that if $f(x_{n-1}, \ldots, x_1, x_0)$ is described by the characteristic number $B_n = (a_{2^n-1}a_{2^n-2} \cdots a_2 a_1 a_0)_2$ then $B_n^d = (\bar{a}_0 \bar{a}_1 \bar{a}_2 \cdots \bar{a}_{2^n-2} \bar{a}_{2^n-1})$ is the characteristic number for the dual function.

Example 3.15 $f(x, y) = x + y$ has the characteristic number $D_2 = 14$. What is the characteristic number of $f^d(x, y)$?

Solution. Write out the characteristic number in binary form. $(a_3 a_2 a_1 a_0)_2 = (1110)_2$. Reverse the order of the coefficients $(a_0 a_1 a_2 a_3)_2 = (0111)_2$. Now complement each coefficient. $(\bar{a}_0 \bar{a}_1 \bar{a}_2 \bar{a}_3)_2 = (1000)_2$. Finally convert this number back to decimal $D_2^d = 8$ and $f^d(x, y) = x \cdot y$.

3.9 FUNCTION COMPLEMENTATION

In this section, we will show how to form the complement of a Boolean function. Before doing so we will need the definition of a literal.

Definition 3.11 A *literal* is any binary variable or the complement of a binary variable. A function of n variables has $2n$ distinct literals.

Definition 3.12 Shannon's Generalized DeMorgan's Law.
   The complement of any Boolean expression or function can be formed by complementing all *literals* in the *dual* expression. Thus,

$$\bar{f}(x_{n-1}, \ldots, x_2, x_1, x_0) = f^d(\bar{x}_{n-1}, \ldots, \bar{x}_2, \bar{x}_1, \bar{x}_0).$$

The reader at this point should be cautioned that 0 and 1 are constants and not literals. Let us verify this simple procedure by using Property 3.9 of a Boolean algebra which states that if $f + g = 1$ and $f \cdot g = 0$, then $f = \bar{g}$, that is,

if

$$f(x_{n-1}, x_{n-2}, \ldots, x_1, x_0) = \bigvee_{J=0}^{2^n-1} a_J P_J^n$$

then

$$g(x_{n-1}, x_{n-2}, \ldots, x_1, x_0) = \bigvee_{J=0}^{2^n-1} \bar{a}_J P_J^n \qquad (3.21)$$

is its complement.

$$\bigvee_{J=0}^{2^n-1} a_J P_J^n + \bigvee_{J=0}^{2^n-1} \bar{a}_J P_J^n = \bigvee_{J=0}^{2^n-1} P_J^n = 1.$$

Since

$$a_J P_J^n \cdot \bar{a}_K P_K^n = 0$$

for all J and K

$$\bigvee_{J=0}^{2^n-1} a_J P_J^n \cdot \bigvee_{K=0}^{2^n-1} \bar{a}_K P_K^n = 0.$$

Hence, Boolean Property 3.9 is satisfied.
   Complementing the literals in a Boolean expression is equivalent to replacing each fundamental product P_J^n with $P_{2^n-1-J}^n$. $2^n - 1 - J$ is the diminished-radix complement of J with Radix 2^n. Thus if

$$f(x_{n-1}, \ldots, x_1, x_0) = \bigvee_{J=0}^{2^n-1} a_J P_J^n,$$

then

$$f(\bar{x}_{n-1}, \ldots, \bar{x}_1, \bar{x}_0) = \bigvee_{J=0}^{2^n-1} a_J P_{2^n-1-J}^n. \qquad (3.22)$$

From Eqs. 3.15 and 3.18 the dual function can be written as:

$$f^d(x_{n-1}, \ldots, x_1, x_0) = \bigwedge_{J=0}^{2^n-1} (\bar{a}_j + S_j^n) = \bigvee_{J=0}^{2^n-1} \bar{a}_J P_{2^n-1-J}^n$$

Complementing the literals by using Eq. 3.22 yields

$$f^d(\bar{x}_{n-1}, \ldots, \bar{x}_1, \bar{x}_0) = \bigvee_{I=0}^{2^n-1} \bar{a}_I P_I^n$$

$$= \bar{f}(x_{n-1}, \ldots, x_1, x_0).$$

Thus we have shown that the complement of a function can be obtained by complementing the literals in the dual function.

Example 3.16

$$f(x_2, x_1, x_0) = (\bar{x}_2 + \bar{x}_1)x_0$$
$$f^d(x_2, x_1, x_0) = (\bar{x}_2 \cdot \bar{x}_1) + x_0$$
$$\bar{f}(x_2, x_1, x_0) = f^d(\bar{x}_2, \bar{x}_1, \bar{x}_0) = (x_2 \cdot x_1) + \bar{x}_0$$

It should be noted again that if one is given a truth table for a Boolean function, then the truth table for the complement function is obtained by *only* complementing the function column

$$f(x_{n-1}, \ldots, x_1, x_0).$$

Example 3.17   Given $f(x_1, x_0) = \bar{x}_1 x_0$ find $\bar{f}(x_1, x_0)$

x_1	x_0	$f(x_1,x_0)$
0	0	0
0	1	1
1	0	0
1	1	0

$$f(x_1,x_0) = \bar{x}_1 x_0$$

x_1	x_0	$\bar{f}(x_1,x_0)$
0	0	1
0	1	0
1	0	1
1	1	1

$$\bar{f}(x_1,x_0) = x_1 + \bar{x}_0$$

x_1	x_0	$f^d(x_1,x_0)$
1	1	1
1	0	0
0	1	1
0	0	1

$$f^d(x_1,x_0) = \bar{x}_1 + x_0$$

$\bar{x}_1$	$\bar{x}_0$	$f^d(\bar{x}_1,\bar{x}_0)$
0	0	1
0	1	0
1	0	1
1	1	1

$$\bar{f}(x_1,x_0) = f^d(\bar{x}_1,\bar{x}_0)$$
$$= x_1 + \bar{x}_0$$

Given a circuit realization for a Boolean function, one has a choice of either adding an inverter to the output of the existing function or of replacing the existing function with its dual and complementing each literal (input lines in this case). Both yield the same result, however one realization may require more circuit elements. Even for those cases where both the true and complements of the variables are available for use as input lines, changing the input variables to their complement values may lead to overloading of some of the input lines. Also, as we shall see a little later, the dual gates may not be available in which case either a new circuit realization for the complement must be formed or an inverter added.

Example 3.18 Find the circuits which realize the complement of circuit (a).

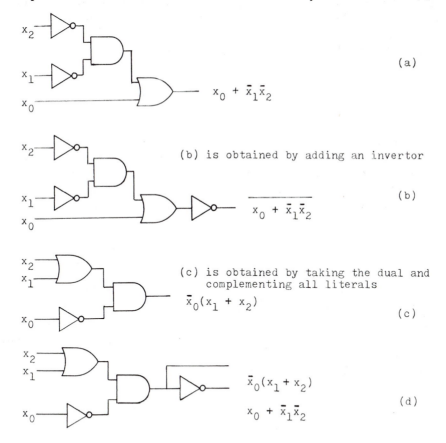

(a)

$x_0 + \bar{x}_1\bar{x}_2$

(b) is obtained by adding an invertor

(b)

$\overline{x_0 + \bar{x}_1\bar{x}_2}$

(c) is obtained by taking the dual and complementing all literals

$\bar{x}_0(x_1 + x_2)$

(c)

$\bar{x}_0(x_1 + x_2)$

$x_0 + \bar{x}_1\bar{x}_2$

(d)

It should be noted that circuit (d) realizes both the function and its complement and uses the same number of elements as circuit (a), which realizes only the function itself.

3.10 STANDARD FORMS FOR BOOLEAN FUNCTIONS

In this section we will discuss the difference between functions expressed in fundamental and nonfundamental forms. Rules will be given for converting a function in fundamental sum-of-product form into fundamental product-of-sum form. A set of rules will also be given for quickly converting between the two forms. Finally a set of rules will be given for writing a function by inspection, in either sum-of-product or product-of-sum form, from either a truth table or a K-map.

Definition 3.13 A Boolean function which is written as logical products ORed together will be referred to as being in *sum-of-product* form (SP). If all of the product terms are fundamental product terms (minterms) then the expression will be referred to as being in *fundamental-sum-of-product* form (FSP). A function in fundamental-sum-of-product form is expressed as

$$f(x_{n-1}, x_{n-2}, \ldots, x_1, x_0) = \bigvee_{I=0}^{2^n-1} a_I P_I^n$$

Definition 3.14 A Boolean function which is written as logical sums ANDed together will be referred to as being in *product-of-sum* form (PS). If all of the sum terms are fundamental sum terms (maxterms) then the expression will be referred to as being in *fundamental-product-of-sum* form (FPS). A function in fundamental-product-of-sum form is expressed as

$$f(x_{n-1}, x_{n-2}, \ldots, x_1, x_0) = \bigwedge_{I=0}^{2^n-1} (a_I + S_{2^n-1-I}^n)$$

Definition 3.15 Two Boolean functions will be called *equal* if they have the same fundamental form, i.e., they both have the same characteristic number, CN_n.

In general, it is difficult to determine by inspection if the sum-of-product form of two Boolean functions are equal unless they happen to be in fundamental form.

For example, let us consider the following two functions:

$$f_1(x, y, z) = xy + xz + yz$$
$$f_2(x, y, z) = xy + x\bar{y}z + \bar{x}yz.$$

By inspection, it is not obvious that these two functions are equal. We can expand these functions into fundamental-sum-of-product form by first multiplying each product term by $(b + \bar{b}) = 1$, where b is a variable not in the product term, and then eliminating the duplicate terms by using the idempotency law.

$$f_1(x, y, z) = xy(z + \bar{z}) + x(y + \bar{y})z + (x + \bar{x})y\bar{z}$$
$$= xyz + xy\bar{z} + xyz + x\bar{y}z + xyz + \bar{x}yz$$
$$= xyz + xy\bar{z} + x\bar{y}z + \bar{x}yz$$
$$f_2(x, y, z) = xy(z + \bar{z}) + x\bar{y}z + \bar{x}yz$$
$$= xyz + xy\bar{z} + x\bar{y}z + \bar{x}yz$$

Since both of these functions have the same fundamental form, they are equal.

Rather than expanding the functions into fundamental form, the two functions can also be mapped onto K-maps.

Example 3.19 Is $f_1(x, y, z) = xy + \bar{x}\bar{y} + xz$ equal to $f_2(x, y, z) = xy + yz + \bar{x}\bar{z}$? Let us map both functions.

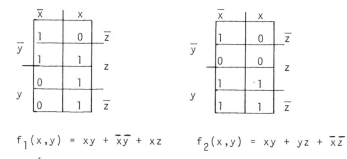

$$f_1(x,y) = xy + \bar{x}\bar{y} + xz \qquad f_2(x,y) = xy + yz + \bar{x}\bar{z}$$

Since the two maps are not identical, the functions are not equal.

ALGORITHM 3.1

For generating the fundamental-sum-of-product form (FSP)

STEP   OPERATION

1.   Express the function f in terms of the operators "·", "+", and "−".
2.   Expand f into sum-of-product form.
3.   Use the idempotency laws to remove any repeated products.
4.   Examine each term. If it is already a fundamental-product term, continue to the next term. If it is not a fundamental-product term, multiply the product by $(x_i + \bar{x}_i)$ for each missing variable x_i.
5.   Multiply out all terms and use the idempotency law to remove any repeated products.
6.   STOP. The resulting sum-of-product form will be the desired fundamental-sum-of-product form.

Example 3.20 Expand: $f(x_2, x_1, x_0) = (x_2 + \bar{x}_1)x_0 + \bar{x}_0$ into (FSP) form.

$$f(x_2, x_1, x_0) = x_2 x_0 + \bar{x}_1 x_0 + \bar{x}_0$$

$$f(x_2, x_1, x_0) = x_2(x_1 + \bar{x}_1)x_0 + (x_2 + \bar{x}_2)(\bar{x}_1 x_0) + (x_2 + \bar{x}_2)(x_1 + \bar{x}_1)\bar{x}_0$$

$$= x_2 x_1 x_0 + x_2 \bar{x}_1 x_0 + x_2 \bar{x}_1 x_0 + \bar{x}_2 \bar{x}_1 x_0$$

$$+ x_2 x_1 \bar{x}_0 + \bar{x}_2 x_1 \bar{x}_0 + x_2 \bar{x}_1 \bar{x}_0 + \bar{x}_2 \bar{x}_1 \bar{x}_0$$

$$= x_2 x_1 x_0 + x_2 \bar{x}_1 x_0 + \bar{x}_2 \bar{x}_1 x_0 + x_2 x_1 \bar{x}_0$$

$$+ \bar{x}_2 x_1 \bar{x}_0 + x_2 \bar{x}_1 \bar{x}_0 + \bar{x}_2 \bar{x}_1 \bar{x}_0$$

It is difficult to convert a function into a fundamental-product-of-sum form, using only the properties of a Boolean algebra. Therefore, in order to convert an arbitrarily written Boolean function into fundamental-product-of-sum form we will resort to the following procedure. First the dual of the function will be taken, then the dual function will be converted into the fundamental-sum-of-product form using Algorithm 3.1. Finally the dual of the resulting FSP form will be taken, resulting in the desired fundamental-product-of-sum form.

ALGORITHM 3.2

For generating the fundamental-product-of-sum form (FPS), use Algorithm 3.1 after inserting the following steps.

STEP   OPERATION

1.5.   Take the dual of the function f.

7.   Take the dual of the fundamental sum-of-product form (FSP) and the result will be the required fundamental product-of-sum form.

Example 3.21 Expand $f(x, y, z) = x\bar{y} + \bar{x}z$ into fundamental-product-of-sum form.

$$f^d(x, y, z) = (x + \bar{y})(\bar{x} + z)$$

$$= xz + \bar{x}\bar{y} + \bar{y}z$$

$$= x(y + \bar{y})z + \bar{x}\bar{y}(z + \bar{z}) + (x + \bar{x})\bar{y}z$$

$$= xyz + x\bar{y}z + \bar{x}\bar{y}z + \bar{x}\bar{y}\bar{z} + x\bar{y}z + \bar{x}\bar{y}z$$

$$= xyz + x\bar{y}z + \bar{x}\bar{y}z + \bar{x}\bar{y}\bar{z}$$

$$f^{dd}(x, y, z) = f(x, y, z) = (x + y + z)(x + \bar{y} + z)(\bar{x} + \bar{y} + z)(\bar{x} + \bar{y} + \bar{z})$$

The reader should attempt to expand this function into fundamental-product-of-sum form without using the principle of duality.

3.11 FUNDAMENTAL FORMS FROM TRUTH TABLES OR *K*-MAPS

Using Eqs. 3.8 and 3.11, it is an easy matter to write out the fundamental-sum-of-product and product-of-sum form of Boolean functions from a truth table or *K*-map.

For the sum-of-product form it is only necessary to scan the output column of the truth table for output a_I's, equal to one; or search the squares of a *K*-map for 1's and then sum the corresponding fundamental-product terms, P_I^n.

To obtain the correct fundamental-product-of-sum form, it is only necessary to scan the truth table output column for zeroes, or search the squares of the *K*-map for zeroes, and then product the corresponding fundamental-sum terms, $S_{2^n-1-I}^n$. Thus if $f(i_{n-1}, i_{n-2}, \cdots, i_1, i_0) = 0$ then the sum $\bigvee_{j=0}^{n-1} x_j^{\bar{i_j}}$ is a product term.

Example 3.22 Write out the FSP and FPS forms for the function described by the truth table below.

x	y	z	f(x,y,z)
0	0	0	1
0	0	1	0
0	1	0	1
0	1	1	1
1	0	0	1
1	0	1	0
1	1	0	1
1	1	1	0

I	a_I	$a_I P_I^3$	$a_I + S_{7-I}^3$
0	1	$\bar{x}\bar{y}\bar{z}$	1
1	0	0	$x+y+\bar{z}$
2	1	$\bar{x}y\bar{z}$	1
3	1	$\bar{x}yz$	1
4	1	$x\bar{y}\bar{z}$	1
5	0	0	$\bar{x}+y+\bar{z}$
6	1	$xy\bar{z}$	1
7	0	0	$\bar{x}+\bar{y}+z$

FSP form

$$f(x, y, z) = \bigvee_{I=0}^{7} a_I P_I^3$$

$$= \bar{x}\bar{y}\bar{z} + \bar{x}y\bar{z} + \bar{x}yz + x\bar{y}\bar{z} + xy\bar{z}$$

FPS form

$$f(x, y, z) = \bigwedge_{I=0}^{7} (a_I + S_{7-I}^3)$$

$$= (x + y + \bar{z})(\bar{x} + y + \bar{z})(\bar{x} + \bar{y} + z)$$

Although the fundamental-product-of-sum form of a Boolean expression appears a little more cumbersome to write, it is really only a little more difficult to obtain by inspection of the truth table.

For those readers who have difficulty with the above one-step procedure, the following two-step procedure may be helpful.

1. Write out the FSP form for the complement of the function.

2. Complement the resulting FSP expression.

Using the data from Example 3.22,

$$\bar{f}(x, y, z) = \bigvee_{I=0}^{7} \bar{a}_I P_I^3$$

$$= \bar{x}\bar{y}z + x\bar{y}z + xyz.$$

Taking the complement yields the desired result:

$$f(x, y, z) = (x + y + \bar{z})(\bar{x} + y + \bar{z})(\bar{x} + \bar{y} + \bar{z}).$$

Another alternative procedure for those readers still finding the previous two methods troublesome uses the principle of duality since these are dual forms.

ALGORITHM 3.3

RULE for writing fundamental-sum-of-product form from a truth table.

Scan each line of the truth table. If the output value is a 1, logically OR to the result obtained so far, the logical product of all n variables complementing each variable in the product whose input value is a 0.

The dual of this Rule becomes

ALGORITHM 3.4

RULE for writing fundamental-product-of-sum form from a truth table.

Scan each line of the truth table. If the output value is a 0, logically AND to the result obtained so far the logical sum of all n variables, complementing each variable in the sum whose input value is a 1.

This last set of rules is the easiest to use and remember. The easiest way to map a Boolean function written in a product-of-sum form is to first complement the function. This will yield a function in sum-of-product form which we already know how to map. An important point to remember when mapping the complement of a function is that zeroes are mapped instead of ones.

Example 3.23 Map the function

$$f(w, x, y, z) = (x + y)(x + \bar{z})(w + \bar{x} + y + z) \text{ on a K-map.}$$

Taking the complement yields

$$\bar{f}(w, x, y, z) = \bar{x}\bar{y} + \bar{x}z + \bar{w}x\bar{y}\bar{z}.$$

Zeroes are now placed on the squares which each of these product terms cover, the remaining squares are then set to ones.

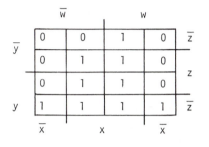

$$f(w,x,y,z) = (x+y)(x+\bar{z})(w+\bar{x}+y+z)$$

An alternate but time-consuming method would be to draw separate K-maps for each term and then AND each of these maps together to obtain the final K-map. Using the data from Example 3.23 this approach yields:

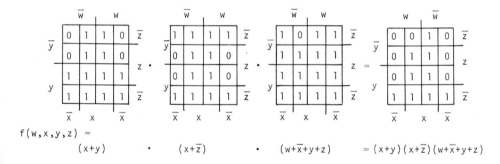

$$f(w,x,y,z) =$$
$$(x+y) \qquad \cdot \qquad (x+\bar{z}) \qquad \cdot \qquad (w+\bar{x}+y+z) \qquad = (x+y)(x+\bar{z})(w+\bar{x}+y+z)$$

When writing a function in product-of-sum form from a K-map, the easiest method is to write out the sum-of-product expression for the complement of the function, grouping zeroes instead of ones. Then to obtain the desired product-of-sum form it is only necessary to complement the product-of-sum expression.

Example 3.24 Write the output function for the K-map below in product-of-sum form.

Using the zeroes,

$$\bar{f}(w, x, y, z) = \bar{w}z + w\bar{x}y + wx\bar{y}\bar{z}.$$
$$\therefore \quad f(w, x, y, z) = (w + \bar{z})(\bar{w} + x + \bar{y})(\bar{w} + \bar{x} + y + z).$$

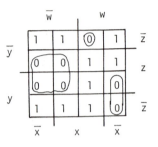

3.12 CONVERSION BETWEEN BOOLEAN FORMS

There are many times when it is necessary to convert a Boolean expression from sum-of-product to product-of-sum and vice versa. In Section 4.3 we shall see one reason for this, the fact that circuits with two levels of NAND gates can be used to realize functions expressed in sum-of-product form; and that circuits with two levels of NOR gates can be used to realize functions expressed in product-of-sum form.

ALGORITHM 3.5

RULE. For conversion from product-of-sum form to sum-of-product form
    Logically multiply out the expression using idempotency to remove any redundant terms.

Example 3.25 Convert $f(x, y, z) = (x + y)(\bar{x} + z)$ into sum of product form.

$$f(x, y, z) = (x + y)(\bar{x} + z)$$
$$= xz + \bar{x}y + yz.$$

ALGORITHM 3.6

RULE. Conversion from sum-of-product to product-of-sum form.

STEP   OPERATION

1.   Take the dual of the function or expression.
2.   Logically multiply out the resulting expression into sum-of-product form, using idempotency to remove any redundant terms.
3.   Take the dual of the result obtained in step 2.

Example 3.26   Convert $f(x, y, z) = x\bar{z} + \bar{x}y + \bar{y}z$ into product-of-sum form.

$$f(x, y, z) = x\bar{z} + \bar{x}y + \bar{y}z.$$
$$f^d(x, y, z) = (x + \bar{z})(\bar{x} + y)(\bar{y} + z)$$
$$= (xy + \bar{x}\bar{z} + y\bar{z})(\bar{y} + z)$$
$$= \bar{x}\bar{y}\bar{z} + xyz.$$
$$f^{dd}(x, y, z) = (\bar{x} + \bar{y} + \bar{z})(x + y + z).$$

3.13 PROBLEMS

3.1 Using only the basic postulates and properties given in Section 3.2, establish the validity of the following statements:
  a) $xy + yz + \bar{x}z = xy + \bar{x}z$
  b) $(x + y)(\bar{x} + \bar{y}) = \bar{x}y + x\bar{y}$
  c) $(\bar{x} + \bar{y})(\bar{x} + y)(x + \bar{y}) = \bar{x}\bar{y}$

3.2 Given that $x + y = x + z$ and that $xy = xz$, prove that $y = z$. (Note: division and subtraction are not valid operations.)

3.3 Verify the validity or nonvalidity of the following statements using the method of perfect induction:
  a) $x + \bar{x}y = x + y$
  b) $\overline{xy + xz + yz} = \bar{x}\bar{y} + \bar{x}\bar{z} + \bar{y}\bar{z}$
  c) $\bar{x}\bar{y} + y\bar{z} + x\bar{y} = \overline{yz}$
  d) $xy + \bar{x}\bar{y} = (x + y)(\bar{x} + \bar{y})$

3.4 Prove that the following Boolean statements are valid:
  a) If $x + y = y$, then $\bar{x} + y = 1$
  b) If $x + y = y$, then $x\bar{y} = 0$

3.5 Starting with Eqs. 3.1 and 3.3, use mathematical induction to prove that
$$f(x_{n-1}, x_{n-2}, \ldots, x_1, x_0) = \bigvee_{j=0}^{2^n-1} a_j \cdot P_j^n$$

3.6 Given the following sets, write out the corresponding Boolean functions in fundamental sum-of-product form.
  a) $A_3 = [a_0, a_2, a_5]$
  b) $A_4 = [a_1, a_3, a_{10}, a_{12}, a_{15}]$
  c) $A_5 = [a_7, a_{12}, a_{15}, a_{21}, a_{24}, a_{25}, a_{30}]$

3.7 Given the following sets, write out the corresponding Boolean functions in fundamental product-of-sum form.
  a) $\bar{A}_5 = [a_1, a_2, a_4, a_{20}, a_{28}, a_{31}]$
  b) $\bar{A}_3 = [a_2, a_3, a_6, a_7]$
  c) $\bar{A}_4 = [a_0, a_5, a_9, a_{12}, a_{13}, a_{14}]$

3.8 Given the following characteristic numbers, express the corresponding Boolean functions in fundamental-sum-of-product form.
   a) $D_2 = 3$ b) $H_3 = A0$
   c) $O_4 = 123456$ d) $D_3 = 129$
   e) $H_4 = A64B$ f) $B_3 = 01011001$

3.9 Given the following characteristic numbers, express the corresponding functions in fundamental-product-of-sum form.
   a) $B_2 = 1001$ b) $D_3 = 254$
   c) $H_4 = EFC7$ d) $D_3 = 143$

3.10 Construct truth tables for the following Boolean functions:
   a) $f(x, y) = (x + y)(x + \bar{y})(\bar{x} + y)$
   b) $f(x, y, z) = xz + y(\bar{x} + z)$
   c) $f(w, x, y, z) = wxy + \bar{x}\bar{y} + w(x + \bar{y}\bar{z})$
   d) $f(x, y, z) = (x + y)(z + \bar{x}\bar{y})$

3.11 From the truth tables for the functions in Problem 3.10, determine the characteristic numbers for each of the functions expressing the result as:
   a) B_N b) D_N
   c) O_N d) H_N

3.12 Map the following Boolean functions on K-maps.
   a) $f(x, y) = \bar{x}\bar{y}$
   b) $f(x, y, z) = xy + \bar{x}\bar{y} + xz$

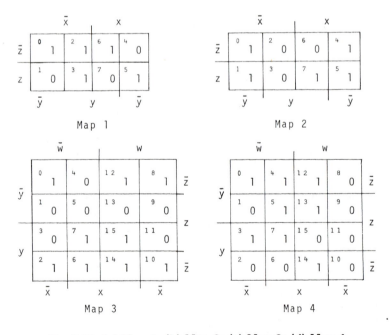

Fig. 3.10 (a) Map 1; (b) Map 2; (c) Map 3; (d) Map 4

c) $f(w, x, y, z) = wx + \bar{w}x\bar{y}z + wy\bar{z}$
d) $f(w, x, y, z) = \bar{w}\bar{y} + wxy + \bar{w}x\bar{z} + w\bar{x}y\bar{z}$
e) $f(w, x, y, z) = (w + x)(x + \bar{z})(\bar{w} + \bar{y} + z)$
f) $f(x, y, z) = (x + y)(x + z)(y + \bar{z})$
g) $f(w, x, y, z) = (wz + yz)(x + \bar{y}\bar{z})$

3.13 Map the Boolean functions corresponding to the following characteristic numbers on the appropriate K-maps:
a) $B_3 = 10110011$ b) $D_4 = 31296$
c) $H_4 = $ A07B   d) $H_3 = 56$
e) $O_4 = 247$ f) $H_4 = $ 16AF

3.14 Write out the Boolean function represented on the K-maps in Fig. 3.10
a) in sum-of-product form.
b) in product-of-sum form.

3.15 Determine the characteristic numbers for the functions shown on the K-maps in Problem 3.14. Express your answers as:
a) B_N b) H_N
c) O_N d) D_N

3.16 Write out the dual of the following Boolean functions, expressing your result in sum-of-product form.
a) $f(x, y) = xy + \bar{x}\bar{y}$
b) $f(x, y, z) = \bar{x}\bar{y} + y\bar{z} + \bar{y}z$
c) $f(x, y, z) = \bar{x}(y + z) + \bar{x}(\bar{y} + \bar{z})$
d) $f(w, x, y, z) = (w + \bar{x}\bar{y}) \cdot (\bar{w} + \bar{z}(x + y))$
e) $f(w, x, y, z) = (x + y)(\bar{w} + yz)$

3.17 A Boolean function is said to be *self dual* if and only if $f^d = f$. For example $f(x) = x$ is a self-dual function. If $f(x_{n-2}, \ldots, x_1, x_0)$ is an arbitrary function of $n - 1$ variables, $n \geq 1$, then show that the function $h(x_{n-1}, x_{n-2}, \ldots, x_1, x_0) = \bar{x}_{n-1}f(x_{n-2}, \ldots, x_1, x_0) + x_{n-1}\bar{f}(\bar{x}_{n-2}, \ldots, \bar{x}_1, \bar{x}_0)$ is a self-dual function.

3.18 Show that there is a one-to-one relationship between functions of $n - 1$ variables and the self-dual functions of n variables.

3.19 Show that there are $2^{2^{n-1}}$ self-dual functions of n variables.

3.20 Devise simple test procedures (algorithms) which can be applied to:
a) a Boolean function   b) a truth table
c) a characteristic number   d) a K-map
to determine if the corresponding Boolean function is self dual.

3.21 Determine the complements of the following Boolean functions expressing your answers in sum-of-product form.
a) $f(x, y) = \bar{x}\bar{y} + xy$
b) $f(x, y, z) = x + yz$
c) $f(w, x, y, z) = (x + y)(\bar{w} + \bar{x}(y + z))$
d) $f(w, x, y, z) = (xy(w + z) + \bar{x}\bar{y}(\bar{w} + z)$
e) $f(w, x, y, z) = (\bar{x}\bar{y})(\overline{w + y\bar{z}})$

3.22 Determine the corresponding characteristic numbers for the dual and complement functions for the Boolean functions whose characteristic numbers are given below
a) $D_3 = 119$ b) $B_3 = 10111001$
c) $O_3 = 235$ d) $H_3 = A7$
e) $H_4 = 62AC$ f) $H_4 = 12E0$

3.23 Construct the dual and complement K-maps for the functions represented by the K-maps in Problem 3.14.

3.24 Construct the dual and complement truth tables for the functions represented by the truth tables in Problem 3.28.

3.25 For the circuit shown below, construct:
a) the dual circuit
b) the complement circuit.

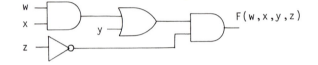

3.26 Expand the following Boolean functions into fundamental sum-of-product form.
a) $f(x, y) = x + y$
b) $f(x, y, z) = x + yz$
c) $f(w, x, y, z) = wx + \bar{y}z + x(y + \bar{w})$
d) $f(w, x, y, z) = (x + y)(\bar{w} + z)(x + \bar{z})$

	Table 1		
x	y	z	F(x,y,z)
0	0	0	0
0	0	1	1
0	1	0	1
0	1	1	0
1	0	0	1
1	0	1	1
1	1	0	1
1	1	1	0

	Table 2		
x	y	z	F(x,y,z)
0	0	0	1
0	0	1	0
0	1	0	0
0	1	1	1
1	0	0	0
1	0	1	1
1	1	0	0
1	1	1	1

Figure 3.11

3.27 Expand the following Boolean functions into fundamental product of sum form.
    a) $f(x, y) = xy$
    b) $f(x, y, z) = \bar{x} + \bar{z}y$
    c) $f(w, x, y, z) = xy + w(x + \bar{z}) + \bar{y}z$
    d) $f(w, x, y, z) = (x + y)(w + z)(y + \bar{x})$

3.28 Express the functions represented by the truth tables (Fig. 3.11) in fundamental-sum-of-product form and also in fundamental-product-of-sum form.

3.29 Express the following Boolean functions in sum-of-product form.
    a) $f(x, y, z) = (x + y)z$
    b) $f(w, x, y, z) = w(x + \overline{yz})$
    c) $f(w, x, y, z) = wx + yz(x + \bar{w})$

3.30 Express the following Boolean functions in product of sum form.
    a) $f(x, y, z) = x + \bar{y}z$
    b) $f(w, x, y, z) = wx + yz(\bar{w} + \overline{xy})$
    c) $f(w, x, y, z) = (w + x)(y + \bar{w}z)$

3.14 BIBLIOGRAPHY

Bartree, T. C., I. L. Lebow, I. S. Reed, *Theory and Design of Digital Machines.* New York: McGraw-Hill, 1962.

Boole, G., *An Investigation of the Laws of Thought.* New York: Dover Publications, 1954.

Friedman, A. D., *Logical Design of Digital Systems.* Woodland Hills, Cal.: Computer Science Press, 1975.

Harrison, M. A., *Introduction to Switching and Automata Theory.* New York: McGraw-Hill, 1965.

Hill, F. J., G. R. Peterson, *Introduction to Switching Theory and Logical Design.* 2nd ed. New York: Wiley, 1974.

Hoernes, G. E., M. F. Heilweil, *Introduction to Boolean Algebra and Logic Design.* New York: McGraw-Hill, 1964.

Hohn, F. E., *Applied Boolean Algebra.* 2nd ed. New York: Macmillan, 1966.

Huntington, E. V., "Sets of independent postulates for the algebra of logic," *Trans. American Mathematical Society*, Vol. 5, pp. 288–309, 1904.

McCluskey, E. J., *Introduction to the Theory of Switching Circuits.* New York: McGraw-Hill, 1965.

Miller, R. E., *Switching Theory Vol. 1.* New York: Wiley, 1965.

Torng, H. C., *Introduction to the Logical Design of Switching Systems.* Reading, Mass.: Addison-Wesley, 1964.

Whitesitt, J. E., *Boolean Algebra and Its Applications.* Reading, Mass.: Addison, Wesley, 1961.

Chapter 4

Special Boolean
Functions and
Basic Logic
Conventions

4.1 INTRODUCTION

In this chapter we will consider a few of the more commonly used two-variable
Boolean functions. In particular we will investigate the properties of the following
functions:

$$
\begin{array}{lll}
\text{NAND} & f(x, y) = \overline{x \cdot y} = x \uparrow y \\
\text{NOR} & f(x, y) = \overline{x + y} = x \downarrow y \\
\text{eXclusive OR} & f(x, y) = x\bar{y} + \bar{x}y = x \oplus y \\
\text{eXclusive NOR} & f(x, y) = xy + \bar{x}\bar{y} = x \odot y
\end{array}
$$

Each of these functions has a generalized multivariable meaning which will also
be discussed. We will also include in Section 4.10 an introduction to digital
multiplexers and their use as a basic digital building block.

In this chapter the logic-symbol notation which is used in this book will also
be introduced. Realizing that there is no national standard for logic symbols,
military standard symbols will be stressed as much as possible in this book, since
all vendors which sell components to the government must adhere to these
symbols or provide a table of symbol correspondence.

With the introduction of new electronic components for logic design and the
rate at which the circuit technology is advancing, we feel that this book should not
attempt to go beyond the logic symbol and discuss the electronic circuit which
realizes its truth table. This is reasonable, since it will allow the reader to free his
thinking of any particular logic building block, be it Resistor Transistor Logic

(RTL), Diode Transistor Logic (DTL), Transistor-Transistor Logic (TTL), Emitter Coupled Logic (ECL), relays, fluidic logic, etc. To be sure, at any given time, and for any given design, one set of logic components is better than the others. However, this introductory course is not the place to learn how to design with today's best components, for, with the innovations of the future, today's best design may well be tomorrow's worst.

Section 4.6 will introduce the concept of positive and negative logic assignments. Section 4.8 will present the logic symbols and conventions to be used in this book. In Section 4.9, we will consider the problem of converting an existing logic design from one set of logical building blocks to another set of different building blocks. A set of transformation rules is also presented.

4.2 THE NAND AND NOR FUNCTIONS

The two most-used functions are NAND and NOR. Their use comes as a direct result of current technology. The electronic circuit elements used to implement these logic functions automatically invert the logical result. Thus it is technically easier to build an electronic logic gate to realize $\overline{x \cdot y}$ or $\overline{x + y}$ than to realize $x \cdot y$ or $x + y$. As we will see shortly, there are certain other advantages to using NAND or NOR gates in our designs. The binary-operation symbols which are assigned to these two functions are "↑" for NAND, and "↓" for NOR. The functions these symbols represent are easy to determine if we rely on our knowledge of set theory, where ∧ is used to denote set intersection or AND, and ∨ is used to denote set union or OR. The vertical line tells us to invert the result.

It should be noted that the two input NAND and NOR gates also carry the name Shaffer stroke (↑) and Pierce arrow (↓) for the men who originally proposed the use of these functions. NAND and NOR tables are shown in Fig. 4.1. Symbolically, the NAND function realizes NOT–AND in one operation. Thus, we

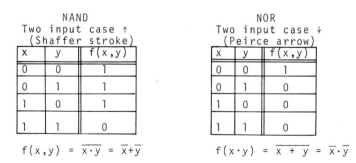

Figure 4.1

will reduce the symbols of AND followed by NOT into one combined symbol, NAND, using the circle on the output line to denote a complement operation.

The NOR function realizes <u>NOT</u>–<u>OR</u> in one operation. Thus we will reduce the symbols of OR followed by NOT into one combined symbol, NOR, using the circle on the output to denote a complement operation.

Inspection of the two truth tables of these functions reveals that the NAND and NOR operators are duals of each other.

SPECIAL PROPERTIES

NAND	NOR
1. $x \uparrow 1 = \overline{x \cdot 1} = \bar{x}$	$x \downarrow 0 = \overline{x + 0} = \bar{x}$
2. $x \uparrow 0 = \overline{x \cdot 0} = 1$	$x \downarrow 1 = \overline{x + 1} = 0$
3. $x \uparrow x = \overline{x \cdot x} = \bar{x}$	$x \downarrow x = \overline{x + x} = \bar{x}$
4. $x \uparrow y = \overline{x \cdot y} = \bar{x} + \bar{y}$	$x \downarrow y = \overline{x + y} = \bar{x} \cdot \bar{y}$
5. $\bar{x} \uparrow \bar{y} = \overline{\bar{x} \cdot \bar{y}} = x + y$	$\bar{x} \downarrow \bar{y} = \overline{\bar{x} + \bar{y}} = x \cdot y$
6. $\overline{x \uparrow y} = \overline{\overline{x \cdot y}} = x \cdot y$	$\overline{x \downarrow y} = \overline{\overline{x + y}} = x + y$

Although both of these operators satisfy the commutative law, neither satisfies either the distributive laws or the associative laws.

NAND	NOR
$x \uparrow (y + z) \neq (x \uparrow y) + (x \uparrow z)$	$x \downarrow (y + z) \neq (x \downarrow y) + (x \downarrow z)$
$x \uparrow (y \cdot z) \neq (x \uparrow y) \cdot (x \uparrow z)$	$x \downarrow (y \cdot z) \neq (x \downarrow y) \cdot (x \downarrow z)$
$(x \uparrow y) \uparrow z \neq x \uparrow (y \uparrow z)$	$(x \downarrow y) \downarrow z \neq x \downarrow (y \downarrow z)$

These statements can be verified by expanding the functions into AND and OR logic and then using either K-maps or truth tables. Because the distributive and associative laws do not hold, be careful when cascading two input NAND and NOR gates together. One must also be sure to use parentheses when writing expressions with more than two variables.

The operations NAND and NOR can be easily expanded to handle n inputs. However, in the multi-input case the operators $\downarrow$ and $\uparrow$ are not applicable. It is best to write out the product or sum and place a bar over the expression. Thus, the NAND of w, x, and y would be written as $\overline{wxy}$, and the NOR of w, x, and y would be written as $\overline{w + x + y}$. The n input NAND operator yields a zero only if all inputs are equal to one, and the n input NOR operator yields a one only if all inputs are zero. NAND and NOR symbols are shown in Fig. 4.2.

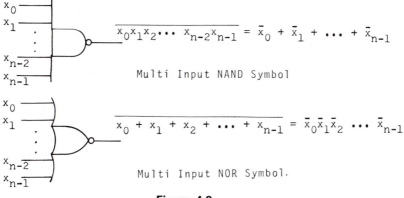

$$\overline{x_0 x_1 x_2 \cdots x_{n-2} x_{n-1}} = \bar{x}_0 + \bar{x}_1 + \cdots + \bar{x}_{n-1}$$

Multi Input NAND Symbol

$$\overline{x_0 + x_1 + x_2 + \cdots + x_{n-1}} = \bar{x}_0 \bar{x}_1 \bar{x}_2 \cdots \bar{x}_{n-1}$$

Multi Input NOR Symbol.

Figure 4.2

4.3 LOGIC DESIGN USING NAND AND NOR GATES

Let us assume that the only logic gates available for our use are an assortment of multi-input NAND and NOR gates. Using the basic properties of these gates listed in Section 4.2, we can easily generate the three basic building blocks required for a Boolean algebra, namely NOT, AND, and OR. (See Fig. 4.3.)

Restricting the number of inputs to our gates will not cause any serious problem in the realization of large sum or product terms. For example, a five input AND gate (Fig. 4.4) can be easily constructed using a simple tree structure with only 2- and 3-input NAND gates. In this case four levels of NAND gates are required.

An alternate approach would be to use the basic postulates and properties instead of brute force. We can break the expression ABCDE into a product of product terms with 2 or 3 variables each. Thus ABCDE becomes (ABC)(DE).

Now, by using the law of involution, we obtain: (ABC)(DE) = $\overline{\overline{(ABC)(DE)}}$.

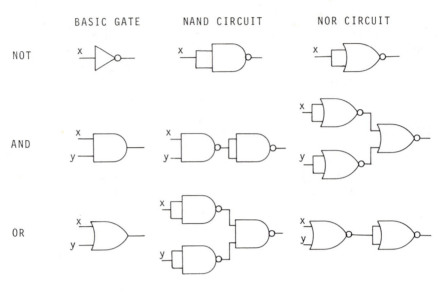

Figure 4.3

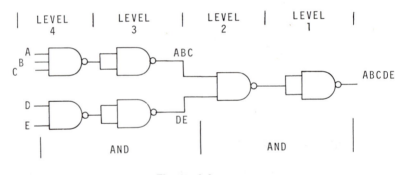

Figure 4.4

By applying the generalized De Morgan's Law we obtain:

 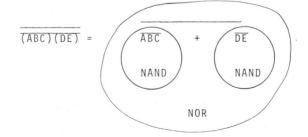

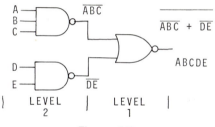

Figure 4.5

The inner expressions are now easily realized by NAND gates, while the outer expression is easily realized by a NOR gate (Fig. 4.5). This result could have been anticipated by examining the four-level NAND circuit and recognizing the middle two-level circuit as an OR gate configuration (Fig. 4.6).

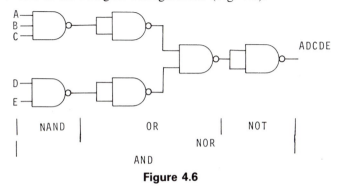

Figure 4.6

Generalizing the preceding observation: *A level of NAND gates followed by a level of NOR gates is logically equivalent to a single multi-input AND gate.* The dual of this observation is also true; i.e., *A level of NOR gates followed by a level of NAND gates is logically equivalent to a single multi-input OR gate* (Fig. 4.7).

Boolean expressions in sum-of-product form can be implemented using 4 levels of NAND gates if the brute force method is employed. Such a circuit for the function $f(x, y, z) = yz + x + \bar{y}\bar{z}$ is shown in Fig. 4.8. However, the law of involution states that $\bar{\bar{x}} = x$, and hence the inverters in levels 2 and 3 are obviously redundant. Their elimination reduces the circuit to only two levels of NAND gates (Fig. 4.9).

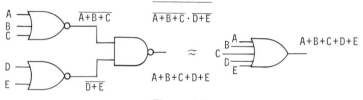

Figure 4.7

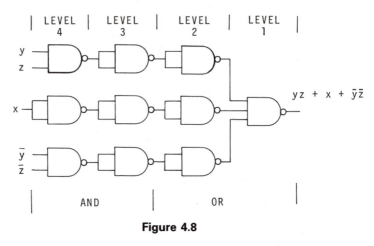

Figure 4.8

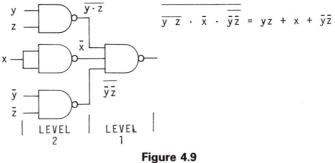

Figure 4.9

We could obtain the same result by applying the law of involution directly on the original expression.

$$f(x,y,z) = \overline{\overline{yz + x + \overline{y}\overline{z}}}$$

$$= \overline{\overline{yz + x + \overline{y}\overline{z}}}$$

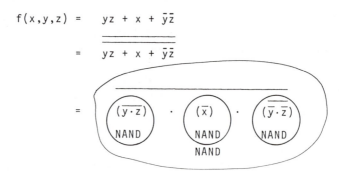

The preceding remarks lead to the following conclusion. *A logic expression in sum-of-product form can be realized by two levels of logic gates: A level of AND gates followed by a level of OR gates, or by two levels of NAND gates.*

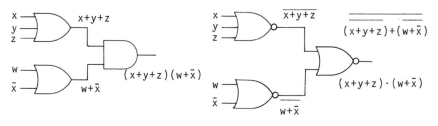

Figure 4.10

The dual statement is also true. *A logic expression in product-of-sum form can be realized by two levels of logic gates: a level of OR gates followed by a level of AND gates, or by two levels of NOR gates.* Thus the function

$$f(w, x, y, z) = (x + y + z)(w + \bar{x})$$

can be realized by either of the circuits in Fig. 4.10.

It was tacitly assumed in the preceding paragraphs that both the true and complements of all of the variables were available. An additional level of NAND or NOR gate inverters can always be used to supply any of the missing complemented variables. If we are given a mixed Boolean function, such as $f(w, x, y, z) = (w\bar{x} + y)(\bar{x} + y\bar{z})(x\bar{z})$, to be realized using NAND or NOR gates, we can either expand it into sum-of-product form or product-of-sum form and then implement the resulting expanded expressions; or we can repeatedly use the law of involution to modify the expression into a form suitable for direct implementation. Using the latter approach we have:

$$(w\bar{x}+y)(\bar{x}+y\bar{z})(x\bar{z}) \quad = \quad \overline{(w\bar{x}+y)(\bar{x}+y\bar{z})(x\bar{z})}$$

$$= \quad \overline{\overline{(w\bar{x}+y)} + \overline{(\bar{x}+y\bar{z})} + \overline{x\bar{z}}}$$

Applying the law of involution again yields:

$$= \quad \overline{\overline{\overline{(w\bar{x}+y)}} + \overline{\overline{(\bar{x}+y\bar{z})}} + \overline{\overline{x\bar{z}}}}$$

$$= \quad \overline{\overline{\overline{w\bar{x} \cdot \bar{y}}} + \overline{\overline{\bar{x} \cdot x\bar{z}}} + \overline{x\bar{z}}}$$

The resulting circuit shown in Fig. 4.11, uses two levels of NAND gates to realize the sum-of-product expressions $w\bar{x} + y$ and $x + y\bar{z}$ and then levels of NAND followed by NOR to realize a multi-input AND gate.

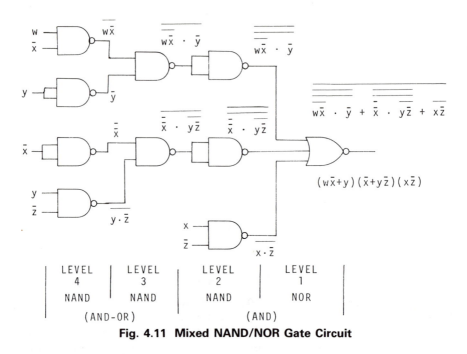

Fig. 4.11 Mixed NAND/NOR Gate Circuit

One last comment on multi-level NAND- and NOR-gate logic (Figs. 4.12 and 4.13). It is clear from the development so far that an *even number* of levels of all NAND or all NOR gates can be used to realize alternate levels of AND followed by OR or OR followed by AND, respectively. If an *odd* number of levels occurs, then an odd level of NAND gates, farthest from the output gate, will realize an

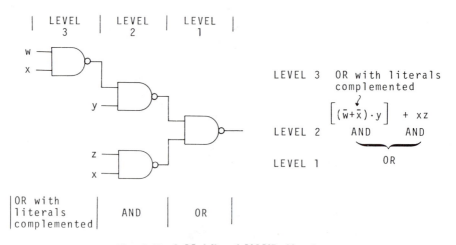

Fig. 4.12 A Multilevel NAND Circuit

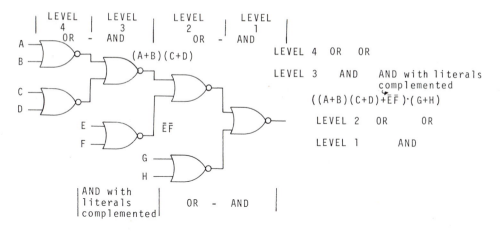

Fig. 4.13 A Multilevel NOR Circuit

OR function with all literals complemented; and an odd level of NOR gates, farthest from the output gate, will realize an AND function with all literals complemented.

The ability to make use of these observations will greatly reduce the amount of time and effort required to analyze NAND and NOR gate logic circuits.

4.4 EXCLUSIVE OR AND EXCLUSIVE NOR

The eXclusive-OR function yields a 1 output only if either input, but not both, are equal to a 1. The Truth Table and K-map, for the two-input function, is shown in Fig. 4.14.

The fundamental sum-of-product expansion for this function is $f(x, y) = \bar{x}y + x\bar{y}$, and the fundamental product-of-sum expansion is $f(x, y) = (\bar{x} + \bar{y})(x + y)$. Because this function is used quite frequently, it has been given

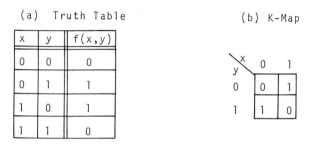

Fig. 4.14 Exclusive OR

the operator symbol, $\oplus$, which is read as either "ring sum" or "XOR."

$$f(x, y) = x \oplus y = x\bar{y} + \bar{x}y = (\bar{x} + \bar{y})(x + y).$$

Because of the wide use of this function in logic circuits, the logic symbol

is used when making logic circuit diagrams. The properties of the operator "$\oplus$" are:

(1) $x \oplus x = x\bar{x} + \bar{x}x = 0$ (6) $\bar{x} \oplus \bar{x} = \bar{x}x + x\bar{x} = 0$

(2) $x \oplus 1 = x \cdot 0 + \bar{x} \cdot 1 = \bar{x}$ (7) $\bar{x} \oplus 1 = \bar{x} \cdot 0 + x \cdot 1 = x$

(3) $x \oplus y \oplus xy = x + y$ (8) $\overline{x \oplus y} = x \oplus y \oplus 1$

(4) $x \oplus y = y \oplus x$ (9) $x \oplus (y \oplus z) = (x \oplus y) \oplus z$

(5) $x \cdot (y \oplus z) = (x \cdot y) \oplus (x \cdot z)$

Properties 4, 9, and 5 reveal that the $\oplus$ operator satisfies the commutative, associative, and distributive laws.

Note: One must be careful when mixing the operators "+" and "$\oplus$" to include brackets around the variables associated with each operator:

$$x + (y \oplus z) \neq (x + y) \oplus z.$$

For if: $x = 1$, $y = 1$, and $z = 1$ then

$$x + (y \oplus z) = 1 \quad \text{while} \quad (x + y) \oplus z = 0.$$

One of the important uses for this function is that of a controlled base-2 one's-complement unit. This is shown in Fig. 4.15.

The exclusive-OR or XOR circuit can be logically constructed using AND, OR, and NOT gates in a variety of ways, two of which are shown in Fig. 4.16.

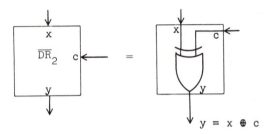

Fig. 4.15 Controlled One's Complement Unit

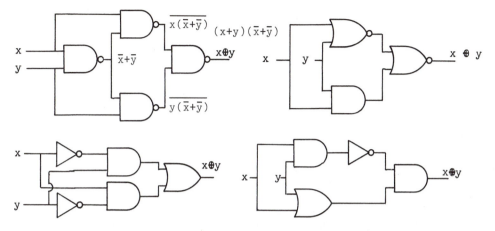

Fig. 4.16 Typical Circuits for Realizing XOR

However in general these units are constructed using NAND and NOR gates as also shown in Fig. 4.16.

If we were to generalize the concept of an exclusive-OR circuit to more than two inputs, we would require that the output be one whenever one and only one input equals a one. However, this type of circuit is seldom used. The more frequently used circuit is obtained by cascading two input exclusive-OR circuits together. In this case all pairs of inputs equal to one will cancel, leading to the interpretation that a multi-level exclusive-OR gate cascade yields a one output only if there is an odd number of ones on the input lines.

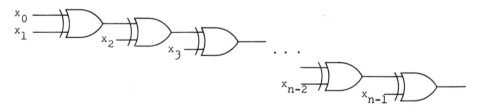

Multi-input XOR cascades are not commonly found as a separate circuit element, and generally are constructed with two input units using either the cascade circuit above, or a tree circuit similar to that shown in Fig. 4.17.

For the three-variable case, $f(x_2, x_1, x_0) = x_2 \oplus x_1 \oplus x_0$, expanding to sum-of-product form we have:

$$f(x_2, x_1, x_0) = x_2\overline{(x_1 \oplus x_0)} + \bar{x}_2(x_1 \oplus x_0)$$
$$= x_2(\bar{x}_1 x_0 + x_1\bar{x}_0) + \bar{x}_2(x_1\bar{x}_0 + \bar{x}_1 x_0)$$
$$= x_2(x_1 + \bar{x}_0)(\bar{x}_1 + x_0) + \bar{x}_2 x_1\bar{x}_0 + \bar{x}_2\bar{x}_1 x_0$$
$$= x_2 x_1 x_0 + x_2\bar{x}_1\bar{x}_0 + \bar{x}_2 x_1\bar{x}_0 + \bar{x}_2\bar{x}_1 x_0.$$

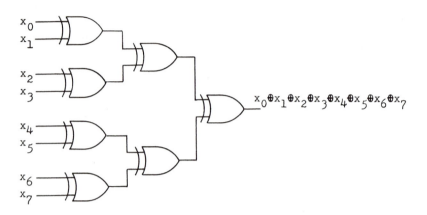

Fig. 4.17 Tree Circuit for Realizing a Multi-Input Exclusive OR

The corresponding product-of-sum form would be:

$$f(x_2, x_1, x_0) = (x_2 + x_1 + x_0)(x_2 + \bar{x}_1 + \bar{x}_0)(\bar{x}_2 + x_1 + \bar{x}_0)(\bar{x}_2 + \bar{x}_1 + x_0)$$
$$= f^d(x_2, x_1, x_0).$$

Another reason for the use of the symbol "$\oplus$" is that the sum-of-product and product-of-sum forms cannot be reduced or minimized. Thus, for an n-input exclusive-OR function, a total of 2^{n-1} terms, with n variables each, would be required to write out the function. Using the $\oplus$ symbol, this is reduced to n terms of 1 variable.

Let us determine the dual of the exclusive-OR operator; first by using the truth tables shown in Fig. 4.18, and then by applying the axioms.

$\oplus$

x_1	x_0	$f(x_1, x_0)$
0	0	0
0	1	1
1	0	1
1	1	0

θ

x_1	x_0	$f^d(x_1, x_0)$
1	1	1
1	0	0
0	1	0
0	0	1

$$f(x_1, x_0) = x_1 \oplus x_0$$
$$= x_1 \bar{x}_0 + \bar{x}_1 x_0$$
$$= (x_1 + x_0)(\bar{x}_1 + \bar{x}_0)$$

$$f^d(x_1, x_0) = x_1 \theta x_0$$
$$= x_1 x_0 + \bar{x}_1 \bar{x}_0$$
$$= (x_1 + \bar{x}_0)(\bar{x}_1 + x_0)$$

Figure 4.18

Applying the principle of duality we have:

$$f(x_1, x_0) = (x_1 \bar{x}_0) + (\bar{x}_1 x_0)$$
$$f^d(x_1, x_0) = (x_1 + \bar{x}_0)(\bar{x}_1 + x_0) = \bar{x}_1 \bar{x}_0 + x_1 x_0.$$

Also, the complement of $f(x_1, x_0) = x_1 \oplus x_0$ is

$$\bar{f}(x_1, x_0) = (\bar{x}_1 + x_0)(x_1 + \bar{x}_0) = \bar{x}_1 \bar{x}_0 + x_1 x_0 = x_1 \odot x_0.$$

Thus for the two-input case, the dual and the complement of an exclusive-OR function are the same. Since the dual of $+$ is $\cdot$, we will assign the operator symbol $\odot$, read "ring dot," as the dual symbol to $\oplus$. Noting that $x \odot y = x \oplus y$, and using property 8, we have $x \odot y = x \oplus y \oplus 1$.

The logic symbol which we might expect to use for the operator $\odot$ would be:

However, it is more common to use the complement operator rather than the dual operator yielding the symbol:

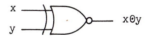

This has led to the name eXclusive NOR (XNOR) for the circuit. We will use both symbols interchangeably in this book.

The properties of the exclusive-NOR function are the duals of those of the exclusive-OR function and will not be listed. Because of the fact that the exclusive-NOR function yields a one output whenever both inputs agree, it is also referred to as a coincidence gate and, as such, is used in comparison circuits.

At first thought, one might expect that since an n input exclusive-OR cascade yields a one for an odd number of inputs equal to one, the dual would be a cascade which yields a one input for an even number of inputs equal to one. Unfortunately, this is not the case. For example:

$$x_2 \odot x_1 \odot x_0 = x_2 \odot (x_1 \odot x_0)$$
$$= x_2 \oplus (x_1 \oplus x_0 \oplus 1) \oplus 1$$
$$= x_2 \oplus x_1 \oplus x_0$$

The problem is not with the concept of duality, but with understanding the concept of a complement; the multi-input cascade which we desire is not the dual function but the complement function.

It is left as an exercise for the reader to show that if

$$f(x_{n-1}, \ldots, x_1, x_0) = x_{n-1} \oplus \cdots x_1 \oplus x_0$$

then

$$\bar{f}(x_{n-1}, x_1, x_0) = x_{n-1} \oplus \cdots x_1 \oplus x_0 \oplus 1$$

and that

$$f^d(x_{n-1}, x_1, x_0) = x_{n-1} \odot \cdots x_1 \odot x_0.$$

It will also be left as an exercise to show that if n is odd,

$$x_{n-1} \oplus x_{n-2} \oplus \cdots \oplus x_1 \oplus x_0 = x_{n-1} \odot x_{n-2} \odot \cdots \odot x_1 \odot x_0.$$

4.5 LOGICAL COMPLETENESS

Definition 4.1 A set of logic operators $(O_1, \ldots, O_k)$ is *functionally complete* if all 2^{2^n} Boolean functions of n variables can be constructed using the variables $x_0, \ldots, x_{n-1}$ and the operators $(O_1, \ldots, O_k)$.

To show that a set of operators is complete, it is sufficient to show that, using this set of operators, the binary operations of "−" (bar) and either "+" or "·" can be obtained, since we know we can realize all the functions with these operators. It is not necessary to obtain both + or ·, since by using De Morgan's law, given the operator "−" and "+" or "·," the missing operator can be always obtained with $x + y = \overline{\bar{x} \cdot \bar{y}}$ and $x \cdot y = \overline{\bar{x} + \bar{y}}$.

Example 4.1 Is the set ($\downarrow$) functionally complete? Yes.

$$\begin{aligned} (x \downarrow x) &= \bar{x} \cdot \bar{x} = \bar{x} && \text{bar can be obtained} \\ (x \downarrow x) \downarrow (y \downarrow y) &= \bar{x} \downarrow \bar{y} = x \cdot y && \text{· can be obtained} \end{aligned}$$

Example 4.2 Is the set ($\uparrow$) functionally complete? Yes.

$$\begin{aligned} (x \uparrow x) &= \bar{x} + \bar{x} = \bar{x} && \text{bar can be obtained} \\ (x \uparrow x) \uparrow (y \uparrow y) &= \bar{x} \uparrow \bar{y} = x + y && \text{+ can be obtained} \end{aligned}$$

Example 4.3 Is the set ($\oplus, \cdot$) functionally complete? No. There is no way of obtaining the NOT operation.

The concept of completeness is important because it reveals that only a few hardware elements are needed in order to realize complex functions. Hence, using only NAND or NOR gates any Boolean function can be realized.

4.6 POSITIVE AND NEGATIVE LOGIC CONVENTIONS

In general, a digital logic circuit can be described in terms of a voltage or current table, which like a logic truth table, specifies the output voltage or current for a fixed combination of input voltages or currents. For example, a three-input single-output device might be described as follows: The output voltage will be low (0 volts) if any input voltage is high (+2.5 volts), and the output voltage will be high (+2.5 volts) if all the input voltages are low (0 volts).

Using the convention of letting H refer to the high voltage or current and L refer to the low voltage or current value, a single device table can be constructed which will adequately specify all devices which satisfy a (high-low) word description similar to it. In doing so, we are ignoring for the moment the actual voltage or current values, whether they are both positive, both negative, or one of each.

There are two choices for assigning a logic "one" to a device table. If the higher value (H) is chosen, then the logic device will be referred to as a positive logic device. If the lower value (L) is chosen, then the logic device will be referred to as a negative logic device. At one time it was hoped that this assignment would correspond to the actual supply voltage used to activate the circuits; i.e.;

positive supply voltage—positive logic

and

negative supply voltage—negative logic.

Voltage Table

x	y	z	f(x,y,z)
L	L	L	H
L	L	H	L
L	H	L	L
L	H	H	L
H	L	L	L
H	L	H	L
H	H	L	L
H	H	H	L

Positive Logic

L=0   H=1
NOR Function

x	y	z	f(x,y,z)
0	0	0	1
0	0	1	0
0	1	0	0
0	1	1	0
1	0	0	0
1	0	1	0
1	1	0	0
1	1	1	0

$$f(x,y,z) = \bar{x}\cdot\bar{y}\cdot\bar{z}$$
$$= \overline{x+y+z}$$

Negative Logic

L=1   H=0
NAND Function

x	y	z	f(x,y,z)
1	1	1	0
1	1	0	1
1	0	1	1
1	0	0	1
0	1	1	1
0	1	0	1
0	0	1	1
0	0	0	1

$$f(x,y,z) = \bar{x}+\bar{y}+\bar{z}$$
$$= \overline{x\cdot y\cdot z}$$

Fig. 4.19 Logic Convention

Unfortunately, industry has not followed this hoped for convention.

The two possible logic assignments yield:

$$\text{a) positive logic} \quad H = \text{logical one}$$
$$L = \text{logical zero}$$
$$\text{b) negative logic} \quad H = \text{logical zero}$$
$$L = \text{logical one}$$

The Voltage Table and the corresponding Positive and Negative Logic Tables are shown in Fig. 4.19.

The two functions realized by the assignments are a positive NOR gate and a negative NAND gate. Again, from these figures we see that dual assignments of the voltage table lead to dual logic function realization.

4.7 THE WIRED GATES

Depending on the actual circuit implementation, it is sometimes possible to use the gates in a wired-OR or wired-AND configuration.

This is sometimes possible when either the low-logic voltage value or the high-logic voltage value is 0 volts. Note that is neither necessary nor sufficient to allow the use of wired gates. Let us assume that we have a circuit with one of the voltage values equal to zero volts.

CASE 1. $L = 0$ volts and $H > 0$ volts. Any time one of the inputs is at ground, the output is forced to ground. Thus the output will only be high when all inputs are high.

Using a positive logic convention on the inputs and output, this device can be considered a wired-AND gate. Using a negative logic convention on the inputs and on the output, this device can be considered a wired-OR gate.

CASE 2. $L < 0$ volts and $H = 0$ volts. In this instance, only if all the input lines are low will a nonzero output voltage be produced, yielding a wired-OR gate for a positive logic convention.

Wired AND   Wired OR

The reader should be cautioned when using this wired-OR or wired-AND circuit for once an output is connected to one of these wired devices, it cannot be used or considered as a separate output. In Fig. 4.20 the output of each NAND

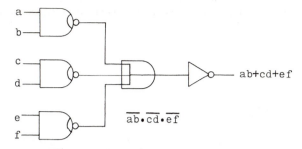

ab+cd+ef

$$\overline{ab} \cdot \overline{cd} \cdot \overline{ef}$$

Fig. 4.20 Use of Wired AND Gates

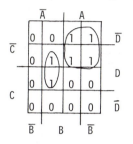

Mapping 1's sample function

$$f(A,B,C,D) = A\overline{C} + \overline{A}BD$$

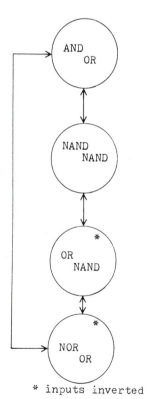

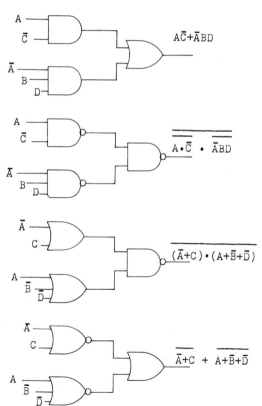

* inputs inverted

Fig. 4.21 Two-Level Logic Circuits from a K-Map Mapping 1's

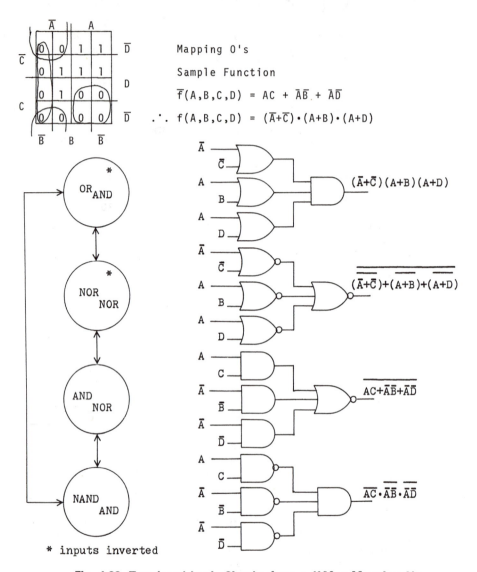

Mapping 0's

Sample Function

$\overline{F}(A,B,C,D) = AC + \overline{A}\overline{B} + \overline{A}\overline{D}$

$\therefore f(A,B,C,D) = (\overline{A}+\overline{C})\cdot(A+B)\cdot(A+D)$

* inputs inverted

Fig. 4.22 Two-Level Logic Circuits from a K-Map Mapping 0's

gate is $\overline{a \cdot b \cdot c \cdot d \cdot e \cdot f}$. A semicircle is added to the logic symbol output in order to denote a logic gate which can be used in a wired configuration.

These devices have a big advantage in lowering parts and packaging costs by eliminating in many cases the need for extra gates. However, circuit devices which can be used as wired gates tend to be slower and somewhat more expensive than circuits without this feature.

The reader is cautioned NOT to adopt the practice of omitting the overlaid AND or OR symbol when using wired gates. Unless one knows the actual logic convention, the symbol:

is ambiguous and hence meaningless.

In Figs. 4.21 and 4.22 the eight two-level logic circuits which can be obtained using NAND, NOR, AND, and OR gates are shown. The realization assumes that both true and complement inputs are available. The reader should follow each loop on these two diagrams and verify the transformations.

4.8 LOGIC SYMBOL CONVENTIONS

Since there is no set of symbols universally agreeable to industry, government, universities, and authors of books on logic design, it is necessary to state the conventions to be used in this book. The personal bias of the author is that the logic symbol should be valid regardless of the logic convention or physical device used to build the actual circuit. For this reason we will adopt the following conventions and logic symbols.

LOGIC CONVENTIONS

a) The basic logic symbols are:

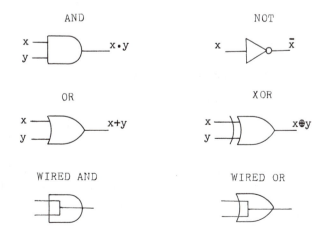

b) A circle on an input line means complement the input variable:

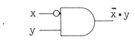

c) A circle on an output line means complement the output function:

d) An equivalent or alternate logic symbol can be obtained by using the dual symbol and complementing all input and output lines:

e) Special functions will be denoted with labeled boxes:

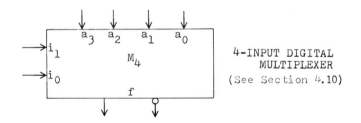

4-INPUT DIGITAL MULTIPLEXER

(See Section 4.10)

The output expression for the circuit in Fig. 4.23 is $z = \overline{x \cdot \overline{y}} = \overline{x} + y$, irrespective of whether the logic convention is positive or negative. Thus as far as writing the logic equations of a logic circuit, *the symbols tell all that is needed.* We only need to know the voltage-logic assignment and the voltage levels when it is actually necessary either to take measurements on a device which has been implemented with hardware or to assemble a new circuit.

Table 4.1 shows the logic symbols which are most commonly used in this book. Although there is no theoretical limit to the number of inputs which a gate can have, there are practical limits. Commercial logic gates are usually packaged as follows: six 1-input gates, four 2-input gates, three 3-input gates, two 4-input gates and one 8-input gate. The cost for each of these packages is generally the

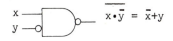

Fig. 4.23 Logic Symbol for $\bar{x}+y$

TABLE 4.1
Commonly Used Logic Symbols

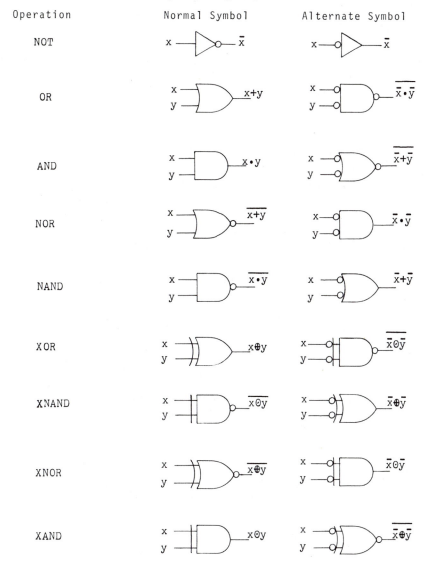

same for a given logic element. Special orders are usually available with odd numbers of inputs, but these generally cost more. When figuring the cost of a logic circuit using these prepackaged units, the designer should also consider the cost of any unused gates.

One should not intermix logic devices which have different voltage levels without providing conversion circuits which yield compatible voltage and current levels. For example, you would not be successful if you attempted to directly connect a logic circuit with H corresponding to zero volts and L corresponding to -3 volts to another set of logic devices with H corresponding to $+5$ volts and L corresponding to 0 volts.

In general, a good design should use as few level converters as possible since these devices increase both the cost, by requiring additional power supplies, and the time required for the circuits to reach a steady-state value. Level converters are, however, commonly used for interfacing existing digital components with different voltage levels.

4.9 LOGIC EQUIVALENT CIRCUITS

There are times when we would like to simplify or modify our logic circuits either for purposes of analysis or because the specific circuit was designed using one type of logic gate and the hardware available is of another type. For example, a NAND circuit diagram is drawn and only NOR gates are available as hardware components. In this section, we will set up some simple rules which will allow us to convert a logic circuit from one set of logic symbols to another set of logic symbols. It should be understood that the conversion is in general not optimum. It is possible that by starting over and redesigning the circuit a better final logic circuit with fewer parts may be obtained. In many cases, in order to preserve the overall input-output equations, inverters will have to be inserted or removed from the new logic circuit.

The following set of equivalence relationships allow one to modify existing logic circuit diagrams while preserving the overall input-output relationship.

1. On Inputs to Gates: The input-output equations will not be changed if you:
   a) remove a circle on an input line and complement the input variable:

$$x \!-\!\!\circ\!| \qquad \approx \qquad \bar{x} \!-\!|$$

   b) insert a circle on an input line and complement the input variable:

2. On Outputs from Gates: The input-output equations will not be changed if you:
   a) remove a circle on an output line and complement the output function:

   b) insert a circle on an output line and complement the output function:

3. The input-output equations will not be changed if a circuit symbol is replaced with its alternate circuit symbol (the dual symbol with all input and output lines complemented), for example:

4. Single-Line Connections Between Gates: The overall input-output equations will not be changed if both ends of the line are complemented and the line variable is complemented:

5. Connections Between Several Gates: The overall input-output equations will not be changed if at any time one end of a line is complemented, all of the other terminals connected to the line and the line variable are also complemented.

Example 4.4 In Fig. 4.24, the four logic diagrams are equivalent. Circuit (b) is obtained from circuit (a) by using rule 3. Circuit (c) is obtained from circuit (b) using rule 4, and finally, circuit (d) is obtained from circuit (c) using rule 3 again.

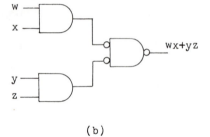

(a)

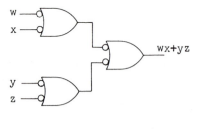

(b)

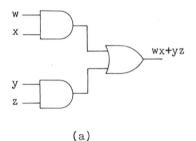

(c)

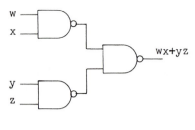

(d)

Figure 4.24

Example 4.5 Realize the following circuit using only NOR gates.

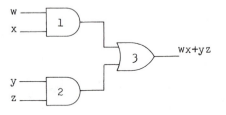

Step 1. Complement the input and output variables using rules 1 and 2. (It will be necessary to correct this later.)

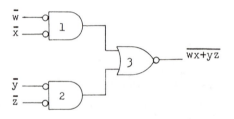

Step 2. Convert gates 1 and 2 to their alternate symbols.

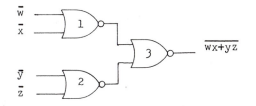

Step 3. Add inverters on the input and output lines in order to yield the correct input-output equation.

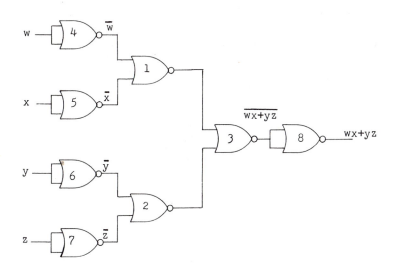

In this case the equivalent circuit requires five additional gates, more than twice the number required originally.

If we had redesigned the circuit, we could have converted the function into product-of-sum form, and then realized the function directly using two levels of NOR gates.

$$f(w, x, y, z) = wx + yz$$

$$f^d(w, x, y, z) = (w + x)(y + z) = wy + wz + xy + xz$$

$$f^{dd}(w, x, y, z) = (w + y)(w + z)(x + y)(x + z)$$

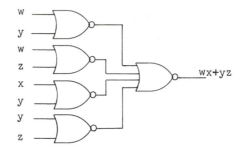

4.10 DIGITAL MULTIPLEXERS AS LOGIC DEVICES

A digital multiplexer is the electronic equivalent of a rotary switch. The unit has 2^n input lines for each section and n binary-selection lines which determine the input line to be transferred to the output of each section. Figure 4.25 shows a logic circuit for a typical 4-input, 2-selection-line, single-stage multiplexer. This unit is the equivalent of a four-position, single-pole, rotary switch. Figure 4.26 shows a two-stage unit with 2 inputs per stage and a single selection line. This unit is equivalent to a double-pole double-throw switch.

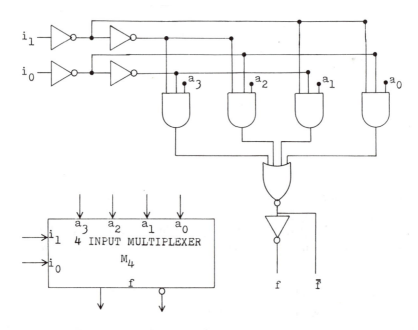

Fig. 4.25 Logic Circuit for a 4-Input Multiplexer

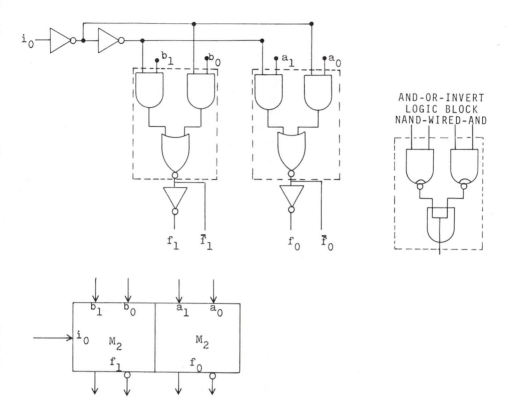

Fig. 4.26 Logic Circuit for a 2-Stage, 2-Input, Single-Line-Control Multiplexer

In general, the selection lines are coded in binary, so that input line a_j is chosen if

$$j = \sum_{k=0}^{n-1} i_k 2^k,$$

where n is the number of selection lines. Thus in Fig. 4.25 if $i_0 = 1$ and $i_1 = 0$ then $j = 1$, and the input line a_1 would be switched to the output.

Digital multiplexer units are quite inexpensive and easy to fabricate as integrated circuit units. By using an AND-OR-INVERT logic unit as its main component, we can reduce the actual number of transistors needed to less than that required for two levels of NAND or NOR logic. The selection lines are buffered by means of two back-to-back inverters, thereby reducing the circuit loading on the selection lines. Finally, both f and $\bar{f}$ are available at the output of the unit. The reader should have no trouble extending this circuit design to include units with 2^n inputs and n selection lines. A second look at the multiplexer unit in Fig. 4.25 reveals that its output, if written out completely with x_1 and x_0 applied to the i_1 and i_0 inputs, is

$$f(x_1, x_0) = a_0\bar{x}_1\bar{x}_0 + a_1\bar{x}_1x_0 + a_2x_1\bar{x}_0 + a_3x_1x_0$$

$$= \bigvee_{I=0}^{3} a_I P_I^2.$$

This is the fundamental sum-of-product form for a general two-variable Boolean function. Thus a 4-input multiplexer with its two selection lines can be wired to realize any two-variable Boolean function simply by converting its characteristic number into binary and applying the corresponding a_i's to the multiplexer inputs. For example, a multiplexer for realizing $D_2 = 14$ would be wired as shown in Fig. 4.27.

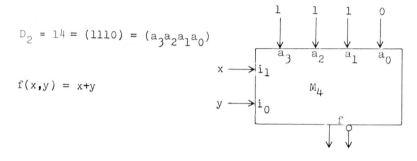

$$D_2 = 14 = (1110) = (a_3a_2a_1a_0)$$

$$f(x,y) = x+y$$

Fig. 4.27 Multiplexer Circuit for $D_2=14$

TABLE 4.2

x y z	$i = 4x + 2y + z$	$T(i) = i^2+2i$	Coefficients $(i^2 + 2i)_2$ $f_i\,e_i\,d_i\,c_i\,b_i\,a_i$
0 0 0	0	0	0 0 0 0 0 0
0 0 1	1	3	0 0 0 0 1 1
0 1 0	2	8	0 0 1 0 0 0
0 1 1	3	15	0 0 1 1 1 1
1 0 0	4	24	0 1 1 0 0 0
1 0 1	5	35	1 0 0 0 1 1
1 1 0	6	48	1 1 0 0 0 0
1 1 1	7	63	1 1 1 1 1 1

These units are often used to build simple read-only memories for computers. For example, if we needed to calculate the function $T(x) = x^2 + 2x$, for any number $0 \leq x \leq 7$, one solution would be to use a table-lookup procedure. By using six 8-input multiplexers wired in parallel, with the a_i inputs wired to yield the binary value $i^2 + 2i$, this function can be easily implemented without a lot of messy logic-equation manipulation or circuit wiring of separate logic gates. See Fig. 4.28 for the required circuit implementation. If $x\,y\,z$ equal 0 1 1, then the output $t_5 t_4 t_3 t_2 t_1 t_0$ is 0 0 1 1 1 1.

There is, however, another use for these multiplexer units. A digital multiplexer unit with 2^n input lines and n selection lines can be wired to realize any Boolean function of $n + 1$ variables. Thus, an 8-input multiplexer unit can be

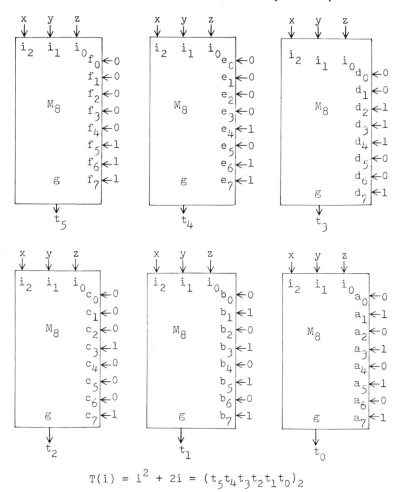

$$T(i) = i^2 + 2i = (t_5 t_4 t_3 t_2 t_1 t_0)_2$$

Fig. 4.28 Multiplexer Realization for Table 4.2

easily wired to realize any of the $2^{2^4} = 65,536$ distinct Boolean functions of four variables.

Let us examine the truth table for a three variable function $f(x, y, z) = x\bar{z} + yz$. See Table 4.3. If we were to expand this function in terms of the binary variables x and y, we would obtain:

$$f(x, y, z) = \bar{x}\bar{y}f(0, 0, z) + \bar{x}yf(0, 1, z) + x\bar{y}f(1, 0, z) + xyf(1, 1, z)$$
$$= \bar{x}\bar{y} \cdot 0 + \bar{x}y \cdot z + x\bar{y} \cdot \bar{z} + xy \cdot 1$$
$$= \bar{x}\bar{y} \cdot a_0 + \bar{x}y \cdot a_1 + x\bar{y} \cdot a_2 + xy \cdot a_3.$$

If we assign the function $f(i_1, i_0, z)$ to input line a_I, $I = 2i_1 + i_0$, and the variables x to y to the selection inputs i_1 and i_0 respectively, then this function can be realized with a digital multiplexer. (See Table 4.3.) Each of the coefficients, a_I's, is a function of one variable, and therefore must be equal to one of the four binary functions of one variable; 0, 1, z, or $\bar{z}$. By comparing the output column, $f(i_1, i_0, z)$, with the z input column for each pair of rows with the same values of i_1 and i_0, the input coefficients a_I's can be quickly and easily determined.

TABLE 4.3

$x = i_1$	$y = i_0$	z	$f(i_1, i_0, z) = f(x,y,z)$	a_i's
0	0	0	0	
0	0	1	0	$a_0 = 0$
0	1	0	0	
0	1	1	1	$a_1 = z$
1	0	0	1	
1	0	1	0	$a_2 = \bar{z}$
1	1	0	1	
1	1	1	1	$a_3 = 1$

In order to determine a_0, we set $i_1 = 0$ and $i_0 = 0$, thereby selecting rows $f(0, 0, 0)$ and $f(0, 0, 1)$. Comparing the output column with column z shows that regardless of the value of z, the output is equal to zero. Hence, $f(0, 0, z) = a_0 = 0$.

To determine a_1, we set $i_1 = 0$ and $i_0 = 1$, thereby selecting rows $f(0, 1, 0)$ and $f(0, 1, 1)$. Comparing the output column with column z shows that the value of z agrees in both cases with the output. Hence, $f(0, 1, z) = a_1 = z$.

To determine a_2, we set $i_1 = 1$ and $i_1 = 0$, thereby selecting rows $f(1, 0, 0)$ and $f(1, 0, 1)$. Comparing the output column with column z shows that the output is always the complement of z. Hence $f(1, 0, z) = a_2 = \bar{z}$.

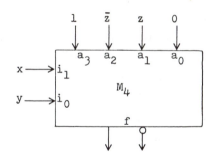

Fig. 4.29 Multiplexer Realization of $f(x, y, z) = xy + x\bar{y}\bar{z} + \bar{x}yz$

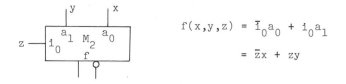

$$f(x,y,z) = \bar{i}_0 a_0 + i_0 a_1$$
$$= \bar{z}x + zy$$

Fig. 4.30 Multiplexer Realization of $f(x, y, z) = x\bar{z} + yz$

To determine a_3, we set $i_1 = 1$ and $i_0 = 1$, thereby selecting rows $f(1, 1, 0)$ and $f(1, 1, 1)$. Comparing the output column with column z shows that regardless of the value of z, the output is equal to 1. Hence $f(1, 1, z) = a_3 = 1$.

The multiplexer circuit in Fig. 4.29 shows how this function can be wired. Writing out the function realized by the multiplexer in Fig. 4.29 yields:

$$f(x, y, z) = a_3xy + a_2x\bar{y} + a_1\bar{x}y + a_0\bar{x}\bar{y}$$
$$= xy + x\bar{y}\bar{z} + \bar{x}yz.$$

Further logical manipulation of this function reveals that $f(x, y, z) = x\bar{z} + yz$. This can be easily determined from a K-map. Using z as the selection line input, this function can also be realized by the multiplexer circuit shown in Fig. 4.30.

Example 4.6 Given $f(x, y, z) = x\bar{y} + yz + \bar{x}\bar{z}$, realize this function with a 4-input multiplexer.

In order to determine the a_i's using x and y as selection inputs, it is only necessary to expand the function so that all terms include the variables x and y. Doing this yields:

$$f(x, y, z) = x\bar{y} + (x + \bar{x})yz + \bar{x}(y + \bar{y})\bar{z}$$
$$= x\bar{y} + xyz + \bar{x}yz + \bar{x}y\bar{z} + \bar{x}\bar{y}\bar{z}$$
$$= xyz + x\bar{y} + \bar{x}y + \bar{x}\bar{y}\bar{z}$$
$$= xya_3 + x\bar{y}a_2 + \bar{x}ya_1 + \bar{x}\bar{y}a_0$$

From this expanded form we see that $a_3 = z$, $a_2 = 1$, $a_1 = 1$, and $a_0 = \bar{z}$. The same set of coefficients could have been obtained by evaluating the function for all values of x and y. Doing this would yield directly:

$$a_0 = f(0, 0, z) = \bar{z}$$
$$a_1 = f(0, 1, z) = 1$$
$$a_2 = f(1, 0, z) = 1$$
$$a_3 = f(1, 1, z) = z$$

The final circuit for this function is:

Let us now consider a four-variable problem, using Truth Table A shown in Fig. 4.31. Using the same procedure as in the last example yields the circuit shown in Fig. 4.32. Again, this function can be quickly implemented using the 8-bit multiplexer by expanding the function about the variables w, x, and y.

w x y z	f (w , x , y , z)	a_i
0 0 0 0	0	
0 0 0 1	0	$a_0 = "0"$
0 0 1 0	1	
0 0 1 1	0	$a_1 = \bar{z}$
0 1 0 0	0	
0 1 0 1	1	$a_2 = z$
0 1 1 0	0	
0 1 1 1	0	$a_3 = "0"$
1 0 0 0	1	
1 0 0 1	1	$a_4 = "1"$
1 0 1 0	0	
1 0 1 1	1	$a_5 = z$
1 1 0 0	1	
1 1 0 1	0	$a_6 = \bar{z}$
1 1 1 0	1	
1 1 1 1	1	$a_7 = "1"$

Fig. 4.31 Truth Table A

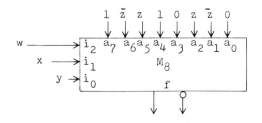

Fig. 4.32 Circuit for Truth Table A

The same function could also be implemented with four 2-input multiplexer units at level one and a 4-input multiplexer unit at the second level as shown in Fig. 4.33. In this case Truth Table A is partitioned into four smaller tables shown in Fig. 4.34.

The second-level unit then has the outputs of each of the first level-units as inputs.

$$f(w, x, y, z) = \bar{w}\bar{x}f(0, 0, y, z) + \bar{w}xf(0, 1, y, z) + w\bar{x}f(1, 0, y, z)$$
$$+ wxf(1, 1, y, z)$$

with

$$f(0, 0, y, z) = y\bar{z}$$
$$f(0, 1, y, z) = \bar{y}z$$
$$f(1, 0, y, z) = \bar{y} + yz = \bar{y} + z$$
$$f(1, 1, y, z) = \bar{y}\bar{z} + y = \bar{z} + y.$$

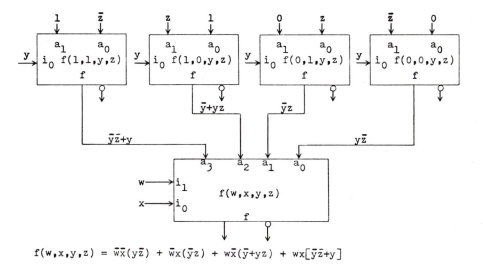

$$f(w,x,y,z) = \bar{w}\bar{x}(y\bar{z}) + \bar{w}x(\bar{y}z) + w\bar{x}(\bar{y}+yz) + wx[\bar{y}\bar{z}+y]$$

Fig. 4.33 Two-Level Multiplexer Circuit

	w = 0	x = 0		
y	z	$f(0,0,y,z)$		a_i
0	0	0	$a_0 = 0$	
0	1	0		
1	0	1	$a_1 = \bar{z}$	
1	1	0		

	w = 0	x = 1		
y	z	$f(0,1,y,z)$		a_i
0	0	0	$a_0 = z$	
0	1	1		
1	0	0	$a_1 = 0$	
1	1	0		

	w=1	x = 0		
y	z	$f(1,0,y,z)$		a_i
0	0	1	$a_0 = 1$	
0	1	1		
1	0	0	$a_1 = z$	
1	1	1		

	w=1	x = 1		
y	z	$f(1,1,y,z)$		a_i
0	0	1	$a_0 = \bar{z}$	
0	1	0		
1	0	1	$a_1 = 1$	
1	1	1		

Figure 4.34

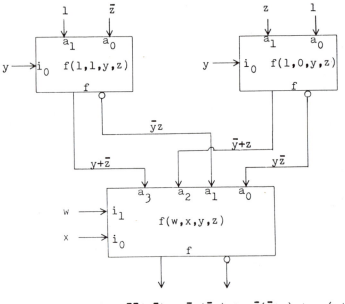

$$f(w,x,y,z) = \bar{w}\bar{x}(y\bar{z}) + \bar{w}x(\bar{y}z) + w\bar{x}(\bar{y}+z) + wx(y+\bar{z})$$

Fig. 4.35 Reduced Two-Level Multiplexer Circuit

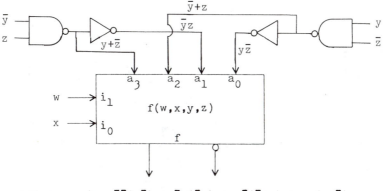

$$f(w,x,y,z) = \bar{w}\bar{x}(y\bar{z}) + \bar{w}x(\bar{y}z) + w\bar{x}(\bar{y}+z) + wx(y+\bar{z})$$

Fig. 4.36 Mixed Logic and Multiplexer Solution

If we examine the functions $f(i_1, i_0, y, z)$ more closely, we observe that:

$$f(0, 0, y, z) = y\bar{z} = \overline{\bar{y} + z} = \bar{f}(1, 0, y, z)$$
$$f(0, 1, y, z) = \bar{y}z = \overline{y + \bar{z}} = \bar{f}(1, 1, y, z)$$
$$f(1, 0, y, z) = \bar{y} + yz = \bar{y} + z$$
$$f(1, 1, y, z) = \bar{y}\bar{z} + y = y + \bar{z}.$$

Since the complement output is available on these multiplexer units, two of the first-level units are not needed. Sharing the multiplexer outputs leads to the circuit shown in Fig. 4.35.

A final alternative for realizing this function is to replace the 2-input multiplexers with simple logic gates, and only use the single 4-input multiplexer. The resulting circuit realization is shown in Fig. 4.36.

We can use several levels of multi-input multiplexers to build a universal logic unit. Such a unit can be used to realize all functions of n variables. Figure 4.37 shows the structure of a unit which uses two levels of 4-input multiplexers to realize a universal 5-variable logic unit. You should have no trouble generalizing this circuit to handle either more variables or larger multiplexer units.

Example 4.7 Given the characteristic number $H_4 = 75C9$, realize the corresponding function using a single 4-input multiplexer and a few additional logic gates.

We can quickly implement the corresponding function using Fig. 4.38 as an aid.

The original characteristic number is written on the first row, with its corresponding binary characteristic number on the second row. The corresponding y and z entries from a truth table are written on the next two rows. Since we

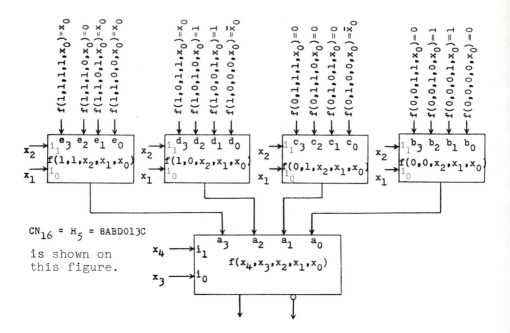

$CN_{16} = H_5 = 8ABD013C$

is shown on
this figure.

Fig. 4.37 5-Variable Multiplexer Unit

must use a 4-input multiplexer, the table is divided into four sections corresponding to the 4-input functions needed for the multiplexer. The required functions are
easily obtained by inspection of the table. The B_4 row specifies the output
function and rows y and z the input variables.

The resulting multiplexer circuit is

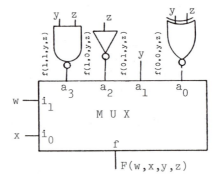

Given the characteristic number of a Boolean function, it is a relatively easy
matter to determine the required input coefficients for a digital multiplexer. For
example, in Fig. 4.39, let $H_5 = 8ABD013C$.

H_4	7	5	C	9
B_4	0111	0101	1100	1001
y	1100	1100	1100	1100
z	1010	1010	1010	1010
	$\overline{y \cdot z}$	$\bar z$	y	$\overline{y \oplus z}$
a_I's	a_3	a_2	a_1	a_0
	$f(1,1,y,z)$	$f(1,0,y,z)$	$f(0,1,y,z)$	$f(0,0,y,z)$

Figure 4.38

	8		A		B		D		0		1		3		C	
$H_5 =$	8		A		B		D		0		1		3		C	
$B_5 =$	1 0	0 0	1 0	1 0	1 0	1 1	1 1	0 1	0 0	0 0	0 0	0 1	0 0	1 1	1 1	0 0
x_0	1 0	1 0	1 0	1 0	1 0	1 0	1 0	1 0	1 0	1 0	1 0	1 0	1 0	1 0	1 0	1 0
a_I's	a_{15}	a_{14}	a_{13}	a_{12}	a_{11}	a_{10}	a_9	a_8	a_7	a_6	a_5	a_4	a_3	a_2	a_1	a_0
$f(i_3,i_2,i_1,i_0,x_0)$	x_0	0	x_0	x_0	x_0	1	1	$\bar x_0$	0	0	0	$\bar x_0$	0	1	1	0
Inputs for Figure 4.37	e_3	e_2	e_1	e_0	d_3	d_2	d_1	d_0	c_3	c_2	c_1	c_0	b_3	b_2	b_1	b_0

Figure 4.39

The characteristic number is first converted into binary. Below these rows of numbers is written a row of alternating ones and zeroes corresponding to the x_0 column of a truth table. The B_5 and x_0 rows are now compared pairwise to determine the correct a_I coefficients. If both values being examined in row B_5 are zero, a_I is zero; if both values being examined are ones, then a_I is set to one. If both values being examined agree with the entries in the x_0 row, the value of a_I is set to x_0; and if both values being examined differ from those in x_0, then the value of a_I is set to $\bar x_0$. This function is superimposed on Fig. 4.37.

4.11 BINARY DEMULTIPLEXERS OR DECODERS

The opposite of a multiplexer is a demultiplexer. This circuit has a single input, n selection lines, and 2^n output terminals. The purpose of the unit is to pass the data contained on the input line to a single output terminal (I) selected by selection lines.

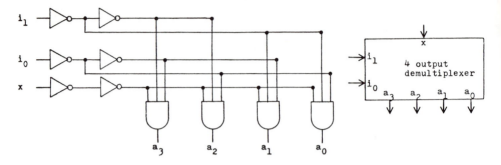

Fig. 4.40 Logic Circuit for a Binary Demultiplexer

A zero level is maintained on the unselected outputs.

Figure 4.40 shows the logic symbol and circuit for a 4-output demultiplexer. In many cases, NAND gates are used instead of AND gates, yielding inverted outputs with ones on the unselected outputs.

One of the main uses for these devices is in binary-to-octal and binary-to-decimal decoders. By placing a logical one on the input x only the selected output line a_I will equal one.

4.12 PROBLEMS

4.1   Verify the following statements:

a) $x \uparrow (x + z) \neq (x \uparrow y) + (x \uparrow z)$

b) $x \downarrow (y \cdot z) \neq (x \downarrow y) \cdot (x \downarrow z)$

c) $x \uparrow (y \uparrow z) \neq (x \uparrow y) \uparrow z$

4.2   Construct a table showing the implementation of the sixteen Boolean functions of two variables. Order your functions according to their characteristic numbers. You may assume that the inputs 0, 1, x, $\bar{x}$, y, and $\bar{y}$ are available for your use.

a) Use only NAND gates.

b) Use only NOR gates.

4.3   Implement the following Boolean functions:

a) Directly using AND, OR, and NOT gates;

b) Directly using NAND and NOR gates (use law of involution);

   c) By converting the function into sum-of-product form and then using two levels of NAND gates;

   d) By converting the function into product-of-sum form and then using two levels of NOR gates;

You may assume that the true and complements of the variables are available for your use.

    1. $F(w, x, y, z) = \overline{wx} + y(\bar{z} + x)$
    2. $F(w, x, y, z) = (wy)[x + \bar{y}(z + \bar{w})]$
    3. $F(w, x, y, z) = \overline{(x + y)(\bar{w}z)}$

4.4 Implement the following Boolean functions using only two input gates and inverters. Assume that only the true inputs are available.

   a) $F(x, y, z) = x + y\bar{z}$

   b) $F(w, x, y, z) = wxyz + w\bar{x}y + x\bar{y}z + \bar{w}\bar{z}$

   c) $F(w, x, y, z) = (x + y)(w + \bar{z} + y)(\bar{x} + \bar{z} + \bar{y})$

4.5 Implement the Boolean functions corresponding to the following characteristic numbers using:

   a) only NAND gates,

   b) only NOR gates,

   c) only 1, 2, 3, or 4 input gates (AND, OR, NAND, NOR, NOT).

    1. $H_3 = A7$ 4. $H_3 = 34$
    2. $H_4 = 0B4D$ 5. $H_4 = 71E2$
    3. $H_4 = 6317$ 6. $H_4 = F321$

You may assume that both the true and the complements of the input variables are available for your use. (Hint: try to use a K-map to help reduce the complexity of your circuit realizations.)

4.6 Show that it is possible to express all sixteen functions of two variables using only AND gates and XOR gates, with $F(x, y) = b_0 \oplus b_1 x \oplus b_2 y \oplus b_3 xy$.

   a) Make a table listing the set of coefficients $[b_3, b_2, b_1, b_0]$ required for each Boolean function along with the function.

   b) Given that $f(x, y)$ can also be expressed as $f(x, y) = a_3 xy + a_2 x\bar{y} + a_1 \bar{x}y + a_0 \bar{x}\bar{y}$, determine a relationship between the b_i's and the a_i's in the two functions.

4.7 Let us assume we have three light switches which, when all are in the down position, turn the light off. Starting from this state any single change of a switch will result in the light being turned on if it was off or being turned off if it was on.

   a) Let $L = 0$ mean the light is off and $L = 1$ mean the light is on. Write a logic expression for the condition when the light is on $L(x, y, z)$ in terms of the three switches x, y, z. A switch in the up position is assumed to be a one.

   b) Express the function obtained in a) in terms of the $\oplus$ operator.

   c) Generalize the preceding result to handle n switches.

4.8 Show that in general if

$$F(x_{n-1}, \ldots, x_1, x_0) = x_{n-1} \oplus \cdots \oplus x_1 \oplus x_0$$

then

$$\bar{F}(x_{n-1}, \ldots, x_1, x_0) = x_{n-1} \oplus \cdots \oplus x_1 \oplus x_0 \oplus 1$$

and that

$$F^d(x_{n-1}, \ldots, x_1, x_0) = x_{n-1} \odot \cdots \odot x_1 \odot x_0.$$

4.9 Show that if n is odd then

$$x_{n-1} \oplus x_{n-2} \oplus \cdots \oplus x_1 \oplus x_0 = x_{n-1} \odot x_{n-2} \odot \cdots \odot x_1 \odot x_0.$$

4.10 Are the following sets of logic operators complete?

a) $[\oplus, \cdot, 1]$ b) $[+, \text{NOT}]$

c) $[\cdot, \text{NOT}]$ d) $[+, \cdot]$

Note: Unless specified in the set of operators, the constants 0 and 1 must be generated using the set of operators.

4.11 Write the output functions in sum-of-product form for each of the gates in the circuits below.

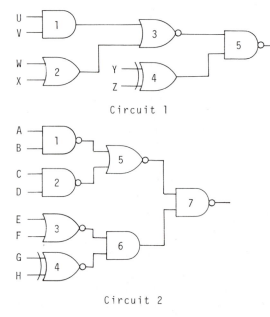

Circuit 1

Circuit 2

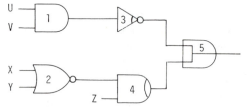

Circuit 3

4.12 What logic functions do the voltage tables below correspond to:
       a) assuming a positive logic convention?
       b) assuming a negative logic convention?

x	y	z	F(x,y,z)
L	L	L	L
L	L	H	L
L	H	L	H
L	H	H	H
H	L	L	H
H	L	H	H
H	H	L	L
H	H	H	H

x	y	z	F(x,y,z)
L	L	L	L
L	L	H	H
L	H	L	H
L	H	H	L
H	L	L	H
H	L	H	L
H	H	L	H
H	H	H	H

4.13 Assuming true and complement input variables are available realize the functions
       shown on the following K-maps using two levels of logic gates:
       a) AND followed by OR,   b) NAND followed by NAND,
       c) OR followed by NAND,   d) NOR followed by OR.

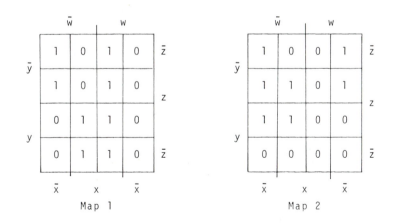

4.14 Assuming true and complement input variables are available, realize the functions
       shown on the following K-maps using two levels of logic gates:
       a) OR followed by AND,   b) NOR followed by NOR,
       c) AND followed by NOR,   d) NAND followed by AND.

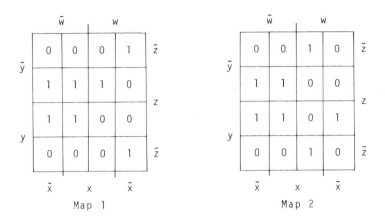

Map 1   Map 2

4.15 Convert the following circuits into equivalent circuits which use only:
   a) NAND gates,
   b) NOR gates,
   c) NAND and NOR gates,
   d) AND, OR, NOT gates.

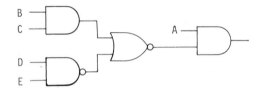

Circuit 1

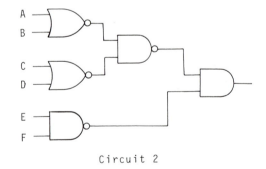

Circuit 2

Hint: Use the guide lines given in Section 4.9.

4.16 Implement the following Boolean functions using digital multiplexers as table look-up units.
   a) $H_2 = A$ b) $H_3 = 71$
   c) $H_3 = BC$ d) $H_4 = A025$

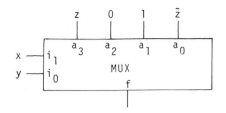

Circuit (1)

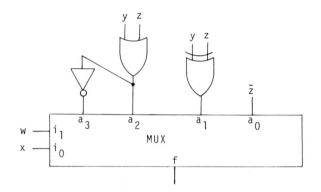

Circuit (2)

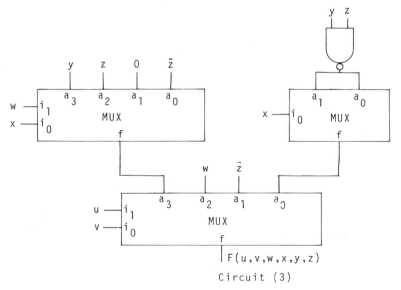

F(u,v,w,x,y,z)

Circuit (3)

Figure 4.41

4.17 Implement the following Boolean functions of n variables using digital multiplexers with only $n - 1$ selection inputs.

a) $H_2 = 7$ b) $H_3 = 1A$

c) $H_4 = AC12$ d) $H_5 = 2B7642E1$

e) $H_3 = 46$ f) $H_4 = 2120$

4.18 Implement the following Boolean functions of n variables using a single digital multiplexer with $n - 2$ selection inputs and a few assorted logic gates.

a) $H_4 = 12E4$ b) $H_3 = 37$

c) $H_4 = 261A$ d) $H_5 = 12345678$

4.19 Implement the following logic functions of n variables using two levels of multiplexers with at most two selection inputs on each unit. A few additional logic gates may be required.

a) $H_4 = A12B$ b) $H_5 = 6374A2F0$

c) $H_5 = 1754B2C3$ d) $H_6 = 631204FE6174265B$

4.20 Write out the corresponding output functions for the following multiplexer circuits. (See Fig. 4.41.)

4.21 Design a special demultiplexer logic circuit which can be used to convert four bits of binary data into decimal data, i.e., the output a_I should be zero unless the control input corresponds to the decimal number I. With

$$(I)_{10} = (i_3 i_2 i_1 i_0)_2.$$

4.22 Determine the circuit configuration of the dual of a multiplexer. Is this a useful circuit?

4.13 BIBLIOGRAPHY

Boyce, J. C., *Digital Logic and Switching Circuits*. Englewood Cliffs, N.J.: Prentice-Hall, 1975.

Carr, W. N., and J. P. Mize, *MOS/LSI Design and Application*. New York: McGraw-Hill, 1972.

Danielsson, P. E., "A Note on Wired-OR Gates," *IEEE Transactions on Computers*, C-19, No. 9, 849–850 (September, 1970).

Ellis, D. T., "A Synthesis of Combinational Logic with NAND or NOR elements," *IEEE Trans. on Electronic Computers*, Vol. EC-14, 701–705, October, 1965.

Harrison, M. A., *Introduction to Switching and Automata Theory*. New York: McGraw-Hill, 1965.

Hill, F. J., and G. R. Peterson, *Introduction to Switching Theory and Logical Design*. 2nd ed. New York: Wiley, 1974.

Larsen, D. G., and P. R. Rony, *The Bugbook I*. Derby, Conn.: E and L Instruments, 1974.

Larsen, D. G., and P. R. Rony, *The Bugbook II*. Derby, Conn.: E and L Instruments, 1974.

Peatman, J. B., *The Design of Digital systems*. New York: McGraw-Hill, 1972.

Shannon, C. E., "Symbolic analysis of Relay and Switching circuits," *Transactions AIEE*, 57, 713–723 (1938).

Chapter 5

Minimization Procedures for Boolean Functions

5.1 INTRODUCTION

The classical Boolean minimization procedures are presented in this chapter. Section 5.2 attempts to answer the question, "What should be minimized and why?" Section 5.3 contains the necessary background material for the classical two-level Boolean minimization procedures. The Map method is explained in Section 5.4, while the Tabular method is explained in Section 5.5. the extension of these two methods to incompletely specified functions is detailed in Section 5.6.

5.2 MINIMIZATION—WHEN AND WHY

With the advent of medium-scale integrated circuits (MSI) and large-scale integrated circuits (LSI) as functional building blocks, the question of gate minimizations has become less important than the problem of minimizing input and output connections from one circuit to another. In many cases nonminimal circuit configurations can be obtained in prepackaged units for less cost than a minimal circuit which must be tailor-made.

The types of building blocks available off-the-shelf for use in design include: NAND, NOR, AND, OR, Inverters, XOR, AND-OR-INVERT, Multiplexers, and Full Adders. In some cases both the true and complement outputs are provided by these devices. The incorporation of these devices in solutions of real design problems often leads to less expensive circuits than can be obtained using classical two-level logic techniques which utilize only NOT, AND, OR, NAND, and NOR gates as basic building blocks.

The basic objective of any two-level logic minimization procedure is to yield the lowest cost two-level logic circuit in either sum-of-product or product-of-sum form subject to the following three constraints.

1. The true and complement values of each variable are available as inputs.

2. The cost criterion for the minimum circuit configuration is chosen such that the addition of a single literal to the optimum circuit will increase the overall circuit cost.

3. The resulting circuit configuration is to be implemented using only the following combinations of logic gates for level 1 and level 2 respectively (AND-OR), (OR-AND), (NAND-NAND), or (NOR-NOR).

Most minimization procedures rely on two levels of logic, AND followed by OR (NAND-NAND) or OR followed by AND (NOR-NOR). They invariably depend on the availability of both the true and the complement of all variables as well as on an unrestricted number of inputs per gate (fan-in) and unlimited output-driving capability (fan-out). Unfortunately, real design problems tend to violate these conditions. Generally, fan-in and fan-out are restricted to the values available off the shelf.

Physical restrictions such as input loading on literals (the number of gates a given input literal can drive) and fan-in (the number of inputs allowed for each gate) are not taken into account in these basic methods. However, for functions of up to four variables, input loading and gate fan-in will not be a problem. For those cases where input loading is a problem, buffered inverters can be used. For example, if the input variable x is capable of driving only 10 gate inputs, and 15 inputs are required, then assuming each gate is also capable of driving 10 gate inputs, the circuit below will provide the necessary input driving capability at a cost of two gates and an additional gate delay. x can now drive 20 gate inputs.

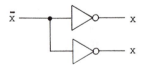

The fan-in problem can be handled by using two or more levels of logic to replace the troublesome gate. For those circuits in which gate fan-in is extremely bothersome, the theory of tree circuits is often applied. In many cases, the

Boolean functions are decomposed into sums and products of functions of fewer variables. For example, if we were to restrict ourselves to only 2-input AND and OR gates, a function such as $f(w, x, y, z) = wxyz + w\bar{x}y\bar{z} + \bar{w}x\bar{y}z + \bar{w}\bar{x}\bar{y}\bar{z}$ could be realized with the four-level tree circuit shown in Fig. 5.1.

Note that most minimization procedures can only handle multiple input-output functions of six variables or less before the use of a computer becomes an absolute necessity. In the case of multiple output functions, if the sum of the number of input and output variables is greater than eight a computer is usually required.

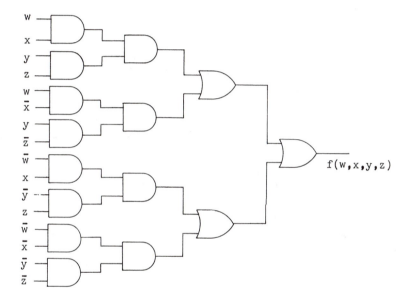

Fig. 5.1 A Tree-Circuit Realization of a 4-Variable Boolean Function

One must also be cautioned that the minimum logic circuit may not be the most appropriate due to other design criteria, such as low cost, low power, and elimination of dynamic hazards (unwanted output excursions), when the input variables are changing from one logic value to the other. Finally, one must consider whether the saving in components which might be obtained is worth the engineering cost of circuit minimization. To determine this, experience or common sense should be applied. The saving of only twenty gates from a 500-gate circuit of which only a few dozen copies are to be manufactured may not be worth the added engineering cost. However, if the circuit is to be mass produced with hundreds of thousands of copies manufactured, then the cost saving in parts is easily justified.

One of the many tasks required of a logic designer is that of selecting a family of electronic logic devices which can be used to implement the logic equations obtained as the solution to a design problem. The selection process is complicated by the tremendous technological developments constantly being made in the fabrication of logic devices. The costs, propagation delays, and power requirements for electronic logic devices are decreasing so rapidly that the components used in a logic circuit design may be obsolete by the time the circuit reaches the mass-production stage.

The chief reason for using two-level logic is to keep the overall delay from the input to the output as low as possible. However, with the present state of the art in electronic logic devices, the gate delays are approaching the point where the wires which interconnect the unit have greater delay than the gates themselves. In circuits in which either wiring is long or the gate delay does not limit the problem solution, cheaper multi-level circuit solutions should be sought. For example, in building an electronic adding machine where it takes the user a quarter of a second to press a key, it makes little difference if the circuit responds to that key press in 5 microseconds or even 1 millisecond. However, where mechanical logic circuits are being used such as fluidic switches with delays in the millisecond region, then two-level logic may be the only way to attain a circuit which can respond quickly enough to input changes.

The design engineer tries to make the best choice at a given time using the device characteristics supplied by the manufacturer. In many cases, it will be necessary for the engineer to design and run special tests in order to determine more accurately the behavior of a logic gate under adverse environmental conditions.

Five device characteristics which can adversely affect the performance of final circuit designs are: high power dissipation, long gate-propagation delays, low output-driving capability, narrow operating-temperature range, and poor noise immunity.

These five parameters to a large extent dictate the cost of a logic circuit. Unfortunately these parameters tend to interact. When a family of logic devices is needed for a particular application, one or more of these characteristics may dominate the selection process. For example, in a logic circuit designed for use as a part of a deep-space probe, where the time required for tracking data and command signals is several seconds (round trip from Earth to the spacecraft to a ground tracking station), high-speed (low propagation delay) is not nearly as important as low gate-power dissipation, high reliability, and resistance to radiation damage.

Logic units to be used as subsystems for portable digital instrumentation packages require low power dissipation and in many cases low gate-propagation time if they are to be useful for high-frequency applications.

Digital wrist watches run by small power cells require extremely low power dissipation and a low power-supply voltage to reduce the size of the unit. They do not require extremely fast logic.

Logic units to be employed as control and monitoring devices for future automobile engines, steering, and braking systems must be capable of reliably operating over a wide range of temperatures and physical stress. High speed and low power dissipation would probably not be critical to the performance of these devices.

Because of the wide variety of logic devices available, it is often difficult to select which type of logic gate should be used in a particular design situation. Some of the trade-offs which should be considered are: gate cost, power dissipation, propagation-delay time, wiring cost, and maintainability of the circuit.

When implementing multiple output logic circuits with propagation-delay times differing from the various input lines, a mixture of low-, medium-, and high-speed logic gates may result in economical circuits with balanced propagation delays.

Sometimes, however, due to quantity price discounts, using the same type of logic gate (such as only NAND or only NOR gates) will result in considerable cost savings.

Because of the wide variety of logic building blocks available to the circuit designer, the comparison of circuit implementations for a given function is extremely difficult. The design engineer must somehow weigh the importance of gate power dissipation, gate cost, gate propagation delay, noise immunity, etc., and then establish a cost criterion against which each circuit configuration can be matched. The selection task is further complicated by the constant changes in the technology.

Gate prices tend to fluctuate depending upon the demand. Constant improvements in circuit fabrication and packaging cause the relative cost between different types of gates and fabrication techniques to vary from time to time.

When using standard packaged units containing several of the same type circuit, the final circuit cost must include the cost of any unused gates as well as the cost of interconnecting the gates on several packages to form the final device.

For example, wiring together four 2-input NAND gates to form an XOR gate may prove to be false economy, since these devices are available internally wired with a lower average propagation delay than can be obtained using separate components.

Due to the rapid changes being made in circuit fabrication and device technology, we will not attempt to compare or discuss the various idiosyncrasies of the major circuit techniques for realizing electronic logic devices. To do so might tend to bias the reader and at the same time detract from the main purpose of this book, which is to teach the basic fundamentals of logic design. The proper place to obtain information on specific device characteristics is from the latest reference material available.

To confront the reader with some design decisions, we will present more than one solution to most of the examples and problems in this book. The reader can then use his or her own cost criterion to select the best circuit based on the variety of devices available and their current costs.

There are many classical techniques for systematically reducing Boolean expressions. Among the better known are the Karnaugh Map procedures and the Quine-McCluskey minimization techniques. The interested reader is referred to the books listed at the end of this chapter for a thorough discussion of these methods.

The reader is also encouraged to search the more recent journals for new multilevel logic minimization procedures. To date, there is no known procedure for guaranteeing minimum multi-input multi-output Boolean functions other than exhaustive trial-and-error methods. There are, however, many procedures which yield reasonably good circuits for only a short amount of time and effort.

In this book, no real attempt will be made to produce optimum circuits. This will allow us greater freedom in our initial designs. It will also provide the enterprising reader with the opportunity to redesign and optimize many of the circuits discussed. Our primary goal is to teach engineers to design circuits which work. In the competitive manufacturing world, it is often necessary to get a prototype unit design as quickly as possible in order to obtain a marketing advantage. Circuit minimization, if cost justified, can then be applied to the prototype in order to obtain an economical production unit. The refinement of the production unit may be continued throughout the production life of the product.

In this chapter we will show the reader how to use the classical minimization procedures. The techniques will not be covered to any great depth and, in many cases, special tricks will not be shown. By using several examples, we hope to give the reader an understanding of the basic philosophy underlying each procedure. We will also present the various procedures in such a way as to allow the reader to quickly implement the algorithms on a digital computer.

5.3 BACKGROUND MATERIAL

The key to obtaining a minimum function in sum-of-product form or product-of-sum form is the ability to systematically apply the laws of absorption and idempotency to a Boolean function. Before proceeding to a discussion of the classical minimization techniques, a few definitions are needed.

Definition 5.1 A Boolean expression g is said to *imply* another Boolean expression f, written $g \Rightarrow f$ (g implies f), if for each variable assignment for which g is equal to one, f is also equal to one. For assignments which yield g equal to zero, f may equal either zero or one.

Example 5.1

$$f(x, y) = x\bar{y} + \bar{x}y$$

Let $g(x, y) = x\bar{y}$ and $h(x, y) = \bar{x}y$. Then $g \Rightarrow f$ and $h \Rightarrow f$, since clearly whenever either g or h equal one, f must also equal 1.

Example 5.2

$$f(x, y, z) = x(y + z)$$

Let $g(x) = x$ $h(x, z) = xz$. In this case $h(x, z) \Rightarrow f(x, y, z)$, however $g(x) \not\Rightarrow f(x, y, z)$ since $f(1, 0, 0) = 0$ when $g(1) = 1$. A simple test can be used to determine if an expression g implies another expression f.

TEST FOR IMPLICATION. (a) if $\bar{g} + f = 1$ then $g \Rightarrow f$. The validity of this test is left as an exercise for the reader.

Example 5.3 Does $abc \Rightarrow ab + ac$?
Using the test procedure

$$\overline{abc} + ab + ac = \bar{a} + \bar{b} + \bar{c} + ab + ac$$
$$= \bar{a} + \bar{b} + \bar{c} + a + a$$
$$= 1$$
$$\therefore abc \Rightarrow ab + ac$$

The importance of implication to the minimization theory is that it allows us to eliminate terms from Boolean expressions. For if a Boolean expression $f = g + h$ is such that $h \Rightarrow g$ then $f = g$ and the term h can be eliminated.

Example 5.4

$$f(x, y, z) = x\bar{z} + yz + xy$$

In this function $xy \Rightarrow x\bar{z} + yz$;

$$\overline{xy} + x\bar{z} + yz = \bar{x} + \bar{y} + x\bar{z} + yz = \bar{x} + \bar{y} + \bar{z} + z = 1$$

and therefore the term xy can be eliminated from the function leaving $f(x, y, z) = x\bar{z} + yz$.

Definition 5.2 A *prime implicant* of an n-variable Boolean function f is a product term P consisting of m literals, $m \leq n$, such that P implies f, $P \Rightarrow f$, but such that any product term P^* obtained by deleting a literal from P does not imply f, $P^* \not\Rightarrow f$.

Example 5.5 $f(x, y, z) = x + x\bar{y} + x\bar{y}\bar{z}$.

Let $P_1 = x$, $P_2 = x\bar{y}$ and $P_3 = x\bar{y}\bar{z}$. For this function $P_1 \Rightarrow f$, $P_2 \Rightarrow f$, and $P_3 \Rightarrow f$, however, only P_1 is a prime implicant, since $\bar{y}$ and $\bar{y}\bar{z}$ can be deleted from the terms P_2 and P_3 without changing the function.

Prime Implicant Theorem If the cost criterion for minimization is such that the addition of a single literal to any expression results in an increased cost for the overall expression, then any minimal sum-of-product expression will always consist of a sum of prime implicants.

Proof. Suppose that a function consists of all prime implicant terms except one. We will call the missing term g. Letting h equal the logical sum of all the prime implicants, we can write $f = h + g$. Since g is not a prime implicant, a new product term g^* can be found such that, for some variable x_i, either $g = x_i g^*$ or $g = \bar{x}_i g^*$ and $g^* \Rightarrow f$. Clearly, g^* will equal 1, for all arguments which make g equal to 1; i.e., $g \Rightarrow g^*$. Because of this and the fact that $g^* \Rightarrow f$, we can replace the term g with the term g^* without changing the function; $f = h + g^*$. This latter expression for f has one less literal than the original expression and, hence, is of lower cost. We can continue to remove literals from g^* in the same way until only a prime implicant remains. Among all the possible ways of expressing f in terms of sums of prime implicants, at least one expression will be minimum for a given cost criterion.

There are an unlimited number of cost criteria applicable to logic circuits which will satisfy the condition that the addition of a literal to an expression will result in a net increase in cost. Six of the more commonly used cost criteria are:

$C1. Gates and inputs are of equal cost.

$C2. Input costs are negligible compared to the cost of gates; i.e., the cost of a gate is n times the cost of an input with n being a large number ($n \geq 10$).

$C3. Gate costs are negligible compared to the cost of inputs; i.e., the cost of an input is k times the cost of a gate with k being a large number ($k \geq 10$).

$C4. Gates are only slightly more expensive than inputs; i.e., the cost of a gate is n times the cost of an input with $1 < n < 10$.

$C5. Inputs are only slightly more expensive than gates; i.e., the cost of an input is k times the cost of a gate with $1 < k < 10$.

$C6. The cost of a gate is a monotonically increasing function of the number of inputs to the gate J. *Example 1* The cost of a gate is aJ where J is the number of inputs and a is a constant. *Example 2* The cost of a gate is $b(e^{+aJ})$ where a and b are fixed constants. *Example 3* The cost of a gate is $b(1 - e^{-aJ})$ where a and b are fixed constants.

Definition 5.3 A *nonredundant sum-of-product expression* is a logical sum of prime implicants such that if any single product term is removed then the reduced expression no longer is equal to the original expression.

It is only necessary to search the set of nonredundant sums for the minimum cost expression. Thus we can divide the set of prime implicants into two types, essential and nonessential.

Definition 5.4 An *essential prime implicant* is a prime implicant q_i which does not imply the sum of the remaining prime implicants; i.e., q_i is essential if

$$q_i \not\Rightarrow q_1 + q_2 + \cdots q_{i-1} + q_{i+1} + \cdots + q_k.$$

The set of all essential prime implicants will be called the *core*.

Definition 5.5 A *nonessential prime implicant* is a prime implicant q_i which does imply the sum of the remaining prime implicants; i.e., q_i is nonessential if

$$q_i \Rightarrow q_1 + q_2 + \cdots q_{i-1} + q_{i+1} + \cdots q_k.$$

When attempting to minimize a Boolean function, all the essential prime implicants or core prime implicants and possibly a few nonessential prime implicants will be needed to form a nonredundant function.

To determine a minimum cost representation for a given function, we must first find the set of all nonredundant representations for the function and then search this set for the expression or expressions which yield the lowest cost. With this in mind we can write a general algorithm for minimizing a Boolean function.

ALGORITHM 5.1

General procedure for minimizing multivariable single-output Boolean functions

STEP   OPERATION

1.   Find all the prime implicants of the function.

2.   Determine the essential prime implicants or core of the function.

3.   Determine the set of all nonredundant sums for the function.

4.   Using a suitable cost criterion, search the set of nonredundant sums for one which yields the lowest cost.

5.4 THE KARNAUGH-MAP METHOD FOR MINIMIZING MULTIVARIABLE SINGLE-OUTPUT BOOLEAN FUNCTIONS

In this section we will show how the K-maps described in Chapter 3 can be used to minimize Boolean functions. We will see shortly that for functions of up to six variables, the map method will yield answers almost by inspection. However, for functions of more than six variables, the map method becomes too cumbersome to use efficiently. The main advantage of the map method lies in the ability of the user to determine visually the prime implicants. On a Karnaugh Map *a prime implicant is any grouping of terms which can not be included in a larger grouping.* The prime implicants of a function are usually circled for easy reference.

Example 5.6 Circle the prime implicants on the K-maps shown below. The prime implicants of these two functions are

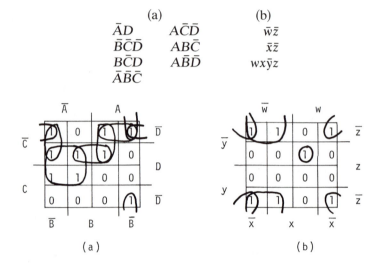

$$\begin{array}{ccc} \text{(a)} & & \text{(b)} \\ \bar{A}D & A\bar{C}\bar{D} & \bar{w}\bar{z} \\ \bar{B}\bar{C}\bar{D} & AB\bar{C} & \bar{x}\bar{z} \\ B\bar{C}D & A\bar{B}\bar{D} & wx\bar{y}z \\ \bar{A}\bar{B}\bar{C} & & \end{array}$$

(a)   (b)

Once the prime implicants have been obtained, either of two procedures can be used to determine the nonredundant sums. Either a visual minimization check can be made by inspection of the K-map, or a prime implicant table can be used in conjunction with a Table of Choices. This alternative will be explained in the discussion of the Quine-McCluskey method in the next section. For functions of six or less variables visual inspection is usually sufficient and much quicker.

The essential prime implicants are easy to spot on a K-map. *If at least one square in a prime implicant grouping is not shared by another prime implicant then the prime implicant is essential.*

Inspection of K-Map (b) in Example 5.6 reveals that all of the prime implicants are essential. The minimum sum-of-product expression for this K-map is $f(w, x, y, z) = \bar{w}\bar{z} + \bar{x}\bar{z} + wx\bar{y}z$. The cost of this function is 4 gates and 11 inputs. Inspection of K-Map (a) in Example 5.6 reveals only two essential

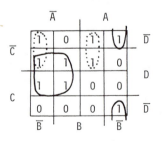

prime implicants, $\bar{A}D$ and $A\bar{B}\bar{D}$. These are shown in solid lines on the following map. It is also clear by inspection that the prime implicants $AB\bar{C}$, and $\bar{A}\bar{B}\bar{C}$ (shown with dotted lines), can be added to yield a lowest-cost function.

Thus we have determined by visual inspection that for the K-map in Example 5.6(a) a minimum sum-of-product expression is

$$f(A, B, C, D) = \bar{A}D + \bar{A}\bar{B}\bar{C} + A\bar{B}\bar{D} + AB\bar{C}.$$

This expression has a cost of 5 gates and 15 inputs. Another minimum solution can be obtained by replacing $\bar{A}\bar{B}\bar{C}$ with $\bar{B}\bar{C}\bar{D}$.

Example 5.7 Find a minimum product-of-sum expression for the K-maps shown in Example 5.6. *Hint:* Find the minimum sum-of-product expression for the complement function and then recomplement the resulting expression.

Solution. To find the minimum sum-of-product expression for the complements of the functions in Example 5.6, it is not necessary to redraw the K-Maps. Instead, circle the zeroes since these clearly identify the complement function. The resulting zero prime implicants for these two maps are circled on the maps below.

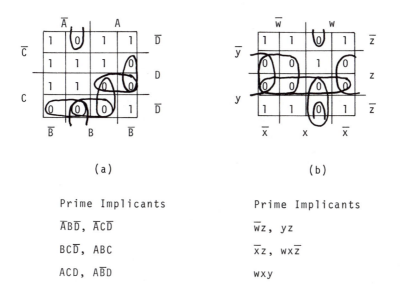

(a)   (b)

Prime Implicants   Prime Implicants

$\bar{A}B\bar{D}$, $\bar{A}C\bar{D}$ $\bar{w}z$, yz

$BC\bar{D}$, ABC $\bar{x}z$, $wx\bar{z}$

$AC\bar{D}$, $A\bar{B}D$ wxy

On the following maps, the essential prime implicants are shown with solid lines. Again, inspection of these maps indicates that the lowest-cost circuit for (a) is obtained by adding the term ABC (shown by dotted lines) and the lowest-cost circuit for (b) is obtained by adding the term yz (also shown by dotted lines).

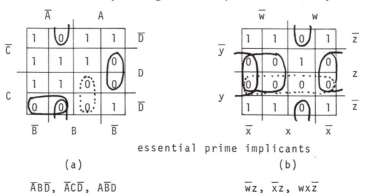

essential prime implicants

(a)   (b)

$\bar{A}B\bar{D}, \ \bar{A}C\bar{D}, \ A\bar{B}D$ $\bar{w}z, \ \bar{x}z, \ wx\bar{z}$

Therefore

$$\bar{f}_a(A, B, C, D) = \bar{A}B\bar{D} + \bar{A}C\bar{D} + A\bar{B}D + ABC$$

and

$$\bar{f}_b(w, x, y, z) = \bar{w}z + \bar{x}z + wx\bar{z} + yz.$$

Complementing these functions yields the desired product-of-sum expressions:

$$f_a(A, B, C, D) = (A + \bar{B} + D)(A + \bar{C} + D)(\bar{A} + B + \bar{D})(\bar{A} + \bar{B} + \bar{C})$$

$$f_b(w, x, y, z) = (w + \bar{z})(x + \bar{z})(\bar{w} + \bar{x} + z)(\bar{y} + \bar{z})$$

with f_a having a cost of 5 gates and 16 inputs and f_b having a cost of 5 gates and 13 inputs. For both of these maps, the sum-of-product expressions are the least costly.

When using K-maps to minimize Boolean functions, a great deal depends on the ability of the user to spot the prime implicants and then select from them a lowest-cost function. The most common error made when using K-maps is the failure to select the largest groups of squares as the prime implicants. Groups which overlap the sides, and the four corners are easily overlooked. With a little practice one can usually overcome this problem. If you find it difficult to make the best selection of prime implicants by inspection, then the table-selection method should be employed until you obtain enough confidence in the visual map method. (See Section 5.5.)

To better illustrate the map method, we will now minimize a series of Boolean functions using K-maps. In each of these examples, both the minimum sum-of-product and the minimum product-of-sum expressions will be obtained. The cost criterion in each case will be the lowest number of inputs plus gates. To aid the reader, two maps are shown for each problem, one showing all the prime implicants and the second showing all the essential prime implicants and nonessential prime implicants used in the minimum expression. The reader is encouraged to attempt to work these examples independently.

Example 5.8 Find a minimum two-level expression for the K-maps shown below.

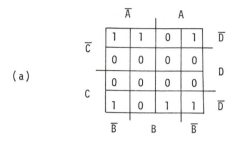

(a)

SP form   PS form

Prime implicant Maps

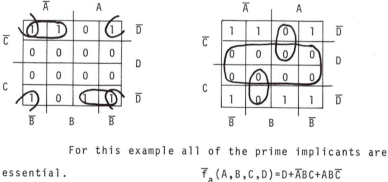

For this example all of the prime implicants are
essential.

$$f_a(A,B,C,D)=\overline{B}\overline{D}+\overline{A}\overline{C}\overline{D}+AC\overline{D}$$

cost: 4 gates, 11 inputs,

total 15.

$$\overline{f}_a(A,B,C,D)=D+\overline{A}BC+AB\overline{C}$$

and

$$f_a(A,B,C,D)=\overline{D}(A+\overline{B}+\overline{C})(\overline{A}+\overline{B}+C)$$

cost: 3 gates, 9 inputs, total 12.

For this example the product-of-sum expression has the lowest cost.

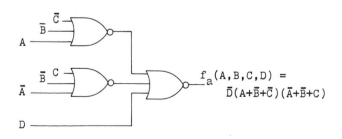

$$f_a(A,B,C,D) = \overline{D}(A+\overline{B}+\overline{C})(\overline{A}+\overline{B}+C)$$

(b)

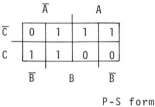

S-P form

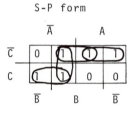

Prime Implicant Map

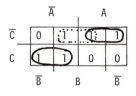

final K-Map. Note there is
another choice.

$f_b(A,B,C) = \overline{A}C + B\overline{C} + A\overline{C}$
cost: 4 gates, 9 inputs; total 13.

P-S form

All Prime implicants are essential.

$\therefore \overline{f}_b(A,B,C) = \overline{A}\,\overline{B}\,\overline{C} + AC$

and

$f_b(A,B,C) = (A+B+C)(\overline{A}+\overline{C})$

cost: 3 gates, 7 inputs; total 10

In this case, the product-of-sum expression yields the lowest cost.

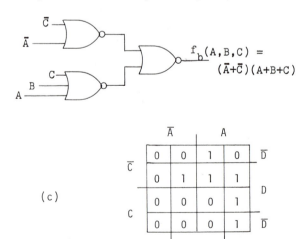

(c)

Prime Implicant Maps

S-P form

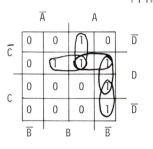

P-S form

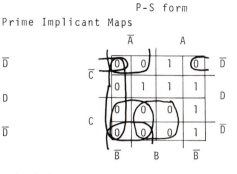

Final Maps

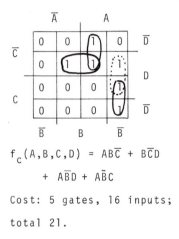

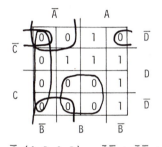

$f_c(A,B,C,D) = AB\bar{C} + B\bar{C}D$

$+ A\bar{B}D + A\bar{B}C$

Cost: 5 gates, 16 inputs;

total 21.

$\bar{f}_c(A,B,C,D) = \bar{A}\bar{B} + \bar{A}D + BC + B\bar{C}\bar{D}$

and

$f_c(A,B,C,D) =$

$(A+B)(A+D)(\bar{B}+\bar{C})(B+C+D)$

Cost: 5 gates, 13 inputs;

total 18.

Thus the product-of-sum expression yields the lowest cost.

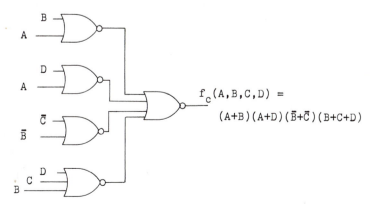

$f_c(A,B,C,D) =$

$(A+B)(A+D)(\bar{B}+\bar{C})(B+C+D)$

Until now we have only used K-maps with no more than four variables. In Example 5.9, we will examine a few of the problems associated with higher order maps.

Example 5.9   Find a minimum two-level logic expression for the K-map below.

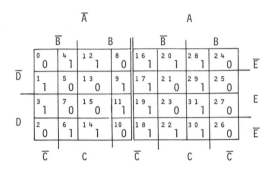

This map should be viewed as being three dimensional rather than as lying on a plane.

The map with $A = 0$ should be viewed as being above the map with $A = 1$. Viewing it in this manner, we see that squares 1, 3, 17, and 19 form the valid group $B\bar{C}E$ and that squares 4, 12, 6, 14, 20, 28, 22, and 30 form the group $C\bar{E}$. It is also clear that squares 9, 11, 17, and 19 do not form a valid group. To prevent such erroneous groups from being formed we have inserted a double line between the two smaller four-variable maps. The prime implicants and the final K-maps for the S-P and P-S solutions are shown in Fig. 5.2.

The difficulties encountered when working with 5-variable maps are at least doubled when six-variable maps are used. Figure 5.3 shows a plane view of a six-variable K-map with the squares labeled with the number corresponding to the fundamental product terms they represent.

Extending these methods to handle functions with more than one output is not easy. In fact, other than trial-and-error techniques, there is no known procedure which will always yield a minimum two-level logic circuit for multi-input multi-output Boolean functions.

However, some general guidelines which yield reasonably economical circuits are given below.

1. Try to minimize each output function separately. Some may end up in sum-of-product form and others in product-of-sum form. Do not try and force all functions to be either sum-of-product or product-of-sum unless you are required to do so by outside constraints.

2. Try to share gates among output functions; i.e., if essential and nonessential prime implicants can be shared, a cost saving should result.

S-P Form

P-S Form

Prime Implicant Map

Prime Implicant Map

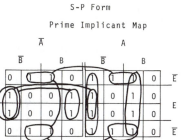

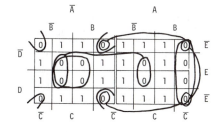

Final K-Map

Final K-Map

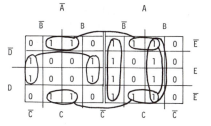

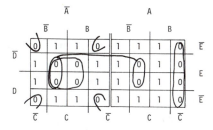

$f(A,B,C,D,E) = C\overline{E} + AB\overline{C} + ABC + \overline{A}\overline{C}E$

Cost: 5 gates, 15 inputs, total 20

$\overline{f}(A,B,C,D,E) = AB\overline{C} + \overline{A}CE + \overline{A}\overline{C}\overline{E} + \overline{B}CE$

and

$f(A,B,C,D,E) = (\overline{A}+\overline{B}+C)(A+\overline{C}+\overline{E})(A+C+E)(B+\overline{C}+\overline{E})$

Cost: 5 gates, 16 inputs; total 21.

∴ the sum of product form is the least costly.

Figure 5.2

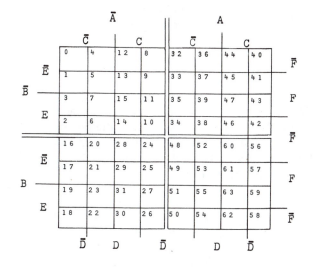

Fig. 5.3 6-Variable *K*-Map

3. Watch for tricky circuits which may use both AND and OR circuits in both levels with several gates being shared.

4. Instead of the above, use a series of programmable read-only memories (P-ROM) with the input variables as the address lines. Store the corresponding output function in each memory location. In the long run, this may be the cheapest solution for multi-input (8-16) and multi-output functions >2.

The type of logic to be used, the number of circuits to be manufactured, and the cost of minimizing the circuit all must be considered during the design of a logic circuit. The design approach to be followed must be selected for each new problem. In the next chapter, several alternate solutions to a series of problems will be presented. These solutions will be tailored to different design constraints.

5.5 QUINE-McCLUSKEY METHOD FOR MINIMIZING MULTIPLE-INPUT SINGLE-OUTPUT FUNCTIONS

One of the main advantages of this tabular minimization procedure is the straightforward, systematic manner in which the prime implicants of a Boolean function are determined. Once the prime implicants have been found the set of nonredundant sums can be quickly generated. The Quine-McCluskey procedure can be used by hand on functions of many variables provided the number of fundamental product terms is less than about thirty. For functions with a larger number of fundamental products, a digital computer should be used. In order to apply Algorithm 5.2, the fundamental product terms which comprise the function are needed:

$$f(x_{n-1}, \ldots, x_1, x_0) = \bigvee_{J=0}^{2^n-1} a_J P_J^n.$$

In many cases, the set of a_J's can be obtained without actually expanding the Boolean function. In order to save time and effort, the binary shorthand notation for representing fundamental product terms will be used. Thus instead of using $P_{11}^4 = w\bar{x}yz$, the binary number 1011 will be employed. Before stating Algorithm 5.2, we must define the weight of a binary number.

Definition 5.6 The *weight* of a binary number x, written $w(x)$, is equal to the number of 1's in its binary representation. Thus the weight of $x = (1011)_2$ is $w(x) = w(1011) = 3$.

Following the statement of Algorithm 5.2 an example will be worked in detail, illustrating each of the steps.

ALGORITHM 5.2

Quine-McCluskey method for finding prime implicants.

$$f(x_{n-1}, \ldots, x_1, x_0) = \bigvee_{J=0}^{2^n-1} a_J P_J^n$$

STEP   OPERATION

1.   Convert the decimal subscript J on each coefficient of the function, a_J, which is equal to one into its corresponding n-digit binary representation $(J)_2$.

2.   Set $K = 1$.

3.   Collect the binary subscripts $(J)_2$'s, into groups according to their binary weight, with Group G_i containing all those $(J)_2$'s with binary weight $w[(J)_2] = i$.

4.   Form Matching Table 1 by placing Group G_i in section S_i^1 of a single column table, each section of the table being separated by a line. The sections are arranged in ascending weight. Within each section, the entries need not be ordered.

5.   If possible, form a new Matching Table $K + 1$. Section S_i^{K+1} of this table is formed by comparing each entry in section i of table K, S_i^K, against each entry of section $i + 1$ of table K, S_{i+1}^K. Two entries in adjacent sections are said to form a match if they agree in all positions except one. If a match occurs a new entry is formed and is placed in section i of table $K + 1$, S_i^{K+1}, which agrees with the matched pair in all positions which match and has a dash, —, in the remaining disagreeing position. If the entry is already in the table, due to a previous match, it is not repeated. A check mark is now placed next to the two entries which formed the match, on the left side of the entries from section i and on the right side of the entries in section $i + 1$.
         If no new entries for table $K + 1$ can be found, go to step 6; otherwise increase K by 1 and repeat step 5.

6.   The prime implicants of the Boolean function correspond to the entries in each table which do not have a check mark beside them. The prime implicant is the product term obtained by barring each variable whose position contains a zero, leaving unbarred those variables whose positions contain ones and deleting those variables whose positions contain a dash.

Let us use this algorithm to determine the prime implicants of the three-variable Boolean function whose characteristic number is $H_3 = 9F$.

We must first convert H_3 into a binary number in order to determine the coefficients in the sum-of-product expansion.

$$(9F)_{16} = (1\ 0\ 0\ 1\ 1\ 1\ 1\ 1)_2$$
$$= (a_7\ a_6\ a_5\ a_4\ a_3\ a_2\ a_1\ a_0)$$

Thus $f(x, y, z) = \bigvee[a_7, a_4, a_3, a_2, a_1, a_0]$.

Step 1 of the algorithm tells us to convert each of the subscripts on the a_j's into binary numbers.

In this example we will carry the corresponding fundamental-product terms along to help the reader gain insight into the procedure.

I	$(I)_2$	$w[(I)_2]$	P_I^3
7	111	3	xyz
4	100	1	$x\bar{y}\bar{z}$
3	011	2	$\bar{x}yz$
2	010	1	$\bar{x}y\bar{z}$
1	001	1	$\bar{x}\bar{y}z$
0	000	0	$\bar{x}\bar{y}\bar{z}$

Note: When converting the subscripts into binary numbers be sure to include n binary positions, one for each variable in the function.

Steps 2, 3, and 4 of the algorithm tell us to reorder these binary numbers placing all terms with the same binary weight in the same section of Table 1, then arranging the sections in ascending order. Although it is not necessary to carry along the fundamental product terms, we are doing so as an aid for the reader.

TABLE 1

I	$(I)_2$	P_I^3
0	000	$\bar{x}\bar{y}\bar{z}$
2	010	$\bar{x}y\bar{z}$
1	001	$\bar{x}\bar{y}z$
4	100	$x\bar{y}\bar{z}$
3	011	$\bar{x}yz$
7	111	xyz

Note that within a given section of the table the terms can be placed in any order.

Step 5 outlines how to construct a new table from the old table.

Starting with the top section we can match term 000 with 010 to form 0–0. This step is just an application of the laws of absorption:

$$\bar{x}\bar{y}\bar{z} + \bar{x}y\bar{z} = \bar{x}\bar{z}$$

The term 0–0 is now placed in section 1 of Table 2. A check mark is placed on the left side of term 000 and on the right side of term 010, in Table 1.

	TABLE 1		**TABLE 2**		
0	√000	$\bar{x}\bar{y}\bar{z}$	(0–2)	0–0	$\bar{x}\bar{z}$
2	010√	$\bar{x}y\bar{z}$			
1	001	$\bar{x}\bar{y}z$			
4	100	$x\bar{y}\bar{z}$			
3	011	$\bar{x}yz$			
7	111	xyz			

The addition of a column to the new table which shows the terms yielding the new terms is not required but is included as aid to the reader.

Continuing with the matching process, the term 000 can also be matched with 001 to yield 00– ($\bar{x}\bar{y}\bar{z} + \bar{x}\bar{y}z = \bar{x}\bar{y}$). The term 00– is then entered into the new table and check marks placed next to 000 and 001. It is not necessary to place a second check mark next to the term. However, this is very helpful in reducing errors since, when you have finished matching the terms in adjacent sections, the check marks on the left side of each section must equal the number of check marks on the right side of the following section. The entry 000 can also be matched with the entry 100 to yield –00 ($\bar{x}\bar{y}\bar{z} + x\bar{y}\bar{z} = \bar{y}\bar{z}$).

Since we have now completed matching the top two sections, a line can be drawn below the entries in the new table to signal the end of a section.

	TABLE 1		**TABLE 2**		
0	√√√000	$\bar{x}\bar{y}\bar{z}$	(0–2)	0–0	$\bar{x}\bar{z}$
2	010√	$\bar{x}y\bar{z}$	(0–1)	00–	$\bar{x}\bar{y}$
1	001√	$\bar{x}\bar{y}z$	(0–4)	–00	$\bar{y}\bar{z}$
4	100√	$x\bar{y}\bar{z}$			
3	011	$\bar{x}yz$			
7	111	$\bar{x}yz$			

Next we must match each term in the 2nd section of Table 1 against each term in the 3rd section. Doing so, we find that 010 matches with 011 yielding 01– ($\bar{x}y\bar{z} + \bar{x}yz = \bar{x}y$); and that 001 also matches 011 yielding 0–1 ($\bar{x}\bar{y}z + \bar{x}yz = \bar{x}z$). The term 100 does not match and hence yields no new entries in Table 2. Since this completes the matching of these two sections, a line is drawn under the entries in this section of Table 2.

TABLE 1			**TABLE 2**		
0	√√√000	$\bar{x}\bar{y}\bar{z}$	(0–2)	0–0	$\bar{x}\bar{z}$
2	√010√	$\bar{x}y\bar{z}$	(0–1)	00–	$\bar{x}\bar{y}$
1	√001√	$\bar{x}\bar{y}z$	(0–4)	–00	$\bar{y}\bar{z}$
4	100√	$x\bar{y}\bar{z}$			
3	011√√	$\bar{x}yz$	(2–3)	01–	$\bar{x}y$
			(1–3)	0–1	$\bar{x}z$
7	111	xyz			

Matching the last two sections, we find that 011 matches with 111 yielding –11, $(\bar{x}yz + xyz = yz)$. This completes the matches for these two sections and also ends the construction of Table 2.

TABLE 1			**TABLE 2**		
0	√√√000	$\bar{x}\bar{y}\bar{z}$	(0–2)	0–0	$\bar{x}\bar{z}$
2	√010√	$\bar{x}y\bar{z}$	(0–1)	00–	$\bar{x}\bar{y}$
1	√001√	$\bar{x}\bar{y}z$	(0–4)	–00	$\bar{y}\bar{z}$
4	100√	$x\bar{y}\bar{z}$			
3	√011√√	$\bar{x}yz$	(2–3)	01–	$\bar{x}y$
			(1–3)	0–1	$\bar{x}z$
7	111√	xyz	(3–7)	–11	yz

We must now proceed to match the entries in each section of Table 2 to see if a third table can be constructed. It should be noted that when looking for matches between adjacent sections, it is only necessary to compare those terms which have dashes in the same positions.

Matching section 1 with section 2, we find that 0–0 matches with 0–1 to form the entry 0––, $(\bar{x}\bar{z} + \bar{x}z = \bar{x})$; and that 00– matches with 01– to also yield the entry 0––, $(\bar{x}\bar{y} + \bar{x}y = \bar{x})$. Since this entry is already included in Table 3, it need not be entered a second time. Since the term –00 does not match any of the terms in Section 2, matching is complete.

TABLE 2			**TABLE 3**		
(0–2)	√0–0	$\bar{x}\bar{z}$	*(0–2–1–3)	0––	$\bar{x}$
(0–1)	√00–	$\bar{x}\bar{y}$			
(0–4)*	–00	$\bar{y}\bar{z}$	* Indicates		
			unchecked terms		
(2–3)	01–√	$\bar{x}y$	corresponding to		
(1–3)	0–1√	$\bar{x}z$	the prime impli-		
			cants.		
(3–7)*	–11	yz			

Since neither term in the second section matches the term −11 we have also completed Table 3. Because Table 3 contains only 1 entry, no new tables can be found and we can proceed to step 6.

Step 6 of Algorithm 5.2 (page 173) tells us to inspect each row of each table for unchecked entries since these terms will be the prime implicants of the function. We find that all rows of Table 1 are checked and that Table 2 yields the two terms −00 and −11. Table 3 yields the entry 0−−. Therefore the function $H_3 = 9F$ has three prime implicants, −00, −11, and 0−−. These prime implicants correspond to the product terms $\bar{y}\bar{z}$, yz, and $\bar{x}$ respectively.

In terms of its prime implicants only

$$f(x, y, z) = \bar{y}\bar{z} + yz + \bar{x}.$$

Example 5.10 Let us determine the prime implicants for the function whose characteristic number is $H_4 = 9BD9$.

Converting H_4 into binary yields the number

$$(1 \quad 0 \quad 0 \quad 1 \quad 1 \quad 0 \quad 1 \quad 1 \quad 1 \quad 1 \quad 0 \quad 1 \quad 1 \quad 0 \quad 0 \quad 1)_2$$
$$(a_{15} \quad a_{14} \quad a_{13} \quad a_{12} \quad a_{11} \quad a_{10} \quad a_9 \quad a_8 \quad a_7 \quad a_6 \quad a_5 \quad a_4 \quad a_3 \quad a_2 \quad a_1 \quad a_0)$$
$$\therefore \quad f(w, x, y, z) = \bigvee[a_{15}, a_{12}, a_{11}, a_9, a_8, a_7, a_6, a_4, a_3, a_0]$$

Applying the Quine-McCluskey algorithm for finding the prime implicants yields:

STEP 1   STEPS 2, 3, and 4

I	(I_2)	$w[(I_2)]$
15	1111	4
12	1100	2
11	1011	3
9	1001	2
8	1000	1
7	0111	3
6	0110	2
4	0100	1
3	0011	2
0	0000	0

These terms must now be reordered according to the number of 1's in each $(I)_2$ to yield Table 1.

STEP   OPERATION

5.   Matching yields two new tables.

TABLE 1	TABLE 2	TABLE 3
√√0000	√–000	*––00
√√1000√	√0–00	*––11
√√0100√	1–00√	
1100√√	*100–	Since a fourth table cannot
√1001√	–100√	be formed, the * terms are
√0110√	*01–0	prime implicants.
√√0011	*10–1	
√1011√√	*011–	
√0111√√	√–011	
1111√√	√0–11	
	1–11√	
	–111√	

STEP   OPERATION

6.   Inspecting the 3 tables for unchecked terms yields the prime implicants of
        $H_4 = 9BD9 = f(w, x, y, z)$

100–	$w\bar{x}\bar{y}$
01–0	$\bar{w}x\bar{z}$
10–1	$w\bar{x}z$
011–	$\bar{w}xy$
––00	$\bar{y}\bar{z}$
––11	yz

Written in terms of the prime implicants

$$f(w, x, y, z) = w\bar{x}\bar{y} + \bar{w}x\bar{z} + w\bar{x}z + \bar{w}xy + \bar{y}\bar{z} + yz$$

Now that we have determined the set of prime implicants for a given Boolean function, all that remains to be done is to select the group or groups of prime implicants which cost the least.

In the series of examples in this section we will use the cost criterion that gates and inputs are of equal cost. Therefore we will be searching the set of nonredundant sums for circuits with the lowest total number of gates and inputs.

Prime Implicants		wxyz 1111	wx$\bar{y}\bar{z}$ 1100	w$\bar{x}$yz 1011	w$\bar{x}\bar{y}$z 1001	w$\bar{x}\bar{y}\bar{z}$ 1000	$\bar{w}$xyz 0111	$\bar{w}$xy$\bar{z}$ 0110	$\bar{w}$x$\bar{y}\bar{z}$ 0100	$\bar{w}\bar{x}\bar{y}$z 0011	$\bar{w}\bar{x}\bar{y}\bar{z}$ 0000
				Fundamental Product Terms							
yz	--11	✓		✓			✓			✓	
$\bar{y}\bar{z}$	--00		✓			✓			✓		✓
$\bar{w}$xy	011-						✓	✓			
w$\bar{x}$z	10-1			✓	✓						
$\bar{w}$x$\bar{z}$	01-0							✓	✓		
w$\bar{x}\bar{y}$	100-				✓	✓					

Fig. 5.4 Prime-Implicant Table for $H_4 = 9BD9$

In order to help us decide which prime implicants to select, a *prime implicant table* will be employed. The prime implicant table for Example 5.10 is shown in Fig. 5.4. This table lists along the side all the prime implicants, usually arranged in order of increasing cost. The fundamental product terms of the function are listed across the top of the table. A check mark is made at each intersection of a row and column if each literal in the prime implicant is also included in the fundamental product. Usually, only the binary numbers are shown on the tops and sides of the table. In this and the next example, however, we will also carry along the variable expressions. From a prime implicant table, it is easy to determine the essential and nonessential prime implicants. To determine the essential prime implicants, each column of the table is scanned from left to right. If a column contains only a single check mark, then the prime implicant at the intersection of the check mark is essential. An asterisk is placed next to all the essential prime implicants found. In Fig. 5.4, columns 1, 2, 9, and 10 have single check marks and hence yz and $\bar{y}\bar{z}$ are essential prime implicants. The next step in our minimization process is to form the set of nonredundant sums. To aid in this process and also to reduce the amount of work required, a new table called a *table of choices* is obtained from the prime implicant table. Each row corresponding to an essential prime implicant is crossed out, and each column with a check mark corresponding to an essential prime implicant is also crossed out. The remaining rows and columns which contain check marks are then condensed to form a new table similar in form to the prime implicant table but now containing only the nonessential or choosable prime implicants. Fig. 5.5 shows the crossed out table for Fig. 5.4 and Fig. 5.6 shows the resulting table of choices.

At this point we will label each row in the table of choices. It is convenient to label each row R_i where i equals the original row assignment in the prime implicant table. Since all the essential prime implicants will appear in all

Prime Implicants		wxyz 1111	wxȳz̄ 1100	wx̄yz 1011	wx̄ȳz 1001	wx̄ȳz̄ 1000	w̄xyz 0111	w̄xyz̄ 0110	w̄xȳz̄ 0100	w̄x̄yz 0011	w̄x̄ȳz̄ 0000
* yz	--11	✓		✓			✓			✓	
* ȳz̄	--00		✓			✓			✓		✓
w̄xy	011-						✓	✓			
wx̄z	10-1			✓	✓						
w̄xz̄	01-0							✓	✓		
wx̄ȳ	100-				✓	✓					

Fundamental Product Terms

Fig. 5.5 Prime-Implicant Table Showing Essential Prime Implicants and Which Rows and Columns Are to be Crossed Out

minimum-cost representations, we need only consider the cost of adding product terms from the remaining nonessential prime implicants.

What we would now like to do is determine all possible ways of combining these nonessential prime implicants so that each grouping contains at least one check mark in the remaining columns of the table of choices. A simple way to accomplish this is by writing a Boolean expression in product-of-sum form with each product corresponding to a column and each sum corresponding to the rows which are checked in that column. For our example, column 1 yields the sum term $(R_4 + R_6)$ and column 2 yields the sum term $(R_3 + R_5)$. Writing out the product of these columns and expanding it into sum-of-product form yields

$$(R_4 + R_6)(R_3 + R_5) = R_3R_4 + R_3R_6 + R_4R_5 + R_5R_6.$$

Nonessential Prime Implicants		wx̄ȳz 1001	w̄xyz̄ 0110	
w̄xy	011-		✓	R3
wx̄z	10-1	✓		R4
w̄x̄z	01-0		✓	R5
wx̄ȳ	100-	✓		R6

Fundamental Product Terms

Fig. 5.6 Table of Choices Obtained from Fig. 5.5

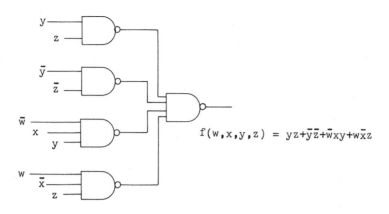

$$f(w,x,y,z) = yz + \bar{y}\bar{z} + \bar{w}xy + w\bar{x}z$$

Fig. 5.7 A Minimum SP Circuit for $H_4 = 9BD9$

Each product term now represents a group of prime implicants with at least one check mark in each column of the table of choices. Since each of these products consists of two terms each having three literals, they are of equal cost. Hence for this function there are four minimum sum-of-product expressions corresponding to the rows.

$$
\begin{array}{cc}
\text{Essential} & \text{Choosable} \\
& \begin{bmatrix} R_3 + R_4 \\ R_3 + R_6 \\ R_4 + R_5 \\ R_5 + R_6 \end{bmatrix} \\
[R_1 + R_2] + &
\end{array}
$$

Therefore

$$
\begin{aligned}
f(w, x, y, z) &= yz + \bar{y}\bar{z} + \bar{w}xy + w\bar{x}z \\
&= yz + \bar{y}\bar{z} + \bar{w}xy \qquad\qquad\quad + w\bar{x}\bar{y} \\
&= yz + \bar{y}\bar{z} \qquad\quad + w\bar{x}z + \bar{w}x\bar{z} \\
&= yz + \bar{y}\bar{z} \qquad\qquad\quad + \bar{w}x\bar{z} + w\bar{x}\bar{y}
\end{aligned}
$$

are all minimum-cost sum-of-product representations. In each case, five gates are required with a total of fourteen inputs needed for these gates. The total number of inputs needed is equal to the sum of the number of literals in each term plus the number of multiple literal terms. The number of gates required is equal to 1 plus the number of terms minus the number of single literal terms. In this example there are ten literals and four multi-literal terms. A typical NAND gate circuit realization for the function is shown in Fig. 5.7.

The prime implicant table for our original example, $H_3 = 9F$, reveals that all three prime implicants are essential and hence there is a unique minimum representation for this function with $f(x, y, z) = \bar{x} + yz + \bar{y}\bar{z}$. This function will require three gates and seven inputs. It should be noted that the term $\bar{x}$ only requires an input to the second level of gates. Figure 5.8 shows a two-level NAND-gate circuit for this function. In this figure, the term $\bar{x}$ appears as an uncomplemented input to the 2nd-level NAND gate.

Prime Implicants		$\overline{x}\,\overline{y}\,\overline{z}$	$\overline{x}\,y\,\overline{z}$	Fundamental Product Terms $\overline{x}\,\overline{y}\,z$	$x\,\overline{y}\,\overline{z}$	$\overline{x}\,y\,z$	$x\,y\,z$
		000	010	001	100	011	111
* $\bar{y}\bar{z}$	-00	✓			✓		
* yz	-11					✓	✓
* $\bar{x}$	0--	✓	✓	✓		✓	

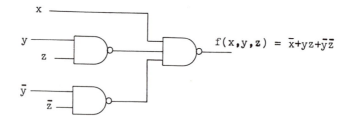

$$f(x,y,z) = \bar{x}+yz+\bar{y}\bar{z}$$

Fig. 5.8 Prime Implicant Table and Minimum SP Circuit for $H_3=9F$

Up until now we have dealt only with minimizing a function in sum-of-product form. It is possible that a product-of-sum expression for the function might yield a lower cost than any sum-of-product expression. *In order to determine a minimum product-of-sum expression, it is only necessary to minimize the complement of the original function in sum-of-product form and then to take the complement of the result.*

In order to determine a minimum gate configuration using only two levels of logic, it is necessary to find both a minimum sum-of-product expression and a minimum product-of-sum expression. The minimum circuit is then selected from these two configurations.

Example 5.11 Find a minimum two-level logic circuit for $H_4 = 5372$.

Assume that the total number of gates plus inputs is to be as low as possible.

Solution. We must find both a minimum product-of-sum and a sum-of-product expression for this function. First let us find the set of fundamental product terms which make up H_4 and $\bar{H}_4$.

$$H_4 = 5372 \quad \Rightarrow B_4 = (0101\ 0011\ 0111\ 0010)_2$$
$$\therefore \quad \bar{H}_4 = AC8D \Rightarrow \bar{B}_4 = (1010\ 1100\ 1000\ 1101)_2$$
$$f(w, x, y, z) = \bigvee [a_{14}, a_{12}, a_9, a_8, a_6, a_5, a_4, a_1]$$

and

$$\bar{f}(w, x, y, z) = \bigvee [a_{15}, a_{13}, a_{11}, a_{10}, a_7, a_3, a_2, a_0]$$

	H_4			$\bar{H}_4$	
i	$(i)_2$	$w((i)_2)$	i	$(i)_2$	$w((i)_2)$
14	1110	3	15	1111	4
12	1100	2	13	1101	3
9	1001	2	11	1011	3
8	1000	1	10	1010	2
6	0110	2	7	0111	3
5	0101	2	3	0011	2
4	0100	1	2	0010	1
1	0001	1	0	0000	0

Let us minimize H_4 in sum-of-product form first. The prime implicants are found using the Quine-McCluskey method.

TABLE 1	TABLE 2	TABLE 3
√√0001	*–001	*–1–0
√√√0100	*0–01	
√√1000	√–100	
√1100√√	√01–0	
1001√√	*010–	
√0110√	*1–00	
0101√√	*100–	
1110√√	11–0√	* Denotes prime implicants.
	–110√	

The prime implicants for $H_4 = 5372$ are given in Fig. 5.9. Choices are tabulated in Fig. 5.10.

We see from inspecting this table that rows R_3 and R_5 will yield the lowest cost cover. All other covers will require at least three rows. Therefore the minimum sum-of-product expression is $f(w, x, y, z) = x\bar{z} + \bar{w}\bar{y}z + w\bar{x}\bar{y}$ with a cost of 4 gates and 11 inputs for a total of 15.

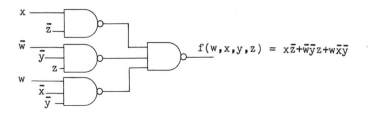

$$f(w,x,y,z) = x\bar{z}+\bar{w}\bar{y}z+w\bar{x}\bar{y}$$

Prime Implicants	Fundamental Product Terms							
	0001	0100	1000	1100	1001	0110	0101	1110
* -1-0		√		√		√		√
-001	√				√			
0-01	√						√	
1-00			√	√				
100-			√		√			
010-		√					√	

The essential prime implicants are marked with an *.

Fig. 5.9 Prime Implicant Table

	Nonessential Prime Implicants	Fundamental Product Terms			
		0001	1000	1001	0101
R_2	-001	√		√	
* R_3	0-01	√			√
R_4	1-00		√		
* R_5	100-		√	√	
R_6	010-				√

Fig. 5.10 Table of Choices

Now we must find a minimum product-of-sum expression. This we will do by first finding the minimum sum-of-product expression for $\bar{H}_4$ and then complementing the resulting expression. The prime implicants will be determined first.

TABLE 1	TABLE 2	TABLE 3
√0000	*00–0	*–01–
√√0010√	√–010	*––11
√1010√	√001–	
√√0011√	101–√	
√1101	√–011√	
√1011√√	√0–1ï	
√0111√	*11–1	* Denotes prime implicants.
1111√√√	1–11√	
	–111√	

The prime-implicant table in Fig. 5.11 will be used to aid in selecting the minimum expression.

Since all the prime implicants are essential, the minimum sum-of-product expression for $\bar{H}_4$ is

$$\bar{f}(w, x, y, z) = yz + \bar{x}y + \bar{w}\bar{x}\bar{z} + wxz.$$

The minimum product-of-sum expression for H_4 is obtained by complementing this expression. Hence

$$f(w, x, y, z) = (\bar{y} + \bar{z})(x + \bar{y})(w + x + z)(\bar{w} + \bar{x} + \bar{z}).$$

This expression has a cost of 5 gates and 14 inputs.

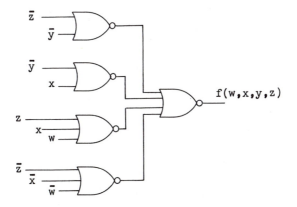

Comparing the two minimum expressions reveals the sum-of-product form to be the more minimal two-level logic expression. Thus the minimum expression for $H_4 = 5372$ is $f(w, x, y, z) = x\bar{z} + \bar{w}\bar{y}z + w\bar{x}\bar{y}$.

Prime Implicants	Fundamental Product Terms							
	0000	0010	1010	0011	1101	1011	0111	1111
* -01-		✓	✓	✓		✓		
* --11				✓		✓	✓	✓
* 00-0	✓	✓						
* 11-1					✓			✓

Fig. 5.11 Prime Implicant Tables for $H_4 = ABCD$

5.6 INCOMPLETELY SPECIFIED FUNCTIONS

Many times a Boolean function is not completely specified; that is, the function is described by telling which product terms must be included and which ones must be excluded with the remaining product terms left as DON'T CARES. Problems of this type generally arise from physical constraints placed on the input variables. Minimizing functions of this type requires only a slight amount of additional effort. The key to minimizing these functions is to set the don't-care terms to 1's when determining the prime implicants of the sum-of-product expression and then set them to zero when trying to find the nonredundant expressions. In general, the resulting final expression will contain some don't-care terms which equal one and some which equal zero, but then you didn't care anyway. In order to illustrate the procedure for handling don't-care terms, two examples will be worked out in detail. The first example uses Quine-McCluskey method and the second uses the Karnaugh map method.

Example 5.12 Design a decimal digit indicator for even numbers $0 \leq N \leq 9$.
Find a minimum two-level logic circuit for this indicator.

Solution. Using four binary positions to represent the decimal integers, we find that the fundamental product terms for this indicator can be placed in three groups:

1. Terms equal to one = [0000, 0010, 0100, 0110, 1000]
2. Terms equal to zero = [0001, 0011, 0101, 0111, 1001]
3. Don't-care terms = [1010, 1011, 1100, 1101, 1110, 1111].

Let us first find a minimum sum of product expression for the indicator. We will include the don't-care states in the matching procedure.

TABLE 1	TABLE 2	TABLE 3	TABLE 4
√√√0000	√√00–0	√0––0	*–––0
————		√––00	
√√0010√	√√0–00	√–0–0	
√√0100√	√√–000	————	
√√1000√	————	––10√	
————	√0–10√	–1–0√	
√0110√√	√–010√	1––0√	
————	————	————	
√√1010√√	√01–0√	*1–1–	
√√1100√	√–100√	*11––	
————	————		
√1011√	√10–0√		
√1101√	√1–00√		
————	————		
√1110√√√	–110√√		
————	————		
1111√√√	√101–		
	√1–10√√		
	√110–		
	√11–0√√		
	————		
	1–11√		
	11–1√		
	111–√√		

Prime implicants.

When we construct the prime-implicant table for this device we will only use those product terms which must equal one.

Inspection of Fig. 5.12 reveals that the two prime implicants 1–1– and 11–– do not cover any terms and hence can be ignored. In fact, these terms result from matching only don't-care terms.

Since there is only one prime implicant, the minimum sum-of-product expression reduces to

$$f(w, x, y, z) = \bar{z}.$$

Prime Implicants	\multicolumn{5}{c}{Fundamental Product Terms}

Prime Implicants	0000	0010	0100	0110	1000
* –––0	√	√	√	√	√
1-1-					
11--					

Figure 5.12

You should satisfy yourself that the minimum sum-of-product expression would have been

$$f(w, x, y, z) = \bar{w}\bar{z} + \bar{x}\bar{y}\bar{z}$$

if the don't-care terms had been set to zero. This latter function is considerably more expensive.

If we were to search for the minimum product-of-sum expression we would minimize the complement of the function. It should be noted that the complement of a don't care is still a don't care.

For this example, the minimum product-of-sum expression will again be $f(w, x, y, z) = \bar{z}$ if the don't-care terms are used in the determination of the prime implicants and $f(w, x, y, z) = (w + \bar{z})(x + y + \bar{z})$ if they are not used. The reader should check these results.

Example 5.13 Find a minimum two-level logic circuit for the K-map below.

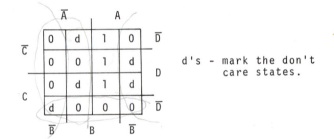

d's - mark the don't care states.

Prime Implicant Maps

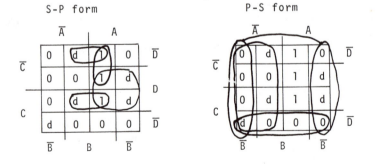

S-P form   P-S form

Remember that on a K-Map an essential prime implicant is determined by a group which has at least one square equal to 1 in f or 0 in $\bar{f}$ which is not covered by another prime implicant. The d's are not considered when selecting essential prime implicants. With this in mind, we see that in the S-P K-map none of the prime implicants are essential, while all of those in the P-S K-map are essential.

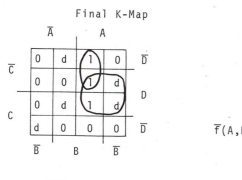

Final K-Map

$$\bar{f}(A,B,C,D) = \bar{A} + C\bar{D} + \bar{B}$$

S-P form

f(A,B,C,D) = AD + AB$\bar{C}$

Cost: 3 gates, 7 inputs;

total 10

P-S form

f(A,B,C,D) = A($\bar{C}$+D)B

Cost: 2 gates, 5 inputs;

total 7.

The product-of-sum circuit yields the lower cost.

If the don't-care terms had not been used then the minimum sum-of-product expression would have been $f(A, B, C, D) = AB\bar{C} + ABD$ with a cost of 3 gates and 8 inputs for a total of 11. A minimum product-of-sum circuit would have been $f(A, B, C, D) = (A + B + \bar{D})(A + C + \bar{D})(\bar{B} + \bar{C} + D)(A + B + C)$ $(\bar{A} + B + D)$ with a cost of 6 gates and 20 inputs for a total of 26. The cost savings is easily seen.

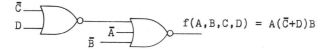

$$f(A,B,C,D) = A(\bar{C}+D)B$$

5.7 PROBLEMS

5.1 Verify the validity of the test for logical implication; i.e., show that if $g \Rightarrow f$ then $\bar{g} + f = 1$.

5.2 Circle all of the prime implicants on the K-maps on page 190. List the essential prime implicants.

5.3 Find a minimum sum-of-product expression for the K-maps on page 190. Assume the cost of gates and inputs is the same. Show your final K-map.

5.4 Find a minimum product-of-sum expression for the K-maps on page 190. Assume the cost of gates and inputs is the same. Show your final K-map.

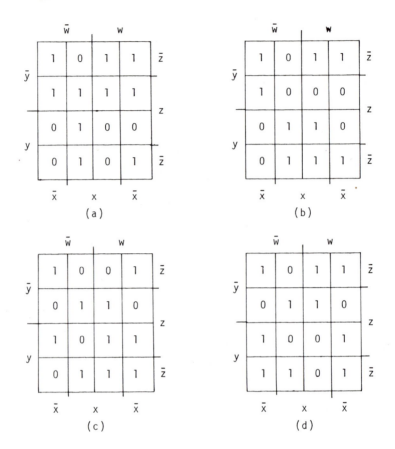

(a) (b) (c) (d)

5.5 Using K-maps find a minimum two-level logic circuit for the following Boolean functions expressed in terms of their characteristic numbers.

a) $H_3 = A6$ b) $H_4 = 13E6$
c) $H_4 = 217F$ d) $H_4 = 1026$
e) $H_4 = 147C$ f) $H_5 = ABCD1234$

Show your final K-map.

Note: You must first find a minimum sum-of-product solution and a minimum product-of-sum solution before you can choose the best two-level solution. You may also have to resort to the use of a table of choices to help determine the minimum expressions.

5.6 Using the Quine-McCluskey tabular procedure, determine the prime-implicants for the Boolean functions listed below.

a) $H_3 = 2C$ b) $H_4 = 16CF$
c) $H_4 = 3579$ d) $H_3 = 72$
e) $H_4 = B76A$ f) $H_4 = 764C$
g) $H_5 = 02FAC167$ h) $H_3 = A6$

5.7 Using the Quine-McCluskey tabular procedure, find a minimum sum-of-product expression for the following Boolean functions.

a) $H_3 = 35$ b) $H_4 = 64A7$
c) $H_4 = 7F62$ d) $H_4 = 3B3D$

5.8 Using the Quine-McCluskey tabular procedure, find a minimum product-of-sum expression for the following Boolean function.

a) $H_3 = 42$ b) $H_4 = C4C2$
c) $H_4 = 35A8$ d) $H_4 = 3B3D$

5.9 Write a computer program which will yield a minimum two-level logic circuit for functions of up to 6 variables.

5.10 Find a minimum sum-of-product and a minimum product-of-sum solution for each of the functions shown on the following K-maps. Show your final K-map.

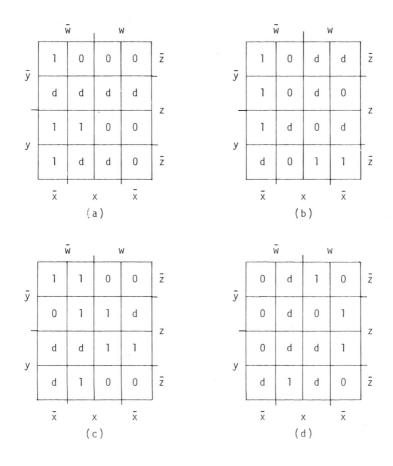

5.8 BIBLIOGRAPHY

Dietmeyer, D. L., and Y. H. Su, "Logic design automation of fan-in limited NAND networks," *IEEE Trans. on Computers*, Vol. C-18, pp. 11–22, Jan. 1969.

Gimpel, J. F., "A reduction technique for prime implicant tables," *IEEE Trans. on Electronic Computers*, Vol. EC-14, pp. 535–541, August, 1965.

Levine, R. I., "Logic minimization beyond the Karnaugh map," *Computer Design*, Vol. 6 No. 3, pp. 40–43, Mar. 1967.

Luccio, F., "A method for the selection of prime implicants," *IEEE Trans. on Electronic Computers*, Vol. EC-15, pp. 205–212, Apr. 1966.

McCluskey, E. J., *Introduction to the Theory of Switching Circuits*. New York: McGraw-Hill, 1965.

McCluskey, E. J., "Minimization of Boolean functions," *Bell System Technical Journal*, Vol. 35, No. 5, pp. 1417–1444, Nov. 1956.

Quine, W. V., *Mathematical Logic*. Cambridge, Mass.: Harvard University Press, 1955.

Quine, W. V., "The problems of simplifying truth functions," *American Mathematical Monthly*, Vol. 59, No. 8, pp. 521–531, Oct. 1952.

Rhyne, V. T., *Fundamentals of Digital Systems Design*. Englewood Cliffs, N.J.: Prentice-Hall, 1973.

Torng, H. C., *Switching Circuits Theory and Logic Design*. Reading, Mass.: Addison-Wesley, 1972.

Veitch, E. W., "A chart method for simplifying truth functions," *Proceedings of a Inf. ACM*, Pittsburgh, Pa., pp. 127–133, May 1952.

Chapter 6

Binary Arithmetic Units

6.1 INTRODUCTION

In the preceding chapters, we have developed the tools and building blocks necessary to assemble practical digital circuits. In this chapter, we will concentrate on designing combinational logic circuits, in particular, binary adding circuits. Binary full and half adder circuits will be developed in Sections 6.2 and 6.3. These units will be assembled into ripple adders in Section 6.4. Sections 6.5, 6.6, and 6.7 will be devoted to high-speed parallel-adding circuitry. The basic concept of a carry lookahead adder will be introduced in Section 6.5. Ripple lookahead adders and first-order lookahead adders will be discussed as a compromise between slow-ripple adders and fast lookahead adders. Finally, Section 6.8 will consider combinational logic circuits for binary multiplication.

6.2 COMBINATIONAL LOGIC CIRCUITS

Before we can design a combinational logic circuit, we must have a clear idea of what constitutes one.

Definition 6.1 A *combinational logic circuit* is any logic circuit whose steady-state output values depend only on the values of its steady-state inputs.
   The amount of time required for the input and output signals to reach steady state depends upon the internal logic-circuit electronics.
   This output time usually consists of the worst-case propagation-delay time between some pair of input and output terminals followed by the worst-case output-transition-time delay.

The input signals are not allowed to change at a rate greater than $1/t_{oss}$ where t_{oss} is the time required for the output to reach steady state.

Logic circuits which do not meet these requirements are called *sequential circuits*.

To help those logic design engineers faced with such tasks as building a single logic circuit from the currently available building blocks or designing an MSI

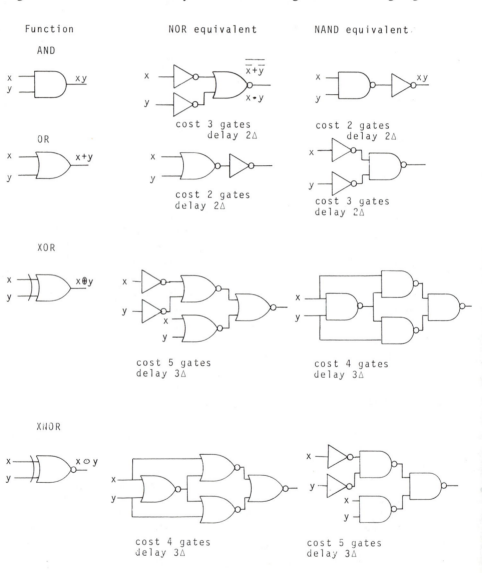

Fig. 6.1 **NAND and NOR Logic Circuits**

(medium scale integrated) or LSI (large scale integrated) circuit package using only NAND or NOR technology, multiple solutions to the design problems in the next section are presented. This serves two goals: the first is to hopefully prevent the design engineer from falling into an "only-one-way-to-build" syndrome; the second is to allow the engineer to roughly estimate both the number of inverting gates needed to fabricate a circuit and the cost savings which can be achieved by using functional building blocks instead of basic logic gates.

In order to accomplish the latter objective and make a fair comparison between logic circuit configurations, a few ground rules must be established.

The basic building blocks will consist of the following three gates: multi-input NORS, multi-input NANDS and inverters. From these we can build all of the functional blocks required. We will assume that these three gates have a nominal propagation delay of one unit (Δ delay). Using these three gates, we can construct AND, OR, XOR, and XNOR gates as shown in Fig. 6.1. Of course, the delay of each of these gates can be reduced if NOR or NAND gates with less than unit delay are utilized in the fabrication process. However, since these faster gates are more expensive, the fairness of the comparison is destroyed unless economic considerations are also included.

It will be up to the design engineer to determine when and where high-speed or even low-speed logic gates will improve the overall cost-performance of his or her final design.

In order to make a valid gate cost and delay comparison, it will be necessary to show an equivalent inverting gate circuit for the functional block. For example, the typical AND-OR-INVERT building block is in fact a set of NAND-WIRED-AND gates and hence a simple gate count can be made. The delay for a wired connection is 0. However the gate propagation delay usually must be modified upward to compensate for the slower basic NAND gate required for wired connections. However, as an MSI circuit, this delay increase is usually small. Thus for comparison purposes, we will assume unit delay and gate cost equal to the number of AND gates.

AND-OR-INVERT

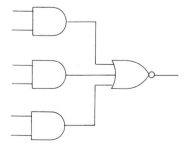

equivalent circuit

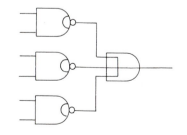

cost 1 gate for each AND function
delay 1Δ

To make a realistic comparison with a gate realization and a multiplexer realization, a valid multiplexer equivalent circuit is required. The model shown in Fig. 6.2 is that for a 4 data-input, 2-selection input unit from which generalizations can be made.

For a multiplexer with k selection inputs, a total of 2^k NAND gates are needed. Each gate will require $(k+1)$ inputs. A single inverter is required at the output to yield the true output function. Finally, $2k$ input isolation inverters

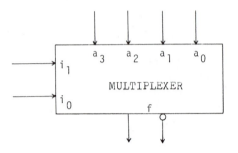

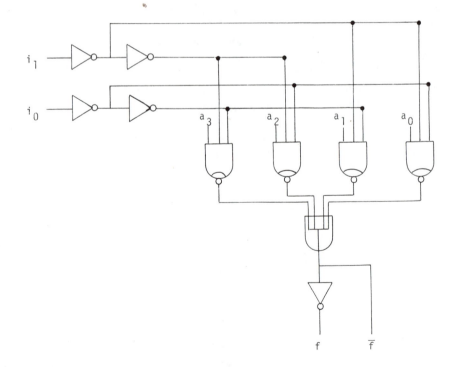

Fig. 6.2 Multiplexer Equivalent Circuit

are required for buffering. Thus a total of $2^k + 2k + 1$ gates are needed. Two delay values are usually stated. The path from data inputs to the output is 2Δ and the delay from a selection input to the output is 4Δ.

When used as a logic element, the inputs to the data points are sometimes externally buffered by double inverters yielding a balanced delay of four units from all input lines.

As the design engineer gains experience implementing logical building blocks, he or she should also develop the equivalent inverting gate circuit to allow for future comparison of circuit designs.

Another design rule is to *assume that either the true or complement of an input signal is available but that both are not available.* The missing inputs can be easily obtained if needed by using inverters. This guideline is used because of the limited number of input and output pins available for external signals on both integrated-circuit chips and the larger printed-circuit cards. In general, the size of the unit is determined not by the amount of integrated-circuit logic but rather by the number of pins or terminals required for external signals. For example, to save input/output terminals, a four-digit binary adder should only use four input lines for the addend and four input lines for the augend; one additional line is needed for the carry in. Five output lines are required for the carry out and sum digits. The addition of a ground and a power connection yields a total of 16 external connections.

6.3 LOGICAL DESIGN OF BINARY ARITHMETIC CIRCUITS

In this section we will concentrate our attention on the design of combinational logic circuits for use in binary arithmetic units. Our goal will be to convert the general word description for the arithmetic device specified in the problem statement into a working functional logic unit using basic logic gates. These logic units will then be used in later chapters as building blocks for larger systems.

Because of the difficulty in assigning realistic cost figures to a particular design solution, multiple solutions will be provided when appropriate. In this way, the design engineer will be able to evaluate cost criterion based on current technology and economic considerations.

Let us start with a simple but important logical element, the design of a controlled one's complement logic box.

Example 6.1 Design a controlled diminished-radix complement logic circuit for base 2 numbers, i.e., a controlled 1's complement logic circuit.

Solution. In Chapter 2, we derived the equations for a controlled diminished-radix logic box. For base 2, the circuit specification and the associated arithmetic table are shown on page 198.

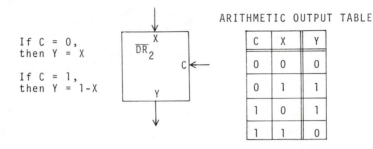

ARITHMETIC OUTPUT TABLE

If C = 0,
then Y = X

If C = 1,
then Y = 1-X

C	X	Y
0	0	0
0	1	1
1	0	1
1	1	0

Close inspection of this table reveals that for binary numbers this arithmetic truth table can be treated as a logic truth table. In this case, the logic output of the device becomes $Y = \bar{C}X + C\bar{X} = C \oplus X$. The logic circuit for this unit is simply an exclusive OR gate.

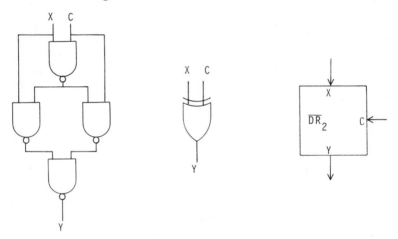

Let us now turn our attention to the design of basic binary adding circuits. First, we will consider the design of simple half-adder cells which accept two binary digits and yield a sum and carry.

Example 6.2 Design a binary half-adder circuit.

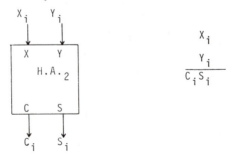

Solution. Below are a truth table and a pair of K-maps for this device.

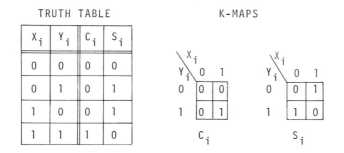

TRUTH TABLE

X_i	Y_i	C_i	S_i
0	0	0	0
0	1	0	1
1	0	0	1
1	1	1	0

From the K-maps we obtain the equations:

$$\text{Carry} = C_i(X_i, Y_i) = X_i Y_i;$$
$$\text{Sum} = S_i(X_i, Y_i) = X_i \bar{Y}_i + \bar{X}_i Y_i = X_i \oplus Y_i.$$

Since both of these output equations are minimum, all that remains is to find an economical circuit configuration. In Fig. 6.3 four logic circuits which realize these

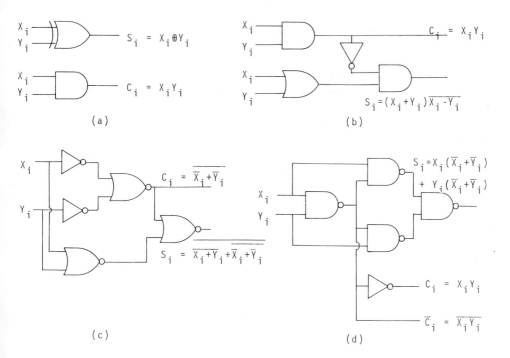

Fig. 6.3 Typical Half-Adder Logic Circuits

TABLE 6.1
Worst-Case Delay and Gate Cost for the Half Adder Circuits in Fig. 6.3

Circuit	INPUT	OUTPUT DELAY		Equivalent Inverting
		Sum	Carry	Gate Cost
a	x_i, y_i	3	2	6
b	x_i, y_i	5	2	7
c	x_i, y_i	3	2	5
d	x_i, y_i	3	2	5

equations are shown: two using a mixture of gates, one using only NOR gates, and one using only NAND gates.

Table 6.1 shows the normalized delay and equivalent gate cost for each of these units.

The final circuit choice will of course depend on the relative cost and availability of logic parts at the time of implementation.

Half adders are also available as building blocks and can be found as integrated-circuit components.

The half-adder circuit can also be assembled using digital multiplexers, with the sum and carry outputs realized by separate stages. Assuming only true variables are available, we can use the two-stage two-data-input single-selection line unit shown in Fig. 6.4.

Remember than an AND-OR-INVERT gate has an equivalent gate cost equal to the number of AND gates and a normalized delay of one. The worst-case delay for the multiplexer half-adder circuit is four units. However, multiplexers, like other logic gates, can be obtained in higher speed (less delay) packages if one is willing to pay the additional price. The higher price for the multiplexer may be more than offset by reduced wiring costs.

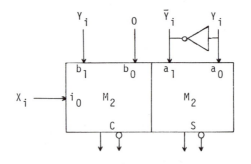

Fig. 6.4 Multiplexer Half Adder

The logical extension to a half-adder circuit is a binary full-adding circuit which utilizes the carry-in from a preceding stage.

Example 6.3 Design a binary full adder cell.

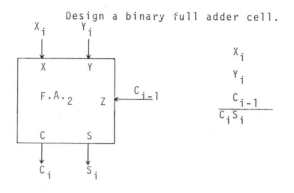

Solution. The 3-input binary adder is referred to as a full adder because the Z input is usually supplying carry information from a previous stage. The truth table for a binary full adder is shown below along with the two K-maps for the sum and carry outputs. Inspection of the K-maps reveals that, although the carry output can be minimized, the sum output cannot.

FULL ADDER TRUTH TABLE

X_i	Y_i	C_{i-1}	C_i	S_i
0	0	0	0	0
0	0	1	0	1
0	1	0	0	1
0	1	1	1	0
1	0	0	0	1
1	0	1	1	0
1	1	0	1	0
1	1	1	1	1

K-MAPS

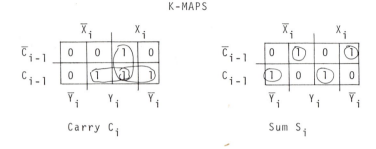

$$S_i(X_i, Y_i, C_{i-1}) = \bar{X}_i\bar{Y}_iC_{i-1} + \bar{X}_iY_i\bar{C}_{i-1} + X_i\bar{Y}_i\bar{C}_{i-1} + X_iY_iC_{i-1},$$
$$= X_1 \oplus Y_i \oplus C_{i-1}. \tag{6.1}$$
$$C_i(X_i, Y_i, C_{i-1}) = X_iY_i\bar{C}_{i-1} + X_i\bar{Y}_iC_{i-1} + \bar{X}_iY_iC_{i-1} + X_iY_iC_{i-1},$$
$$= X_iY_i + X_iC_{i-1} + Y_iC_{i-1}, \tag{6.2}$$
$$= X_iY_i + C_{i-1}(X_1 \oplus Y_i). \tag{6.3}$$

These equations can be implemented in a variety of ways using NAND and NOR logic. However, if one is observant, one can see that the logic equations for a binary full adder are *self dual*; i.e., $S_i^D = S_i$ and $C_i^D = C_i$. This can be shown by taking the dual of Eqs. 6.1 and 6.2.

$$S_i(X_i, Y_i, C_{i-1}) = X_i \oplus Y_i \oplus C_{i-1},$$
$$S_i^D(X_i, Y_i, C_{i-1}) = X_i \odot Y_i \odot C_{i-1},$$
$$= (X_i \oplus 1) \oplus Y_i \oplus (C_{i-1} \oplus 1),$$
$$= X_i \oplus Y_i \oplus C_{i-1},$$
$$= S_i(X_i, Y_i, C_{i-1}).$$
$$C_i(X_i, Y_i, C_{i-1}) = X_iY_i + X_iC_{i-1} + Y_iC_{i-1}.$$
$$C_i^D(X_i, Y_i, C_{i-1}) = (X_i + Y_i)(X_i + C_{i-1})(Y_i + C_{i-1}),$$
$$= (X_i + Y_iC_{i-1})(Y_i + C_{i-1}),$$
$$= X_iY_i + X_iC_{i-1} + Y_iC_{i-1},$$
$$= C_i(X_i, Y_i, C_{i-1}).$$

Therefore any circuit implementation using logic gates yields a second implementation using the dual gates.

Figure 6.5 shows a circuit implementation which can be obtained by interconnecting two binary half-adder circuits. However, the direct implementation of Eqs. 6.1 and 6.2 leads to simpler and less-expensive circuits.

Figures 6.6 and 6.7 show typical full-adder implementations using NAND gates. Because the output equations are self dual, two additional implementations

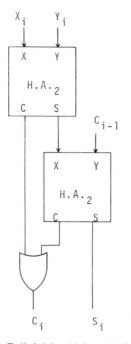

Fig. 6.5 Binary Full-Adder Using Half-Adder Circuits

can be obtained by replacing each NAND gate by a NOR gate. In each of the circuits shown it is assumed that only true variables are available as inputs.

Configuration I is the direct implementation of Eqs. 6.1 and 6.2. Configuration II realizes the sum output with back-to-back XOR circuits and the carry output C_i as $X_i Y_i + C_{i-1}(X_i \oplus Y_i)$. Configuration II only requires two input gates.

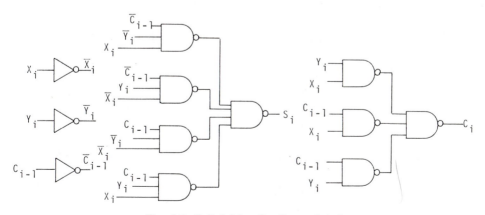

Fig. 6.6 Full-Adder Configuration I

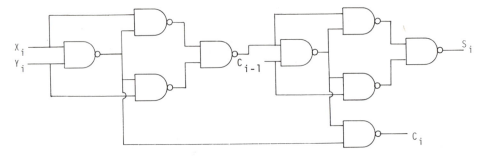

Fig. 6.7 Full-Adder Configuration II

The reader should try to obtain other circuit realizations for these two equations by further manipulation of the Boolean equations.

Table 6.2 lists the gate cost, gate-input cost, and worst-case delay time for each of the circuit configurations.

Figure 6.8 shows the multiplexer implementation for a full adder. The input values for the multiplexers can be obtained by inspection of the truth table for the full adder.

Unless the cost of the multiplexer used is very low or other factors such as reduced package costs or fewer connections enter into the design, the multiplexer-circuit configuration may be uneconomical.

The reader should also be aware of the fact that high-speed full adders can be obtained as integrated-circuit units. One popular circuit configuration suitable for integrated-circuit fabrication utilizes AND-OR-INVERT logic for the chip design. Because this gate is really a set of NAND gates followed by an internally WIRED-AND gate, a unit of delay is saved on the sum output. However, the reader should be cautioned against building the circuit with separate logic gates since the cost and delay saving is realized internally by the AND-OR-INVERT integrated-circuit-fabrication process. A NAND-WIRED-AND configuration is shown in Fig. 6.9.

There are certain applications, not commonly found in many computers, which require only the difference between two binary-arithmetic quantities. For these special applications, binary subtracting units should be considered.

TABLE 6.2
Full Adder Comparison

Circuit Configuration	No. of Gates	No. of Gate Inputs	Input	Output Delay S_i	Output Delay C_i
I	12	28	X_i, Y_i	3	2
			C_{i-1}	3	2
II	9	18	X_i, Y_i	6	5
			C_{i-1}	3	2

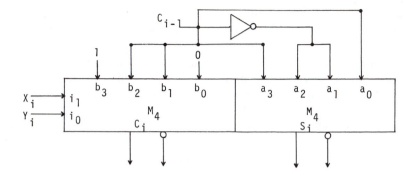

Fig. 6.8 Full-Adder Circuit Using Digital Multiplexers

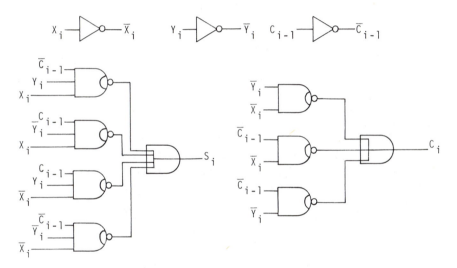

Fig. 6.9 NAND-WIRED-AND Full-Adder Circuit

Example 6.4 Design a binary full subtracter cell.

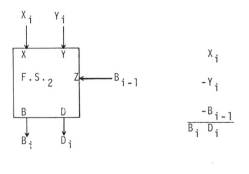

Solution. In general, subtraction is carried out in the binary number system in much the same way as it is in decimal. Thus, the subtrahend is usually complemented and the rules for addition are then used to form the result. However, there may be cases where only subtraction is required, in which case separate subtracting hardware would be more economical. Thus the minuend X_i, the subtrahend Y_i, and a possible borrow B_{i-1} from the previous stage must be combined to yield the difference D_i and borrow B_i.

A truth table and pair of K-maps for a binary full subtracter are shown in Fig. 6.10.

The logic equations for the full subtracter are easily obtained by inspection of the truth table or K-maps.

$$D_i(X_i, Y_i, B_{i-1}) = X_i\bar{Y_i}\bar{B}_{i-1} + \bar{X_i}Y_i\bar{B}_{i-1} + \bar{X_i}\bar{Y_i}B_{i-1} + X_iY_iB_{i-1}$$

$$= X_i \oplus Y_i \oplus B_{i-1}. \tag{6.4}$$

$$B_i(X_i, Y_i, B_{i-1}) = \bar{X_i}\bar{Y_i}B_{i-1} + \bar{X_i}Y_i\bar{B}_{i-1} + \bar{X_i}Y_iB_{i-1} + X_iY_iB_{i-1}$$

$$= \bar{X_i}Y_i + B_{i-1}[\bar{X_i} \oplus Y_i] \tag{6.5}$$

$$= \bar{X_i}Y_i + \bar{X_i}B_{i-1} + Y_iB_{i-1}. \tag{6.6}$$

Inspection of Eqs. 6.4 and 6.6 or the truth table reveals that the Boolean equations for a binary full subtracter are also self dual.

If the output equations for a full adder and a full subtracter are compared, we see that the sum and difference equations are the same and that only the carry and borrow equations differ. If we share gates between the outputs of a full subtracter, a simple logic circuit results. One such circuit is shown in Fig. 6.11.

Minuend X_i	0	0	0	0	1	1	1	1
Subtrahend Y_i	0	0	1	1	0	0	1	1
Borrow B_{i-1}	0	1	0	1	0	1	0	1
Difference D_i	0	1	1	0	1	0	0	1
Borrow B_i	0	1	1	1	0	0	0	1

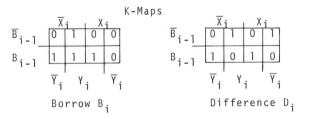

K-Maps

Borrow B_i

Difference D_i

Figure 6.10

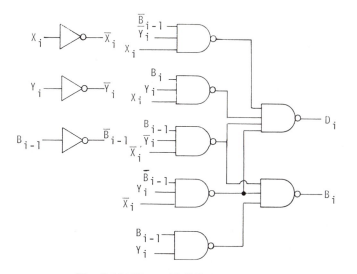

Fig. 6.11 Binary Full-Subtracter Cell

It should be again noted that due to the duality of the output equations, an alternate cell configuration can be obtained simply by replacing each NAND gate with its dual NOR gate.

The reader is encouraged to find other full subtracter configurations by manipulating the output logic equations, as was done with the full-adder output equations.

6.4 BINARY RIPPLE ARITHMETIC UNITS

In this section, the full-adder circuit developed in the previous section will be interconnected to form multi-digit two's complement arithmetic units. Because the carry output of each adder is to be connected to the carry input of the next unit in line, the information on the carry line will appear to ripple down the unit from the least significant position to the most significant position. Thus the ith sum cannot be calculated until after the carry information from the previous stage is verified. Due to this ripple effect on the carry line, arithmetic units with this type of organization are often referred to as ripple adders.

Let us begin by constructing a four-bit ripple-adder configuration, R(4), using the binary full-adder unit shown in Fig. 6.7. The resulting ripple adder is shown in Fig. 6.12. The addition time for this 4-cell unit is 12 units of gate delay, and in general would equal $2n + 4$ for an n bit adder. Using alternate full-adder cells in this ripple configuration will only reduce the overall add time by a few gate

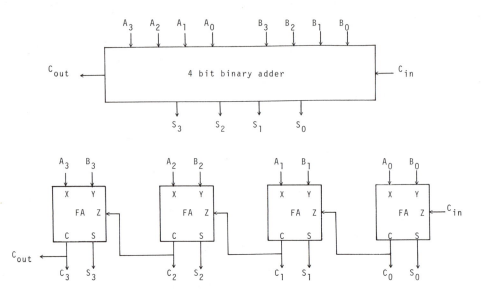

Fig. 6.12 4-Bit Ripple Binary-Arithmetic Unit

delays. The least propagation delay will be $2n$ when AND-OR-INVERT logic (NAND-WIRED-AND) logic is used.

Let us examine the organization of a 4-bit ripple adder subtracter. (See Fig. 6.13.) This unit has been designed in such a way as to enable us to use it, along with a few additional external logic gates, to form two's complement, one's complement, or sign and magnitude binary adder-subtracters.

The complement enable line M allows us to take the one's complement of the subtrahend for subtraction. The carry in line, C_{in}, can be used to pass carry, or borrow information from a previous stage; or it can be used as the end-around-carry input required for one's complement arithmetic.

By connecting the carry input of the first stage to the complement enable line M, two's complement subtraction can be accomplished.

The configuration in Fig. 6.13 uses separate XOR gates as inputs to binary full adders.

6.5 n-BIT CARRY-LOOKAHEAD ADDERS, CLA(n)'s

Let us continue our study of binary arithmetic units. Now, however, we will focus our attention on methods for reducing the total addition time.

We might ask "Can the overall addition time be reduced from a linear function of the number of bits n to a constant independent of the adder length?"

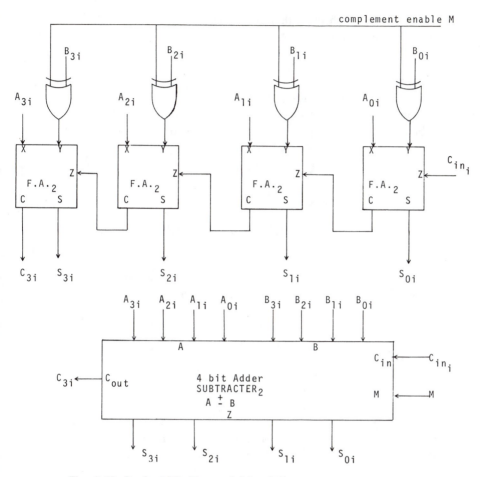

Fig. 6.13 Basic 4-Bit Binary Adder-Subtracter Configuration

The answer is yes. The how is the subject of this section.

We will start by reexamining the basic equations for a full adder.

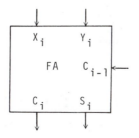

Definition 6.2 G_i, the ith *generate function* will equal one whenever it is known, independent of the carry-in to the full adder, C_{i-1}, that the carry out, C_i, of the full adder is a one. In terms of the inputs to the full adder, $G_i = X_i \cdot Y_i$.

Definition 6.3 P_i, the ith *propagate function* will equal one whenever it is possible for the carry out, C_i, of the full adder to equal one only if the carry in, C_{i-1}, is equal to a one. In terms of the inputs to the full adders $P_i = X_i \oplus Y_i$.

The basic full-adder equations can be rewritten in terms of these two functions with

$$S_i = (X_i \oplus Y_i) \oplus C_{i-1}$$
$$= P_i \oplus C_{i-1} \tag{6.7}$$

and

$$C_i = X_i Y_i + C_{i-1}(X_i \oplus Y_i)$$
$$= G_i + C_{i-1}P_i. \tag{6.8}$$

Using these modified full-adder equations, we can express the sum and carry equations for the four-bit adder in Fig. 6.12 as follows:

$$C_{-1} = C_{in}$$

$$C_0 = G_0 + C_{in}P_0 \qquad S_0 = P_0 \oplus C_{in}$$

$$C_1 = G_1 + C_0P_1 \qquad S_1 = P_1 \oplus C_0$$

$$C_2 = G_2 + C_1P_2 \qquad S_2 = P_2 \oplus C_1$$

$$C_3 = G_3 + C_2P_3 \qquad S_3 = P_3 \oplus C_2$$

The key to the development of a fast arithmetic unit lies in the direct expansion of these carry equations. The carry equation for C_0 does not change:

$$C_0 = G_0 + C_{in}P_0. \tag{6.9}$$

By substituting this equation into the carry equation for C_1 we obtain:

$$C_1 = G_1 + C_0P_1$$
$$= G_1 + (G_0 + C_{in}P_0)P_1$$
$$= G_1 + G_0P_1 + C_{in}P_0P_1. \tag{6.10}$$

The carry equation for C_1 can now be substituted in the carry equation for C_2 yielding:

$$C_2 = G_2 + C_1P_2$$
$$= G_2 + (G_1 + G_0P_1 + C_{in}P_0P_1)P_2$$
$$= G_2 + G_1P_2 + G_0P_1P_2 + C_{in}P_0P_1P_2. \tag{6.11}$$

Finally, the carry equation for C_2 can now be substituted in the carry equation for C_3 yielding:

$$C_3 = G_3 + C_2P_3$$
$$= G_3 + (G_2 + G_1P_2 + G_0P_1P_2 + C_{in}P_0P_1P_2)P_3$$
$$= G_3 + G_2P_3 + G_1P_2P_3 + G_0P_1P_2P_3 + C_{in}P_0P_1P_2P_3. \qquad (6.12)$$

The importance of these expanded carry equations lies in the realization that all four of these carry equations can be generated simultaneously using two levels of logic. Once the four carries are known, the four sums can also be simultaneously generated. This type of lookahead-carry generation is commonly referred to

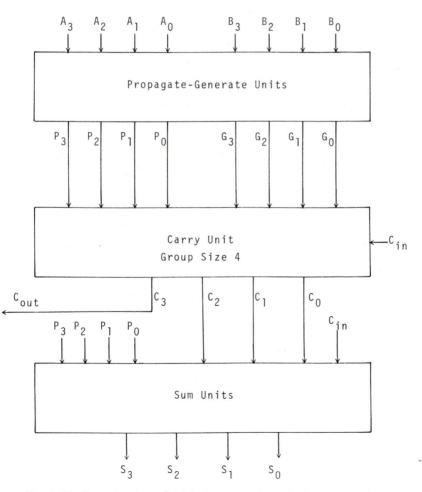

Fig. 6.14 Organization of 4-Bit Carry-Lookahead Adder, CLA(4)

as a *carry-lookahead* technique. The basic organization of a typical four-bit carry-lookahead adder, CLA(4), is shown in Fig. 6.14.

It is clear, from the basic organization of the carry-lookahead adder, CLA(n), that the average propagation delay will be a constant. The overall propagation delay is dependent upon the propagation delay for each of the three major units: the propagate-generate section, the carry section, and the sum section. A detailed examination of each of these three major sections follows.

The propagate and generate units satisfy the equations

$$P_i = A_i \oplus B_i,$$
$$G_i = A_i B_i \qquad 0 \leq i < 3.$$

These equations are also recognized as the equations for the binary half adder which was investigated in Section 6.3. A simple five-gate NAND-logic-circuit realization for these equations is shown in Fig. 6.15.

Using NAND logic, five gates are required with a worst-case delay of three units. The output $\bar{G}_i$ is also available with this circuit configuration.

The carry unit for the four-bit adder can be implemented directly using two levels of NAND gates. The number of carry outputs generated by the unit will be referred to as the Carry *Group Size* (*GS*). In this case, a group-size 4-carry unit is shown in Fig. 6.16.

A total of 18 NAND's are required for the carry unit with a worst case propagation delay of two units. If the complemented inputs $\bar{G}_i$ are available from

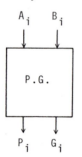

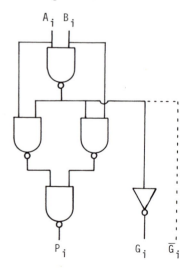

Propagation Delay

Input Output	A_i	B_i
P_i	3	3
G_i	2	2

Fig. 6.15 Typical NAND-Gate Logic Circuit for a Propagate-Generate Unit

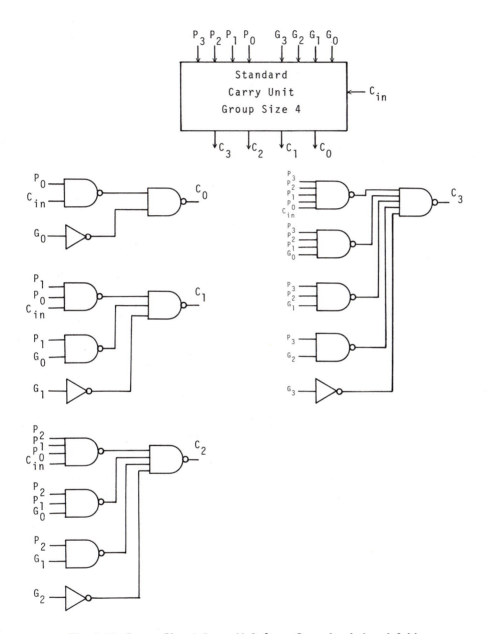

Fig. 6.16 Group-Size-4 Carry Unit for a Carry-Lookahead Adder

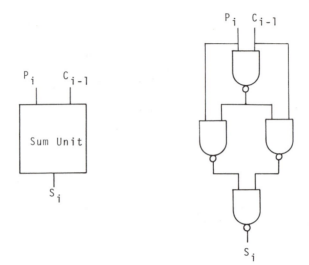

Fig. 6.17 Typical NAND-GATE Sum Unit for a Carry-Lookahead Adder

the propagate-generate units then the four inverters shown in Fig. 6.16 can be eliminated.

Finally, the sum unit consists of a single exclusive OR unit, as shown in Fig. 6.17. The sum unit satisfies the equation $S_i = P_i \oplus C_{i-1}$ and requires four NAND gates with a worst-cast propagation delay of three units. $C_{-1} = C_{in}$.

TABLE 6.3
NAND-Gate Summary for a 4-Bit Carry-Lookahead Adder, CLA(4)

Function	Number of Units	NAND-Gates per Unit	Total Number of NAND Gates	Propagation Delay
Propagate-Generate	4	5	20	3
Carry GS-4	1	18	18	2
Sum	4	4	16	3

Total number of NAND gates - 54
Total propagation delay - 8 units

Table 6.3 on page 214 summarizes the NAND gate requirements for a four-bit carry-lookahead adder.

The total propagation delay for a four-bit binary adder, using the functional units shown in Figs. 6.15 to 6.17, has been reduced to only 8 units of gate delay. This reduction in propagation delay, however, comes at the expense of 18 additional NAND gates.

Using this four-bit adder as a starting point, let us extend the length of our adder to *n* bits (Fig. 6.18). We would like to determine the total parts count required for an *n*-bit carry-lookahead adder, CLA(*n*), as well as information on the fan-in and fan-out requirements on the gates used in the various sections of the adder.

For an *n*-bit unit, the structure of the propagate-generate units and the sum units remains unchanged with these units satisfying the equations:

$$P_i = A_i \oplus B_i,$$
$$G_i = A_i B_i,$$
$$S_i = P_i \oplus C_{i-1}.$$

The carry equations can be extended to *n* bits by continuously expanding the recursively defined carry equation of a ripple adder.

$$C_k = G_k + C_{k-1} P_k,$$

with

$$C_0 = G_0 + C_{in} P_0. \tag{6.13}$$

This equation when expanded becomes

$$C_k = C_{in} \bigwedge_{i=0}^{k} P_i + \bigvee_{j=0}^{k-1} \left\{ G_j \cdot \left[\bigwedge_{i=j+1}^{k} P_i \right] \right\} + G_k. \tag{6.14}$$

Using this equation, carry C_3 would be evaluated as

$$C_3 = C_{in} \bigwedge_{i=0}^{3} P_i + \bigvee_{j=0}^{2} \left\{ G_j \cdot \left[\bigwedge_{i=j+1}^{3} P_i \right] \right\} + G_3,$$

$$= C_{in} P_0 P_1 P_2 P_3 + \qquad \qquad \text{(term 1)}$$

$$G_0 P_1 P_2 P_3 + G_1 P_2 P_3 + G_2 P_3 + \qquad \text{(term 2)}$$

$$G_3. \qquad \qquad \text{(term 3)}$$

Using Eq. 6.14, the expanded carry equations for the first 12 carries, C_0 to C_{11}, are tabulated for future reference in Table 6.4. We will use this table as an aid during our derivation of formulas for the total NAND-gate parts count. These equations will also be used to help us establish the fan-in and fan-out requirements for the gates in both the propagate-generate and the carry units.

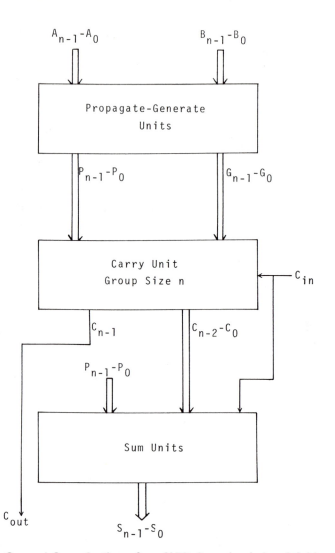

Fig. 6.18 General Organization of an *N*-Bit Carry-Lookahead Adder, CLA(*n*)

An examination of the basic carry equations reveals that carry C_{k+1} has one more term than carry C_k with

$$C_k = G_k + C_{k-1}P_k$$

and

$$C_{k+1} = G_{k+1} + C_k P_{k+1}$$
$$= G_{k+1} + [G_k + C_{k-1}P_k]P_{k+1}$$
$$= G_{k+1} + G_k P_{k+1} + C_{k-1}P_k P_{k+1}.$$

TABLE 6.4
Expanded Carry Equations

$$C_0 = G_0 + C_{in}P_0$$

$$C_1 = G_1 + G_0P_1 + C_{in}P_0P_1$$

$$C_2 = G_2 + G_1P_2 + G_0P_1P_2 + C_{in}P_0P_1P_2$$

$$C_3 = G_3 + G_2P_3 + G_1P_2P_3 + G_0P_1P_2P_3 + C_{in}P_0P_1P_2P_3$$

$$C_4 = G_4 + G_3P_4 + G_2P_3P_4 + G_1P_2P_3P_4 + G_0P_1P_2P_3P_4 + C_{in}P_0P_1P_2P_3P_4$$

$$C_5 = G_5 + G_4P_5 + G_3P_4P_5 + G_2P_3P_4P_5 + G_1P_2P_3P_4P_5 + G_0P_1P_2P_3P_4P_5 + C_{in}P_0P_1P_2P_3P_4P_5$$

$$C_6 = G_6 + G_5P_6 + G_4P_5P_6 + G_3P_4P_5P_6 + G_2P_3P_4P_5P_6 + G_1P_2P_3P_4P_5P_6 + G_0P_1P_2P_3P_4P_5P_6 + C_{in}P_0P_1P_2P_3P_4P_5P_6$$

$$C_7 = G_7 + G_6P_7 + G_5P_6P_7 + G_4P_5P_6P_7 + G_3P_4P_5P_6P_7 + G_2P_3P_4P_5P_6P_7 + G_1P_2P_3P_4P_5P_6P_7 + G_0P_1P_2P_3P_4P_5P_6P_7 + C_{in}P_0P_1P_2P_3P_4P_5P_6P_7$$

$$C_8 = G_8 + G_7P_8 + G_6P_7P_8 + G_5P_6P_7P_8 + G_4P_5P_6P_7P_8 + G_3P_4P_5P_6P_7P_8 + G_2P_3P_4P_5P_6P_7P_8 + G_1P_2P_3P_4P_5P_6P_7P_8 + G_0P_1P_2P_3P_4P_5P_6P_7P_8 + C_{in}P_0P_1P_2P_3P_4P_5P_6P_7P_8$$

$$C_9 = G_9 + G_8P_9 + G_7P_8P_9 + G_6P_7P_8P_9 + G_5P_6P_7P_8P_9 + G_4P_5P_6P_7P_8P_9 + G_3P_4P_5P_6P_7P_8P_9 + G_2P_3P_4P_5P_6P_7P_8P_9 + G_1P_2P_3P_4P_5P_6P_7P_8P_9 + G_0P_1P_2P_3P_4P_5P_6P_7P_8P_9 + C_{in}P_0P_1P_2P_3P_4P_5P_6P_7P_8P_9$$

$$C_{10} = G_{10} + G_9P_{10} + G_8P_9P_{10} + G_7P_8P_9P_{10} + G_6P_7P_8P_9P_{10} + G_5P_6P_7P_8P_9P_{10} + G_4P_5P_6P_7P_8P_9P_{10} + G_3P_4P_5P_6P_7P_8P_9P_{10} + G_2P_3P_4P_5P_6P_7P_8P_9P_{10} + G_1P_2P_3P_4P_5P_6P_7P_8P_9P_{10} + G_0P_1P_2P_3P_4P_5P_6P_7P_8P_9P_{10} + C_{in}P_0P_1P_2P_3P_4P_5P_6P_7P_8P_9P_{10}$$

$$C_{11} = G_{11} + G_{10}P_{11} + G_9P_{10}P_{11} + G_8P_9P_{10}P_{11} + G_7P_8P_9P_{10}P_{11} + G_6P_7P_8P_9P_{10}P_{11} + G_5P_6P_7P_8P_9P_{10}P_{11} + G_4P_5P_6P_7P_8P_9P_{10}P_{11} + G_3P_4P_5P_6P_7P_8P_9P_{10}P_{11} + G_2P_3P_4P_5P_6P_7P_8P_9P_{10}P_{11} + G_1P_2P_3P_4P_5P_6P_7P_8P_9P_{10}P_{11} + G_0P_1P_2P_3P_4P_5P_6P_7P_8P_9P_{10}P_{11} + C_{in}P_0P_1P_2P_3P_4P_5P_6P_7P_8P_9P_{10}P_{11}$$

The additional sum term is G_{k+1}, since the product term P_{k+1} only adds one additional input to each of the first-level NAND gates in the circuit for carry C_k.

Since the NAND implementation for $C_0 = G_0 + C_{in}P_0$ requires only three gates,

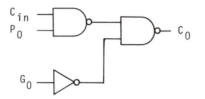

a formula for the number of NAND gates can be easily established as the sum of a simple series.

Carry	Number of Gates
C_0	3
C_1	4
C_2	5
C_3	6
.	.
.	.
.	.
C_{n-1}	$(n-1) + 3 = n + 2$

The total number of gates required by a carry unit which generates n carries is the sum of this series:

$$N_G = \sum_{j=0}^{n-1} (j + 3).$$

This equation can be rewritten as:

$$N_G = \sum_{j=3}^{n+2} j = \sum_{j=0}^{n+2} j - \sum_{j=0}^{2} j.$$

The sum of a sequence of consecutive integers is given by

$$\sum_{j=0}^{k} j = \frac{k(k + 1)}{2}.$$

Therefore

$$N_G = \sum_{j=0}^{n+2} j - \sum_{j=0}^{2} j = \frac{(n + 2)(n + 3)}{2} - \frac{2(3)}{2}$$

$$= \frac{n^2 + 5n}{2}$$

$$= \frac{n(n + 5)}{2}. \tag{6.15}$$

TABLE 6.5
NAND-Gate Parts-Count Summary for an N-Bit Carry-Lookahead Adder, CLA(n)

Functional Unit	Number of Units	NAND Gates per Unit	Total Number of NAND Gates	Propagation Delay per Unit
Propagate-Generate	n	5	$5n$	3
Carry Group Size n	1	$\dfrac{n(n+5)}{2}$	$\dfrac{n(n+5)}{2}$	2
Sum	n	4	$4n$	3

Total number of gates: $\dfrac{n(n+23)}{2} = N_{CLA(n)}$.

Total propagation delay: 8 units $= D_{CLA(n)}$.

Before proceeding to analyze the fan-in and fan-out requirements, let us summarize the NAND parts-count information we have obtained.

The calculations in Table 6.5 were made assuming the standard 3-level NAND-gate propagate-generate units (P-G units) shown in Fig. 6.15 without $\bar{G}_i$, the 2-level NAND gate carry units shown in Fig. 6.16, and the 3-level NAND-gate sum units shown in Fig. 6.17. Thus the general carry-lookahead adder organization, shown in Fig. 6.18, requires $[n(n + 23)]/2$ NAND gates while maintaining a constant propagation delay of eight units of gate delay. An n-bit ripple adder configuration, $R(n)$ using the NAND-gate full adder shown in Fig. 6.7 would have required only $9n$ NAND gates, but would have had a propagation delay of $2n + 4$ units of gate delay.

The use of alternate logic configurations such as AND-OR-INVERT logic will not alter these results significantly, except for small values of n. For example, n gates can be saved in the carry section if $\bar{G}_i$ is made available by the propagate-generate units. This change will reduce the total gate count to

$$[n(n + 21)]/2$$

but will not affect the overall addition time.

A 16-bit carry-lookahead adder will require approximately twice as many gates as a ripple adder while reducing the addition time by a factor of 4.5. A 64-bit carry-lookahead adder will require approximately five times as many gates as a ripple adder while reducing the addition time by a factor of 17.

From a theoretical standpoint, a fast adder, whose add time is independent of the number of bits to be added, is realized by using a carry-lookahead adder. However, there may be a realization problem for large values of n (greater than ≈ 16) due to the requirements for NAND gates with a large number of inputs. Let us reexamine each of the three major units in the carry-lookahead adder, in order to establish a set of formulas which will help us calculate the number of NAND gates which have j inputs.

Examination of Figs. 6.15 and 6.17 reveals that the P-G and sum units require a total of $8n$ 2-input NAND gates and n 1-input NAND gates or inverters. An examination of the carry equations (Table 6.4) reveals that carry output, C_k, requires 2 NAND gates with $k + 2$ inputs and several NAND gates each with j inputs, $1 \le j \le k + 1$.

Combining this information reveals that the n carry equations will require a total of:

n	1-input gates
$n + 1$	2-input gates
n	3-input gates
$n - 1$	4-input gates
$\ldots$	$\ldots$
$n + 3 - j$	j-input gates
$\ldots$	$\ldots$
3	n-input gates
2	$(n + 1)$-input gates

Thus the carry unit requires n inverters and $(n + 3 - j)$ NAND gates with j inputs each.

Combining the gate-input requirements for the three units, we find that an n-bit carry-lookahead adder requires:

PG		Carry		Sum		Total	
n	$+$	n	$+$	0	$=$	$2n$	1-input gates
$4n$	$+$	$n + 1$	$+$	$4n$	$=$	$9n + 1$	2-input gates
0	$+$	$n + 3 - j$	$+$	0	$=$	$n + 3 - j$	j-input gates

$$3 \le j \le n + 1$$
$$(6.16)$$

A second limiting factor on the actual circuit realization of a carry-lookahead adder will be the fan-out requirements of the various external inputs and internal gates. Both inputs A_i and B_i have fan-out requirements of two. This can be easily determined by examining Fig. 6.15. The carry in, C_{in}, must drive one gate in each of the carry equations and also two NAND gates in the S_o sum unit, for a total of $(n + 2)$ gates. If the gate-driving capability on the carry in, C_{in}, is not adequate, it can be easily increased by using an inverter tree without increasing the overall add time.

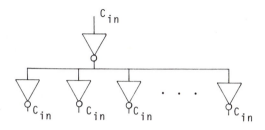

Each of the carry equations will only be required to drive two gates in the sum section and, hence, will not cause any loading difficulty. The propagate-generate functions must drive a varying number of gates in the carry and sum sections of the adder. Upon careful examination of the carry equations in Table 6.4, it is clear that variable G_j appears a total of $(n - j)$ times in the carry equations of an *n*-bit adder. It is also clear that the variable P_j will appear a total of $(j + 1)(n - j)$ times where $0 \le j \le (n - 1)$; since the variable P_j will be used exactly $(j + 1)$ times in the carry equations C_j through C_{n-1}.

If *n* is odd, then the maximum gate loading or fan-out in the carry section will be on variable $P_{(n-1)/2}$, and will be equal to $[(n + 1)^2]/4$. If *n* is even, then the maximum carry-unit loading will be on both $P_{(n/2)-1}$ and $P_{n/2}$, and will be equal to $[n(n + 2)]/4$. In either case, the propagate functions will be required to drive gate loads on the order of $[(n + 1)^2]/4$. Each of the propagate functions will also be required to drive two additional gates in the sum units. Thus the maximum fan-out requirements for the internal gates will be:

$$\frac{n(n + 2)}{4} + 2 \qquad \text{for } n \text{ even}$$

and

$$\frac{(n + 1)^2}{4} + 2 \qquad \text{for } n \text{ odd.} \qquad (6.17)$$

For large values of *n* the internal driving capability of the propagate functions can be extended by duplicating the output gate in the appropriate P-G unit several times.

The extra gates added to meet the fan-out requirement will not increase the add time for the unit and, for large values of *n*, will not significantly increase the total parts count.

From the above discussion, we can conclude that, for small values of *n*, the fan-out requirements for a carry-lookahead adder can be satisfied by adding a few extra gates in the propagate-generate units. However internal gate loading will become a serious problem for large carry-lookahead adders.

The fan-in problem is also serious but can be eliminated in the carry section by using more than two levels of logic. However, for each additional level an additional unit of propagation delay will be required by the adder. Alternate carry sections will be discussed in Sections 6.6 and 6.7.

In summary, we have shown in this section that it is, in fact, possible to build fast parallel adders with an add time independent of the number of bits (n), by trading gates for propagation delay. However, practical carry-lookahead adders will be limited to small values of n due to the severe gate fan-in and fan-out problems.

6.6 RIPPLE-CARRY-LOOKAHEAD ADDERS, RLA($n : a, b$)

In this section our objective will be to investigate an alternate carry-section organization, which will allow for a reduction in the gate fan-in and fan-out requirements at the expense of additional propagation delay. The increased delay time may be tolerable, if a corresponding reduction in overall parts count occurs.

One simple solution to the high fan-in requirement is to break the large single carry unit into a number of smaller carry units and allow the carries to ripple between the various units.

By breaking the carry unit into a units each of group size b with $n/b = a$, we obtain the general organization shown in Fig. 6.19. An additional restriction on a is that a is not equal to 1 or n since, in these cases, either a ripple adder or a complete carry-lookahead will result. We will denote the organization of these adders as being an n-bit ripple-lookahead adder having a ripple-carry units each with group size b,RLA($n:a, b$). Fig. 6.20 shows the organization of an RLA($12:3, 4$) unit.

The effect of this modification on the overall parts count and propagation delay is shown in Table 6.6. This ripple-lookahead organization reduces the parts count by a factor of $\simeq a$ at the expense of adding $2(a-1)$ units of propagation delay. The new formulas for the parts count and propagation delay are

$$\text{Parts count:} \quad N_{\text{RLA}(n:a,b)} = \frac{n(b + 23)}{2}; \qquad (6.18)$$

$$\text{Propagation delay:} \quad D_{\text{RLA}(n:a,b)} = 6 + 2a. \qquad (6.19)$$

The maximum fan-out requirements on the propagate functions are also reduced to

$$\frac{b(b + 2)}{4} + 2, \qquad \text{for } b \text{ even}$$

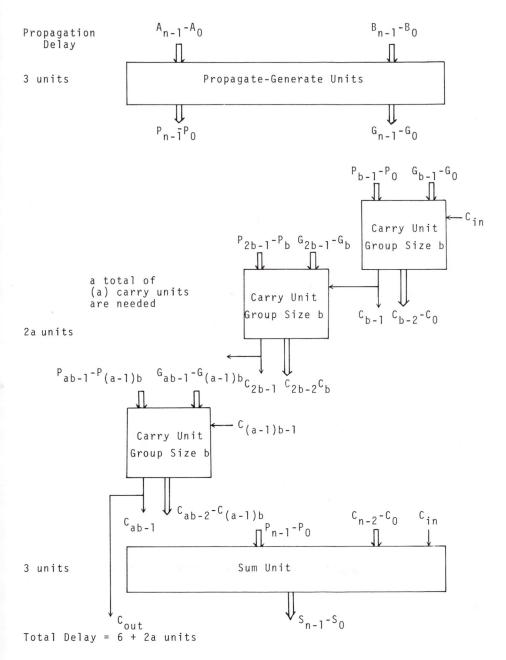

Propagation
Delay

3 units

a total of
(a) carry units
are needed

2a units

3 units

Total Delay = 6 + 2a units

Fig. 6.19 General Organization of a Ripple-Lookahead Adder

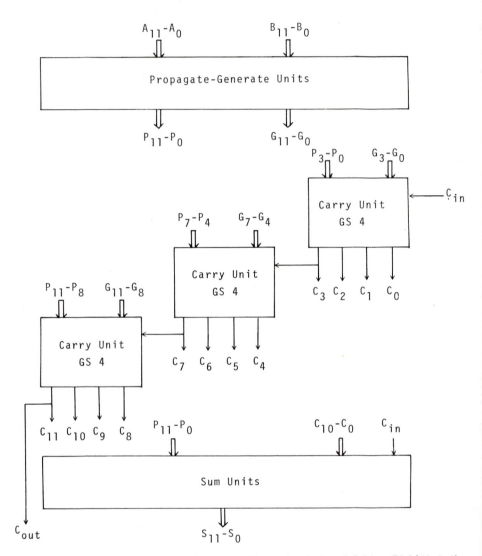

Fig. 6.20 Typical Organization of a 12-Bit Ripple-Lookahead Adder, RLA(12:3,4)

or to

$$\frac{(b + 1)^2}{4} + 2, \qquad \text{for } b \text{ odd.} \tag{6.20}$$

Thus for the RLA(12:3, 4) unit shown in Fig. 6.19, the maximum fan-out requirement on the P_i's is reduced from 44 to 8.

The maximum fan-in requirement of the carry section is also reduced to

TABLE 6.6
Parts Breakdown for an N-Bit Ripple-Lookahead Adder, RLA(n: a, b)

Functional Unit	Number of Units Needed	NAND Gates per Unit	Total Parts	Propagation Delay
Propagate-Generate	n	5	$5n$	3
Carry GS b	$a = \dfrac{n}{b}$	$\dfrac{b(b+5)}{2}$	$\dfrac{n(b+5)}{2}$	$2a$
Sum	n	4	$4n$	3

Restrictions $a \cdot b = n$

Total parts $9n + \dfrac{n(b+5)}{2} = \dfrac{n(b+23)}{2}$

Total propagation delay $6 + 2a = 6 + \dfrac{2n}{b}$

Maximum gate fan in $b + 1$

Maximum gate fan out $\dfrac{b(b+2)}{4} + 2$, $\dfrac{(b+1)^2}{4} + 2$

b even b odd

$(b + 1)$ from $(n + 1)$. In the case of the RLA(12:3, 4) the fan-in is reduced to 5 from 13 for the CLA(12) adder.

Using the ripple-lookahead approach, the carry equations for a RLA(12:3, 4) adder become:

$$C_0 = G_0 + C_{in}P_0$$
$$C_1 = G_1 + G_0P_1 + C_{in}P_0P_1$$
$$C_2 = G_2 + G_1P_2 + G_0P_1P_2 + C_{in}P_0P_1P_2$$

$$C_3 = G_3 + G_2P_3 + G_1P_2P_3 + G_0P_1P_2P_3 + C_{in}P_0P_1P_2P_3$$
$$C_4 = G_4 + C_3P_4$$
$$C_5 = G_5 + G_4P_5 + C_3P_4P_5$$
$$C_6 = G_6 + G_5P_6 + G_4P_5P_6 + C_3P_4P_5P_6$$
$$C_7 = G_7 + G_6P_7 + G_5P_6P_7 + G_4P_5P_6P_7 + C_3P_4P_5P_6P_7$$
$$C_8 = G_8 + C_7P_8$$
$$C_9 = G_9 + G_8P_9 + C_7P_8P_9$$
$$C_{10} = G_{10} + G_9P_{10} + G_8P_9P_{10} + C_7P_8P_9P_{10}$$
$$C_{11} = G_{11} + G_{10}P_{11} + G_9P_{10}P_{11} + G_8P_9P_{10}P_{11} + C_7P_8P_9P_{10}P_{11}. \qquad (6.21)$$

6.7 FIRST-ORDER CARRY-LOOKAHEAD ADDERS, CLA($n : a, b$)

One drawback with the ripple-lookahead adders discussed in the preceding section is the dependence of the propagation time on the size of the adder, n; and the selected group size, b, with $D_{RLA(n:a,b)} = 6 + 2n/b$.

The most advantageous adder configuration would be one which reduced the gate fan-in and fan-out problems and the parts count, while maintaining a propagation delay independent of the adder size.

Such an adder configuration does exist and is the subject of this section. Let us start by reexamining the carry equations for the 12-bit ripple-lookahead adder, Eqs. 6.21. In particular, let us examine the three carry equations C_3, C_7, and C_{11}:

$$C_3 = G_3 + G_2P_3 + G_1P_2P_3 + G_0P_1P_2P_3 + C_{in}P_0P_1P_2P_3$$
$$C_7 = G_7 + G_6P_7 + G_5P_6P_7 + G_4P_5P_6P_7 + C_3P_4P_5P_6P_7$$
$$C_{11} = G_{11} + G_{10}P_{11} + G_9P_{10}P_{11} + G_8P_9P_{10}P_{11} + C_7P_8P_9P_{10}P_{11}.$$

We would like to be able to calculate these three carry equations simultaneously as we did with all the carries in standard carry-lookahead adders.

By defining a new set of first-order propagate-generate functions, P_j^1, G_j^1, we will be able to reduce the logical complexity of these equations. The superscript 1 will be used to denote the order of the propagate-generate functions. Let us assign

$$P_0^1 = P_0P_1P_2P_3.$$
$$G_0^1 = G_3 + G_2P_3 + G_1P_2P_3 + G_0P_1P_2P_3.$$
$$P_1^1 = P_4P_5P_6P_7.$$
$$G_1^1 = G_7 + G_6P_7 + G_5P_6P_7 + G_4P_5P_6P_7.$$
$$P_2^1 = P_8P_9P_{10}P_{11}.$$
$$G_2^1 = G_{11} + G_{10}P_{11} + G_9P_{10}P_{11} + G_8P_9P_{10}P_{11}.$$

Using these new terms, the ripple-carry equations for C_3, C_7, and C_{11} can be rewritten as:

$$C_3 = G_0^1 + C_{in}P_0^1$$

$$C_7 = G_1^1 + C_3P_1^1,$$

$$= G_1^1 + G_0^1P_1^1 + C_{in}P_0^1P_1^1$$

and

$$C_{11} = G_2^1 + C_7P_2^1,$$

$$= G_2^1 + G_1^1P_2^1 + G_0^1P_1^1P_2^1 + C_{in}P_0^1P_2^1P_2^1.$$

The reader should expand these equations and compare the results with the corresponding carries in Table 6.4. The importance of this set of equations lies in the fact that these three carry equations can be simultaneously generated as soon as the values for the first order propagate-generate function become known. The three carries can then be supplied simultaneously to the three standard carry units, thereby allowing the remaining carries to be simultaneously generated. In this case, we are using two steps instead of one for the complete generation of the carries for the adder.

The complete equations for this unit are:

Propagate-generate:

$$P_i = A_i \oplus B_i; \qquad G_i = A_iB_i, \qquad 0 \leq i \leq 11$$

First-order propagate-generate:

$$P_0^1 = P_0P_1P_2P_3, \qquad G_0^1 = G_3 + G_2P_3 + G_1P_2P_3 + G_0P_1P_2P_3,$$

$$P_1^1 = P_4P_5P_6P_7, \qquad G_1^1 = G_7 + G_6P_7 + G_5P_6P_7 + G_4P_5P_6P_7,$$

$$P_2^1 = P_8P_9P_{10}P_{11}. \qquad G_2^1 = G_{11} + G_{10}P_{11} + G_9P_{10}P_{11} + G_8P_9P_{10}P_{11}.$$

First-order carry:

$$C_3 = G_0^1 + C_{in}P_0^1,$$

$$C_7 = G_1^1 + G_0^1P_1^1 + C_{in}P_0^1P_1^1,$$

$$C_{11} = G_2^1 + G_1^1P_2^1 + G_0^1P_1^1P_2^1 + C_{in}P_0^1P_1^1P_2^1.$$

Carry:

$$C_0 = G_0 + C_{in}P_0,$$

$$C_1 = G_1 + G_0P_1 + C_{in}P_0P_1,$$

$$C_2 = G_2 + G_1P_2 + G_0P_1P_2 + C_{in}P_0P_1P_2.$$

$$C_4 = G_4 + C_3P_4,$$

$$C_5 = G_5 + G_4P_5 + C_3P_4P_5,$$

$$C_6 = G_6 + G_5P_6 + G_4P_5P_6 + C_3P_4P_5P_6.$$

$$C_8 = G_8 + C_7P_8,$$

$$C_9 = G_9 + G_8P_9 + C_7P_8P_9,$$

$$C_{10} = G_{10} + G_9P_{10} + G_8P_9P_{10} + C_7P_8P_9P_{10}.$$

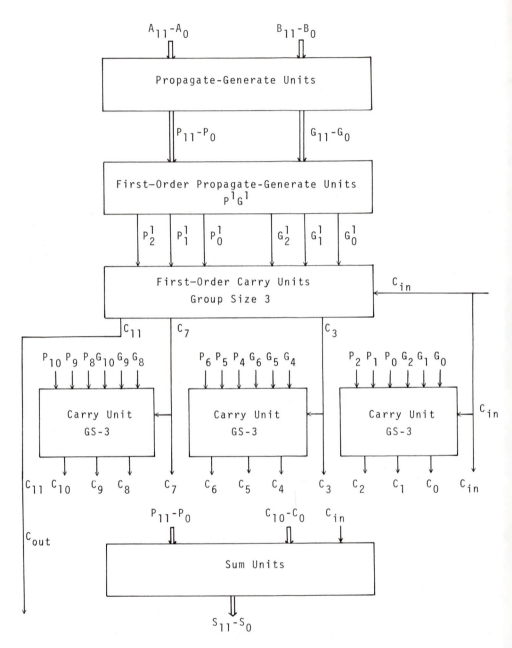

Fig. 6.21 Organization of a Carry-Lookahead Adder, (CLA(12:3,4))

Sum:

$$S_0 = P_0 \oplus C_{in}, \quad S_j = P_j \oplus C_{j-1} \quad \text{for} \quad 1 \le j \le n - 1. \tag{6.22}$$

The adder organization implied by these equations is shown in Fig. 6.21. It is easily seen, from an examination of the first-order carry equations, that the hardware logic circuit for this unit is the same as for a standard carry unit with the exception that the first-order propagate-generate functions are used. It is also clear that first order propagate-generate functions can be realized using two levels of NAND gates.

If we group b of the propagate functions together, then each P_i^1 equation will require two NAND gates and each G_i^1 equation will require $(b + 1)$ NAND gates. By requiring that b divide n, we determine that a total of $a = n/b$ first-order propagate-generate units will be needed with a gate count of $a(b + 3)$ NAND gates. The adder organizations in this section will require that $a \cdot b = n$; however, other adder organizations are possible. These alternate configurations will be left as exercises.

The two additional first-order logic units will add a total of four units of propagation delay to the overall addition time. A summary of the gate counts and propagation delay for the 12-bit adder CLA(12:3, 4) is shown in Table 6.7.

TABLE 6.7

Gate Count and Delay Summary for a CLA(12:3,4) Arithmetic Unit

Function	Number of Units	Parts Per Unit	Total Parts	Delay Per Unit
PG	12	5	60	3
$P^1 G^1$	3	7	21	2
1st order carry GS-3	1	12	12	2
Carry GS-3	3	12	36	2
Sum	12	4	48	3
		Total	177	12

Comparing this unit against the ripple lookahead adder, RLA(12:3,4), designed in the preceding section (see Table 6.6) we see that the CLA(12:3,4) has the same propagation delay. However, it also cost 15 extra gates to implement. These extra gates are required for the first-order propagate-generate functions and the first-order carry unit. For small values of n, the increased delay time for these extra functions is not warranted. However; this parts-count disadvantage will quickly disappear as n increases.

Figure 6.22 shows the basic organization of an n-bit first-order carry lookahead adder, CLA($n:a, b$). The parameter b is used to indicate the group size of the first-order propagate-generate functions with the restriction that b be an integer that divides n. The parameter $a = n/b$ indicates the number of first-order propagate-generate units. The parameters a and b should not be set equal to either 1 or n as these choices will yield either a ripple adder or a full carry lookahead adder. A general summary of the parts count is shown in Table 6.8.

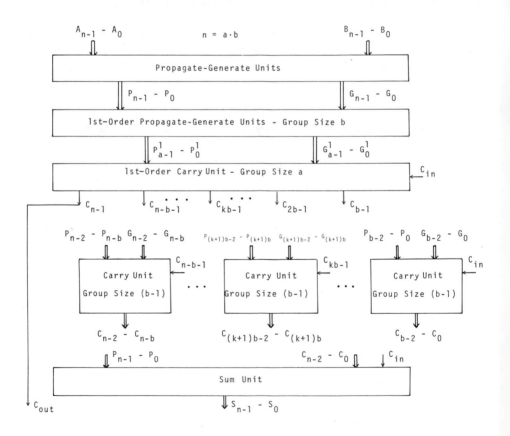

Fig. 6.22 Basic Organization of a First-Order Carry-Lookahead Adder, CLA($n:a, b$)

TABLE 6.8

NAND-Gate Count Summary for a First-Order Carry-Lookahead Adder, CLA(n:a, b)

Function	Number of Units	Parts Per Unit	Total Parts	Propagation Delay
Propagate-Generate PG	n	5	$5n$	3
1st Order Propagate-Generate P^1G^1 Group Size b	a	$b + 3$	$a(b+3)$	2
1st Order Carry Group Size a	1	$\dfrac{a(a+5)}{2}$	$\dfrac{a(a+5)}{2}$	2
Carry Units Group Size $(b-1)$	a	$\dfrac{(b-1)(b+4)}{2}$	$\dfrac{a(b^2+3b-4)}{2}$	2
Sum Units	n	4	$4n$	3

Note $n = a \cdot b$

Total parts $N_{CLA(n:a,b)} = \dfrac{n(b+23) + a(a+7)}{2}$

$$= \frac{n(n+b^3 + 23b^2 + 7b)}{2b^2}$$

Total delay $D_{CLA(n:a,b)} = 12$ units

The maximum gate fan-in will again occur in the carry units; however, since we have two different group sizes (a and $b - 1$) the maximum fan-in will be equal to the larger of $a + 1$ or b. The maximum gate fan-out from the propagate-generate units will also depend on the values of a and b. If a is larger, then the maximum fan-out will be from one of the first-order propagate functions. If b is larger than a, then the maximum fan-out will be from one of the standard propagate functions.

If a is larger than b, then the maximum fan-out will be equal to $a(a + 2)/4$ if a is even, or it will be equal to $(a + 1)^2/4$ if a is odd. If a is equal to or less than

b, then the maximum fan out will be equal to $b(b + 2)/4 + 2$ if b is even or it will be equal to $[(b + 1)^2/4] + 2$ if b is odd. The input variable C_{in} will be required to drive (a) gates in the first-order carry unit, $(b - 1)$ gates in the standard carry units and 2 gates in the sum section for a total of $a + b + 1$.

TABLE 6.9

Comparison of Adder Organizations

Ripple R(n)			Carry Lookahead CLA(n)			Ripple Lookahead RLA(n:a,b)			First Order Carry Lookahead CLA(n:a,b)		
n	Gates	Delay	n	Gates	Delay	n:a,b	Gates	Delay	n:a,b	Gates	Delay
12	108	28	12	210	8	12:3,4	162	12	12:3,4	177	12
16	144	36	16	312	8	16:4,4	216	14	16:4,4	238	12
24	216	62	24	564	8	24:3,8	372	12	24:6,4	363	12
32	288	68	32	880	8	32:4,8	496	14	32:8,4	492	12
48	432	100	48	1704	8	48:6,8	744	18	48:8,6	756	12
64	576	132	64	2784	8	64:8,8	992	22	64:8,8	1052	12

Finally, to summarize our discussion, Table 6.9 shows a selection of typical adder configurations for various values of n for each of the adder organizations examined up until now. The values of a and b for the ripple-lookahead adders were chosen to keep them competitive in terms of overall parts count with the first-order carry-lookahead adders.

From a practical point of view, the straight carry-lookahead adder CLA(n) would be prohibitive to construct for values of n greater than 16 due to the high gate fan-in and fan-out problems. The ripple lookahead and first-order carry-lookahead units are very competitive in terms of parts count, fan-in, fan-out and overall propagation delay, for adders up to 32 bits in length. For large adders with bit lengths longer than 32, the first-order carry-lookahead adder is clearly superior. However, for large adders, gate fan-in and fan-out is again going to be a serious problem.

6.8 PARALLEL MULTIPLICATION TECHNIQUES

In this section combinational logic arrays suitable for binary multiplication will be introduced. We will start by considering arrays for unsigned (positive integers) binary numbers. The logic arrays can then be modified so that signed numbers can also be multiplied.

For unsigned parallel binary multiplication, the binary product for the two binary numbers,

$$A = \sum_{i=0}^{m-1} a_i 2^i$$

and

$$B = \sum_{j=0}^{n-1} b_j 2^j,$$

is given by

$$A \cdot B = \sum_{i=0}^{m-1} a_i 2^i \cdot \sum_{j=0}^{n-1} b_j 2^j$$

$$= \sum_{i=0}^{m-1} \sum_{j=0}^{n-1} a_i b_j \cdot 2^{i+j}. \tag{6.23}$$

To carry out this calculation using combinational logic, let us start by examining in detail the multiplication of two 4-bit binary numbers $A = a_3 a_2 a_1 a_0$ and $B = b_3 b_2 b_1 b_0$. Direct multiplication yields the following array.

			a_3	a_2	a_1	a_0
			b_3	b_2	b_1	b_0
			$a_3 b_0$	$a_2 b_0$	$a_1 b_0$	$a_0 b_0$
		$a_3 b_1$	$a_2 b_1$	$a_1 b_1$	$a_0 b_1$	
	$a_3 b_2$	$a_2 b_2$	$a_1 b_2$	$a_0 b_2$		
$a_3 b_3$	$a_2 b_3$	$a_1 b_3$	$a_0 b_3$			
p_7	p_6	p_5	p_4	p_3	p_2	p_1	p_0

In order to realize this array with hardware, we must form each of the products $a_i b_j$ and, then sum the columns. The products can be easily obtained using AND gates. The sums can be formed by interconnecting a large array of binary adding circuits. One such circuit is shown in Fig. 6.23.

The array in Fig. 6.23 uses a total of 12 full adders and 16 AND gates. Generalizing this unit, an additional diagonal of full adders will be required for each additional bit in the multiplicand A. An additional row will be required for each additional bit in the multiplier B. If A has m bits and B has n bits, then a total of $(m - 1)n$ full adders will be needed, along with mn AND gates.

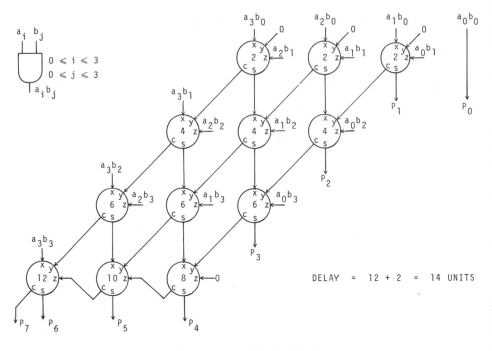

Fig. 6.23 A 4 × 4 Multiplier Array

The time required for a multiplication will depend to a large extent on the choice of the full adder configuration. As we scan down the columns, we notice that the propagation time from any input to the sum output is an important parameter. Scanning the array down the diagonals, we notice that the propagation time from any input to the carry output is also an important parameter. By selecting an adder configuration with low propagation delay from its inputs to its outputs, very fast multiplication times can easily be achieved. For example using NAND-WIRED-AND (AND-OR-INVERT) logic, a propagation delay of approximately two units of gate delay can be obtained. Such an adder configuration was shown in Fig. 6.9.

Using this full adder cell, the worst-case propagation delay will be along the right-most diagonal and across the last row. For the array shown in Fig. 6.23 this delay will be equal to 12 units of gate delay. We must add to this the time required to generate the products $a_i b_j$ which will increase the propagation delay by an additional 2 units. Thus using NAND-WIRED-AND full-adder cells, the average propagation delay for a 4 × 4 multiplier will be 14 units of gate delay.

In general, for an $m \times n$ multiplier array, the propagation delay will be $2(m - 1 + n - 1) + 2 = 2(m + n - 1)$. The multiplier array which we have just developed is suitable for construction as a large-scale integrated circuit. However, these LSI units will require a large number of input and output pins.

6.9 PROBLEMS

6.1 Design a controlled-radix complement cell for binary numbers.

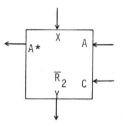

6.2 Design a combinational controlled-radix and diminished-radix logic cell which will operate on four binary digits (a single binary-coded hexadecimal digit). This unit should be capable of being cascaded so that either the radix or diminished-radix complement of numbers with $4k$ digits can be taken

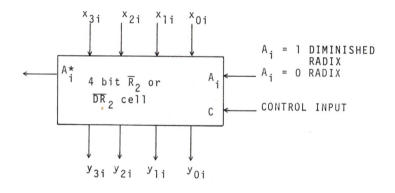

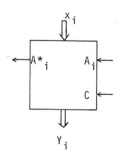

6.3 Verify that the circuits shown below are binary full-adder circuits. Extend Table 6.2 to include these two circuits.

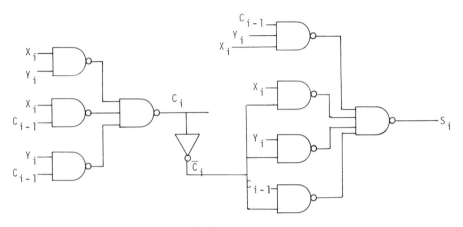

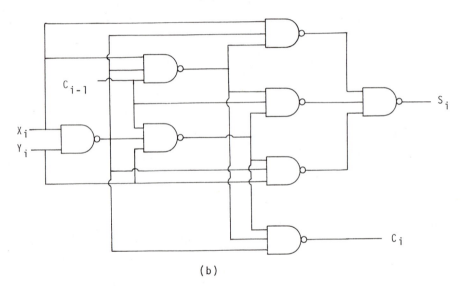

(a)

(b)

6.4   Assuming that the true and complement inputs are available, design a two-level logic circuit for adding 2-bit binary numbers.

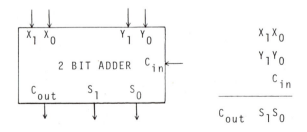

a) Minimize your circuit using K-maps.
b) Write out your output equations in sum-of-product form.
c) Compare your circuit's gate count and delay against the 2-bit ripple adder shown below

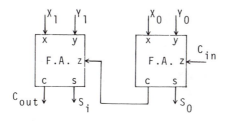

The full adder used is that shown in Fig. 6.7.

6.5 Design a two-level two's complement adder-subtracter cell.

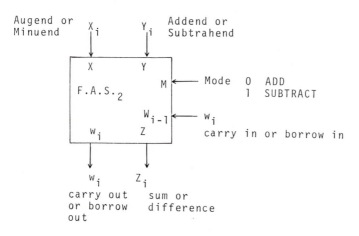

6.6 Design a two's complement adder-subtracter cell using an XOR gate and a full-adder cell. Explain how the unit works.

6.7 You have been asked to design a special-purpose arithmetic unit for adding a column of n single-bit binary numbers where (a) $n = 3$; (b) $n = 7$; (c) $n = 15$. (Hint: Use binary full adders.)

6.8 Using the basic 4-bit adder-subtracter circuit shown in Figure 6.12 as a building block, design a 12-bit adder-subtracter using
a) Two's complement arithmetic;
b) One's complement arithmetic.
Be sure to provide an overflow indicator with your circuit (see Chapter 2).

6.9 Design a 12-bit binary adder-subtracter for sign and magnitude numbers.

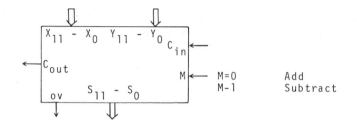

6.10 Design a combinational logic circuit capable of comparing two 8-bit binary integers (no sign bit) A and B. The output signal X should be 1 whenever $A \geq B$.

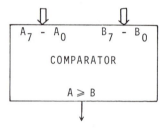

6.11 Repeat Problem 6.10 using 8-bit signed binary integers and
a) Two's complement notation,
b) One's complement notation,
c) Sign-and-magnitude notation.
Note: Consider $+0 \geq -0$ to simplify your work.

6.12 Carefully design a carry-lookahead adder, CLA(n), (a) CLA(6); (b) CLA(8); (c) CLA(9).
1) Show a block diagram for your adder.
2) Write out the basic equations for your design.
3) Include a gate count and delay summary for your unit.

6.13 Design an 8-bit carry-lookahead adder-subtracter using two's complement arithmetic. You will need to add a Mode line ($M = 0$, add; $M = 1$, subtracter) to the basic carry-lookahead adder. (Hint: You need only modify the propagate-generate unit.)

6.14 Carefully design a ripple-lookahead adder (a) RLA(15:3,5); (b) RLA(18:6,3); (c) RLA(24:4,6).
1) Show a block diagram for your adder.
2) Write out the basic equations for your unit.
(Hint: See Eq. 6.21 for the carry section.)
3) Include a gate count and delay summary for your design.

6.15 Carefully design a first-order carry-lookahead adder (a) CLA(24:6, 4); (b) CLA(18:3, 6); (c) CLA(20:4, 5)
1) Show a block diagram for your adder.

2) Write out the basic equations for each of the functional units in your design. (Hint: See Eq. 6.22 for help.)

3) Include a gate count and delay summary for your design.

6.16 Carefully design an n-bit parallel adder subject to the constraints that only gates with up to 10 inputs are used, and that the maximum fan-out for a gate is 10. Try to keep the add time as short as possible.

  a) Let $n = 8$.

  b) Let $n = 12$.

  c) Let $n = 16$.

  d) Let $n = 24$.

  e) Let $n = 32$.

6.17 Design a 6×4 multiplier array forming $P = A \cdot B$ with $A = a_5a_4a_3a_2a_1a_0$ and $B = b_3b_2b_1b_0$.

6.18 Design a 4×2 multiply cell as indicated below.

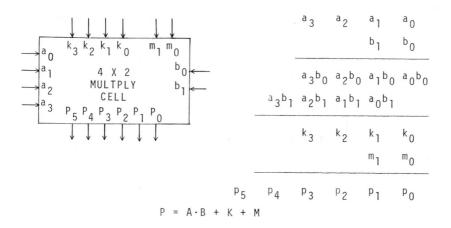

$$P = A \cdot B + K + M$$

Calculate the worst-case multiply time for your cell.

6.19 Using the 4×2 multiply cell designed in Problem 6.18 construct:

  a) a 4×6 multiplication array,

  b) an 8×4 multiplication array,

  c) an 8×8 multiplication array.

  Calculate the worst-case multiply time for each unit.

6.20 Signed binary multiplication can be carried out by (1) pre-complementing any negative inputs; (2) multiplying the resulting positive numbers; and (3) post-complementing the product if the result should be negative. Using this basic approach, design an $n \times n$ binary array multiplication unit using:

  a) Sign-and-magnitude arithmetic,

  b) One's complement arithmetic,

  c) Two's complement arithmetic.

The inputs are

$$A = a_{n-1}a_{n-2} \cdots a_0$$

and

$$B = b_{n-1}b_{n-2} \cdots b_0$$

The resulting signed product is to be a $2n$-bit number

$$P = p_{2n-1}p_{2n-2} \cdots p_n p_{n-1} \cdots p_0$$

6.10 BIBLIOGRAPHY

Baugh, C. R., and B. A. Wooley, "A two's complement parallel array multiplication algorithm," *IEEE Trans. on Computers*, Vol. C-22, pp. 1045–1047, Dec. 1973.

Booth, T. L., *Digital Networks and Computer Systems*. New York: Wiley, 1971.

Chu, Y., *Digital Computer Design Fundamentals*. New York: McGraw-Hill, 1962.

Dietmeyer, D. L., *Logic Design of Digital Systems*. Boston, Mass.: Allyn and Bacon, 1971.

Flores, I., *The Logic of Computer Arithmetic*. Englewood Cliffs, N.J.: Prentice-Hall, 1963.

Lehman, M., and N. Burla, "Skip techniques for high-speed carry propagation in binary arithmetic units," *IRE Trans. Prof. Group on Electronic Computers*, Vol. EC-10, pp. 691–698. Dec. 1961.

Peatman, J. B., *The Design of Digital Systems*. New York: McGraw-Hill, 1972.

Pezaris, S. D., "A 40-ns 17-bit by 17-bit array multiplier," *IEEE Trans. on Computers*, Vol. C-20, pp. 422–447, Apr. 1971.

Stein, M. L., and W. D. Munro, *Introduction to Machine Arithmetic*. Reading, Mass.: Addison-Wesley, 1971.

Chapter 7

Decimal
Arithmetic

7.1 INTRODUCTION

In this chapter, we will concentrate on the special logic problems which confront the design engineer when decimal data is involved.

Section 7.2 discusses a few of the codes commonly used to represent decimal data. Section 7.3 discusses briefly error-control hardware which can be used to protect decimal data during transmission. Section 7.4 discusses code conversion and display circuits. Section 7.5 discusses the design of decimal addition circuits using coded-binary numbers. Section 7.6 combines the material in these three sections to form useful decimal arithmetic units.

7.2 BINARY CODING OF DECIMAL DATA

Up to this point, we have been assuming that all of the numerical data has been expressed in binary arithmetic. However, even though computers are naturally suited to operate efficiently using binary data, people are not normally able to cope with the problems of interpreting numerical and nonnumerical binary data.

If we wish to design a small portable desk calculator and hope to sell the unit, then the keyboard and display panel must use the familiar decimal information symbols.

Information transmitted to a computer from a teletype or similar input device accepts, interprets, and converts the keyboard commands into a string of binary information symbols, which must be unambiguously understood by our digital hardware. For the moment, let us turn our attention to the binary coding of numerical data.

Assuming we are planning to encode our information on a digit-by-digit basis, it is only necessary to establish a one-to-one correspondence between the decimal numbers 0 through 9, and 10 groups of four or more binary digits. At least four binary digits are needed, since with three binary digits we can only distinguish between eight objects. In general, it will take at least $\lceil \log_2 N \rceil$ binary digits to distinguish uniquely N objects.

If we use the minimum number of four binary digits to encode our 10 decimal symbols, then we will be able to choose from a total of $(16!/6!) \approx 3 \times 10^{10}$ different binary codes. This number is arrived at by placing the 16 possible 4-bit codes in a hat and then assigning the code words to the decimal digits by random selection. Thus there are 16 ways of assigning a code to the decimal number 0, 15 ways left for the decimal number 1 and finally, 7 ways left for decimal number 9. Fortunately, certain codes have become widely used for coding decimal numbers.

These codes were selected not for their randomness but rather for their ease in both interpretation and conversion to and from the normal binary-coded-decimal notation. Normal binary-coded-decimal notation, henceforth referred to as BCD or 8-4-2-1 code, uses the four-digit binary representation of the decimal digits as its code format. This code, shown in Table 7.1, is widely used due to its ease of decimal interpretation.

Examining the codes in Table 7.2, we see that the 8-4-2-1 code is the normal BCD code. The excess-3 and excess-6 codes are formed from the normal 8-4-2-1 code by adding the codes for the constants 3 and 6 respectively to it. The next to last code is negatively weighted with the weights +8, +4, −2, and −1. The remaining codes except for the Gray are all positively weighted and do not yield a unique coding correspondence with the decimal digits. This ambiguity can result in coding errors when hand-coding decimal digits. The main feature of the Gray code is the fact that adjacent numbers differ by only one position.

One additional desirable property of binary codes, used for the coding of decimal digits, would be the ability to easily implement the diminished-radix (9's

TABLE 7.1

BCD Code Format

$(N)_{10}$	$(N)_2$	$(N)_{10}$	$(N)_2$
0	0000	5	0101
1	0001	6	0110
2	0010	7	0111
3	0011	8	1000
4	0100	9	1001

TABLE 7.2

Typical 4-Bit Binary Codes for Decimal Digits

$(N)_{10}$	8-4-2-1	Excess 3	Excess 6	5-4-2-1	2-4-2-1	4-3-2-1	5-2-1-1	5-2-2-1	8-4-$\bar{2}$-$\bar{1}$	Gray
0	0 0 0 0	0 0 1 1	0 1 1 0	0 0 0 0	0 0 0 0	0 0 0 0	0 0 0 0	0 0 0 0	0 0 0 0	0 0 1 0
1	0 0 0 1	0 1 0 0	0 1 1 1	0 0 0 1	0 0 0 1	0 0 0 1	0 0 0 1	0 0 0 1	0 1 1 1	0 1 1 0
2	0 0 1 0	0 1 0 1	1 0 0 0	0 0 1 0	0 0 1 0	0 0 1 0	0 0 1 1	0 0 1 0	0 1 1 0	0 1 1 1
3	0 0 1 1	0 1 1 0	1 0 0 1	0 0 1 1	0 0 1 1	0 1 0 0	0 1 0 1	0 0 1 1	0 1 0 1	0 1 0 1
4	0 1 0 0	0 1 1 1	1 0 1 0	0 1 0 0	0 1 0 0	1 0 0 0	0 1 1 1	0 1 1 0	0 1 0 0	0 1 0 0
5	0 1 0 1	1 0 0 0	1 0 1 1	1 0 0 0	1 0 1 1	1 0 0 1	1 0 0 0	1 0 0 0	1 0 1 1	1 1 0 0
6	0 1 1 0	1 0 0 1	1 1 0 0	1 0 0 1	1 1 0 0	1 0 1 0	1 0 1 0	1 0 0 1	1 0 1 0	1 1 0 1
7	0 1 1 1	1 0 1 0	1 1 0 1	1 0 1 0	1 1 0 1	1 1 0 0	1 1 0 0	1 0 1 0	1 0 0 1	1 1 1 1
8	1 0 0 0	1 0 1 1	1 1 1 0	1 0 1 1	1 1 1 0	1 1 0 1	1 1 1 0	1 0 1 1	1 0 0 0	1 1 1 0
9	1 0 0 1	1 1 0 0	1 1 1 1	1 1 0 0	1 1 1 1	1 1 1 0	1 1 1 1	1 1 1 0	1 1 1 1	1 0 1 0

complement) operation. This would allow us to efficiently use complement arithmetic for subtraction operations. It can easily be shown that if the sum of the weights used to code the decimal digits equals 9, then the code words can always be selected to yield a self-complementing code. A *self-complementing* code is one which yields the diminished-radix complement by simply complementing the zeros and ones in the binary code. The 5-2-1-1, 8-4-$\bar{2}$-$\bar{1}$, and 2-4-2-1 codes have this property, as does the excess-3 code whose weight sums to 18, if the excess-3 is considered to be a constant weight or a bias of 3. Thus for these four codes a controlled 9's complement unit can be constructed using four exclusive OR gates as shown in Fig. 7.1.

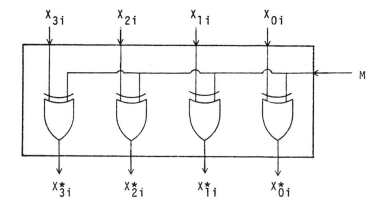

Fig. 7.1 Controlled Diminished-Radix Unit for Self-Complementing 4-Digit Binary Codes for Decimal Digits

However, the design of a controlled 9's complement unit for the remaining codes in Table 7.1 is not a difficult task and requires only a modest amount of hardware.

Example 7.1 Design a controlled 9's complement unit for the normal BCD (8-4-2-1) code.

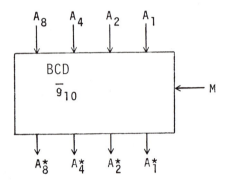

Solution. There are many ways of attacking a problem of this type depending on the availability of functional parts, the number of units to be manufactured, and the time required for the operation to be performed.

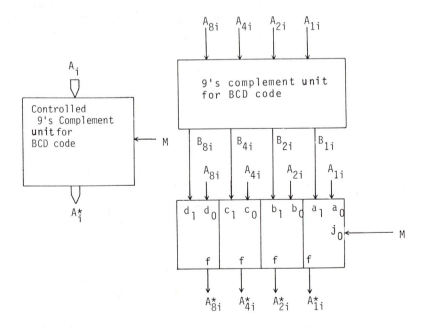

Fig. 7.2 Controlled Nine's Complement BCD Logic Unit

One method of solving this problem is to break the problem into two parts. This can be done by noting that, when the control line $M = 0$, the BCD digit N is transferred directly to the output of the unit and when $M = 1$, the 9's complement of N is passed to the output. This switching action suggests the use of a multiplexer with M as the selection line. Figure 7.2 shows a block diagram for such a solution. For such a unit, it is only necessary to design the 9's complement unit for the BCD code.

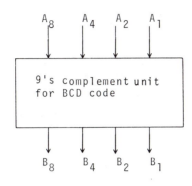

The next step in solving this problem is to convert the problem statement into a form which can be handled with Boolean algebra. For this particular problem, a truth table is easily constructed from the problem statement. Since the numbers of interest are 0 through 9, the numbers 10 through 15 will be don't-care numbers; i.e., we don't care what the output is for these numbers, since they will never be applied to the inputs. Remembering that the diminished-radix complement of a decimal digit N is $9 - N$, the truth table on page 246 becomes a restatement of the problem.

Once we have converted the problem statement into a truth table, all that remains to be done is to determine a reasonable logic circuit which will satisfy this table. In effect we have four output functions to consider:

$$B_8(A_8, A_4, A_2, A_1), \qquad B_2(A_8, A_4, A_2, A_1),$$
$$B_4(A_8, A_4, A_2, A_1), \qquad B_1(A_8, A_4, A_2, A_1).$$

In Table 7.3, d stands for a don't-care output value. The K-maps for each of these functions are shown in Fig. 7.3.

An inspection of these K-maps will reveal that for each function the product-of-sum or sum-of-product minimum expressions are of equal cost.

$$B_8(A_8, A_4, A_2, A_1) = \bar{A}_8\bar{A}_4\bar{A}_2 = \overline{A_8 + A_4 + A_2}.$$

Cost: one NOR gate.

$$B_4(A_8, A_4, A_2, A_1) = A_4\bar{A}_2 + \bar{A}_4 A_2 = A_4 \oplus A_2.$$

Cost one XOR gate.

$$B_2(A_8, A_4, A_2, A_1) = A_2.$$

Cost—nothing—direct feed through.

$$B_1(A_8, A_4, A_1) = \bar{A}_1.$$

Cost—one inverter.

The resulting multigate logic circuit is shown in Fig. 7.4.

TABLE 7.3

N		(N)$_2$				(9-N)$_2$			9-N
	A_8	A_4	A_2	A_1	B_8	B_4	B_2	B_1	
0	0	0	0	0	1	0	0	1	9
1	0	0	0	1	1	0	0	0	8
2	0	0	1	0	0	1	1	1	7
3	0	0	1	1	0	1	1	0	6
4	0	1	0	0	0	1	0	1	5
5	0	1	0	1	0	1	0	0	4
6	0	1	1	0	0	0	1	1	3
7	0	1	1	1	0	0	1	0	2
8	1	0	0	0	0	0	0	1	1
9	1	0	0	1	0	0	0	0	0
10	1	0	1	0	d	d	d	d	d
11	1	0	1	1	d	d	d	d	d
12	1	1	0	0	d	d	d	d	d
13	1	1	0	1	d	d	d	d	d
14	1	1	1	0	d	d	d	d	d
15	1	1	1	1	d	d	d	d	d

The header of the table spans: Input over the (N)$_2$ columns and Output over the (9-N)$_2$ and 9-N columns.

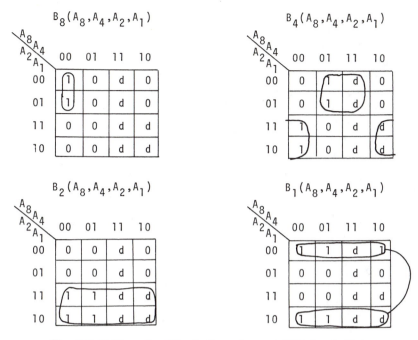

$B_8(A_8,A_4,A_2,A_1)$

$\begin{smallmatrix}A_8 \\ A_2 A_4 \\ A_1\end{smallmatrix}$	00	01	11	10
00	1	0	d	0
01	1	0	d	0
11	0	0	d	d
10	0	0	d	d

$B_4(A_8,A_4,A_2,A_1)$

$\begin{smallmatrix}A_8 \\ A_2 A_4 \\ A_1\end{smallmatrix}$	00	01	11	10
00	0	1	d	0
01	0	1	d	0
11	1	0	d	d
10	1	0	d	d

$B_2(A_8,A_4,A_2,A_1)$

$\begin{smallmatrix}A_8 \\ A_2 A_4 \\ A_1\end{smallmatrix}$	00	01	11	10
00	0	0	d	0
01	0	0	d	0
11	1	1	d	d
10	1	1	d	d

$B_1(A_8,A_4,A_2,A_1)$

$\begin{smallmatrix}A_8 \\ A_2 A_4 \\ A_1\end{smallmatrix}$	00	01	11	10
00	1	1	d	1
01	0	0	d	0
11	0	0	d	d
10	1	1	d	d

Fig. 7.3 *K*-Maps for Nine's Complement BCD Logic Unit

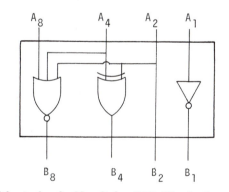

Fig. 7.4 Multigate Logic Circuit for BCD Nine's Complement Unit

7.3 BINARY CODING OF DECIMAL DATA WITH ERROR CONTROL

In the preceding section, a few of the 4-bit binary codes for decimal information were presented. These codes are ideally suited for use on computers or communications equipment where the possibility of a random-noise pulse changing a zero to a one or a one to zero is extremely remote, for example, one error a week, a month, or a year. The reason for this requirement is the inability to detect when these errors occur. In the case of ordinary binary information, any error yields another valid binary number. When working with 4-digit binary-coded-decimal numbers, some errors can be detected since there are six invalid code words.

For a single error to be detected, such an error must convert the valid code word into an invalid code word. One simple way that this can be accomplished is to add an extra bit to each number code. This extra bit is called a *parity* bit. If the parity bit is selected so that there are an even number of ones in the resulting code words, then the code is said to have even parity; if there are an odd number of ones, then the code has odd parity. Table 7.4 shows the resulting odd and even code words which result from adding an extra parity bit to the normal BCD code.

An inspection of Table 7.4 reveals that all single errors will yield an invalid number. Simple combinational logic circuits can be used to calculate and check for single errors. Two such circuits for the normal BCD code with even parity are shown in Figs. 7.5 and 7.6.

The parity bit P_e is equal to $B_8 \oplus B_4 \oplus B_2 \oplus B_1$, since a cascaded exclusive OR circuit yields a one if an odd number of ones exists in the code word. The output of the last exclusive-OR circuit yields the correct parity bit to make the total word contain an even number of ones.

TABLE 7.4
Odd and Even Parity for BCD Code

Decimal number	Normal BCD	BCD with Even Parity	BCD with Odd Parity
N	$B_8 B_4 B_2 B_1$	$B_8 B_4 B_2 B_1 P_e$	$B_8 B_4 B_2 B_1 P_o$
0	0 0 0 0	0 0 0 0 0	0 0 0 0 1
1	0 0 0 1	0 0 0 1 1	0 0 0 1 0
2	0 0 1 0	0 0 1 0 1	0 0 1 0 0
3	0 0 1 1	0 0 1 1 0	0 0 1 1 1
4	0 1 0 0	0 1 0 0 1	0 1 0 0 0
5	0 1 0 1	0 1 0 1 0	0 1 0 1 1
6	0 1 1 0	0 1 1 0 0	0 1 1 0 1
7	0 1 1 1	0 1 1 1 1	0 1 1 1 0
8	1 0 0 0	1 0 0 0 1	1 0 0 0 0
9	1 0 0 1	1 0 0 1 0	1 0 0 1 1

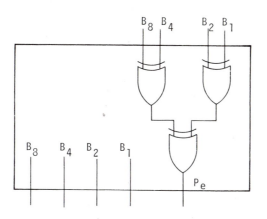

Fig. 7.5 Even-Parity Generator for BCD Code

To detect an error, we need only to check to see if the code word in question has an odd number of ones. Again the cascaded exclusive-OR circuit can be used.

Several codes with more than four bits per word have been used to code the decimal digits. Among the more common are the 2-out-of-5 code and the 2-out-of-7 code or the biquinary code and the Hamming code. These codes are shown in Table 7.5.

The Hamming code is particularly useful, because it has the added capability of correcting single errors. This is because each code word in the Hamming code differs in at least three positions from every other code word. Because of this, at least two errors must occur before there can be any confusion about the digit received. As long as only a single error occurs, the received word will differ from the correct digit in only one place and from all other digits in two or more places. However, if more than one error occurs, then the received code word will equal or differ by one from an incorrect digit. Hence any set of code words with a minimum distance of three between each pair of code words can correct all single errors.

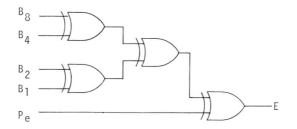

Fig. 7.6 Even-Parity Error Detector for BCD Code with Even Parity

Table 7.5

Error-Detecting Binary Codes for Decimal Digits

N	2 out of 5	biquinary 5-0-4-3-2-1-0	Hamming - BCD $I_4 I_3 I_2 P_3 I_1 P_2 P_1$
0	00011	0 1 0 0 0 0 1	0 0 0 0 0 0 0
1	00101	0 1 0 0 0 1 0	0 0 0 0 1 1 1
2	00110	0 1 0 0 1 0 0	0 0 1 1 0 0 1
3	01001	0 1 0 1 0 0 0	0 0 1 1 1 1 0
4	01010	0 1 1 0 0 0 0	0 1 0 1 0 1 0
5	01100	1 0 0 0 0 0 1	0 1 0 1 1 0 1
6	10001	1 0 0 0 0 1 0	0 1 1 0 0 1 1
7	10010	1 0 0 0 1 0 0	0 1 1 0 1 0 0
8	10100	1 0 0 1 0 0 0	1 0 0 1 0 1 1
9	11000	1 0 1 0 0 0 0	1 0 0 1 1 0 0

The Hamming code is a popular single-error-correcting code because of the ease in which the code words are generated, and also because of the interesting corrective property inherent in its structure. The 7-bit code consists of three parity bits and four information bits arranged in the order $I_4 I_3 I_2 P_3 I_1 P_2 P_1 = X_7 X_6 X_5 X_4 X_3 X_2 X_1$, where $I_4 I_3 I_2 I_1$ represents the four bits of binary information to be transmitted and $P_3 P_2 P_1$ represents the three parity bits. The parity bits are selected and interleaved with the information bits, in order to make the detection and correction of a single error a simple process.

By selecting even parity over transmitted bits $X_7 X_6 X_5 X_4$, $X_7 X_6 X_3 X_2$, and $X_7 X_5 X_3 X_1$; a single-bit error in bit position X_i will identify itself at the receiver, when a parity check is made on the received code word $Y_7 Y_6 Y_5 Y_4 Y_3 Y_2 Y_1$. The parity bits are determined by the equations:

$$X_7 \oplus X_6 \oplus X_5 \oplus X_4 = 0,$$
$$X_7 \oplus X_6 \oplus X_3 \oplus X_2 = 0,$$
$$X_7 \oplus X_5 \oplus X_3 \oplus X_1 = 0.$$

Thus in terms of the information to be transmitted:

$$I_4 \oplus I_3 \oplus I_2 = P_3,$$
$$I_4 \oplus I_3 \oplus I_1 = P_2,$$
$$I_4 \oplus I_2 \oplus I_1 = P_1.$$

A decoder and parity-checking circuit for the received information word is relatively easy to construct. Assuming additive noise can change a zero to a one, or a one to a zero, during the transmission process, the code word received may contain an error. In Fig. 7.7 a simple circuit model for the code generator and transmission lines are shown with the received code word indicated by Y.

The received code word Y can be checked for error by testing for even parity on the received lines. The received bits of the word Y are:

$$Y_7 = e_7 \oplus X_7 = e_7 \oplus I_4,$$
$$Y_6 = e_6 \oplus X_6 = e_6 \oplus I_3,$$
$$Y_5 = e_5 \oplus X_5 = e_5 \oplus I_2,$$
$$Y_4 = e_4 \oplus X_4 = e_4 \oplus P_3 = e_4 \oplus I_4 \oplus I_3 \oplus I_2,$$
$$Y_3 = e_3 \oplus X_3 = e_3 \oplus I_1,$$
$$Y_2 = e_2 \oplus X_2 = e_2 \oplus P_2 = e_2 \oplus I_4 \oplus I_3 \oplus I_1,$$
$$Y_1 = e_1 \oplus X_1 = e_1 \oplus P_1 = e_1 \oplus I_4 \oplus I_2 \oplus I_1.$$

These bits are inputted into a parity checking circuit similar to that shown in Fig. 7.8.

The output equations for the parity checking units become:

$$S_2 = Y_7 \oplus Y_6 \oplus Y_5 \oplus Y_4,$$
$$= (e_7 \oplus I_4) \oplus (e_6 \oplus I_3) \oplus (e_5 \oplus I_2) \oplus (e_4 \oplus I_4 \oplus I_3 \oplus I_2),$$
$$= e_7 \oplus e_6 \oplus e_5 \oplus e_4;$$
$$S_1 = Y_7 \oplus Y_6 \oplus Y_3 \oplus Y_2,$$
$$= (e_7 \oplus I_4) \oplus (e_6 \oplus I_3) \oplus (e_3 \oplus I_1) \oplus (e_2 \oplus I_4 \oplus I_3 \oplus I_1),$$
$$= e_7 \oplus e_6 \oplus e_3 \oplus e_2;$$
$$S_0 = Y_7 \oplus Y_5 \oplus Y_3 \oplus Y_1,$$
$$= (e_7 \oplus I_4) \oplus (e_5 \oplus I_2) \oplus (e_3 \oplus I_1) \oplus (e_1 \oplus I_4 \oplus I_2 \oplus I_1),$$
$$= e_7 \oplus e_5 \oplus e_3 \oplus e_1.$$

An examination of the equations for S_2, S_1, and S_0 reveals that they depend only on the noise errors and not on the information transmitted. Table 7.6 shows the resulting output of the parity checking circuit assuming a single error has taken place.

Thus an inspection of either Table 7.6 or the parity check equations reveals that a single error in bit position X_i reveals itself by forcing $S_2S_1S_0$ to be the binary number i, thus identifying the bit position in error.

All that remains is to correct a bit received in error. This last task can be accomplished by using another set of exclusive-OR circuits, as shown in Fig. 7.9, where one input to the gate is bit X_i, and the other input is $S_2^{i_2}S_1^{i_1}S_0^{i_0}$, where

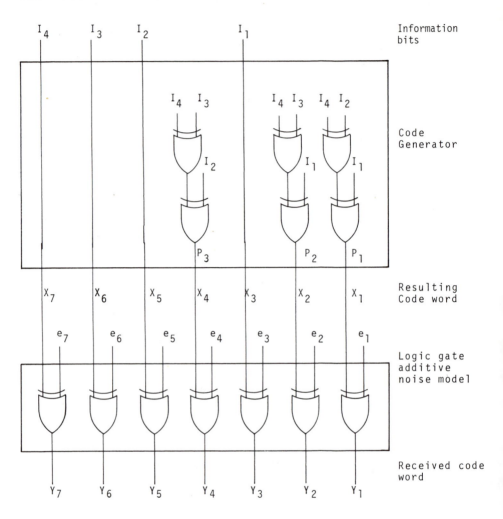

Fig. 7.7 Hamming-Code Generator with Additive Noise

$(i)_{10} = (i_2 i_1 i_0)_2$. Thus if a single error occurs in position Y_i, it will be removed by the XOR gate on the output of line Y_i.

In general, only bit positions Y_7, Y_6, Y_5, and Y_3 need actually be corrected, since only these bits correspond to the information transmitted. Figure 7.10 shows a complete flow chart for this Hamming 7-bit single-error-correcting code. Since the code does not depend on the information sent, any of the 4-bit binary codes for decimal digits can make use of the same coder and decoder.

For additional information on error correcting codes the reader is referred to the books at the end of this chapter.

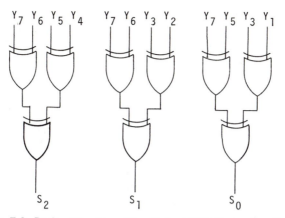

Fig. 7.8 Parity-Checking Circuit for 7-Bit Hamming Code

TABLE 7.6

Output of Parity-Check Circuit

Parity Check Equation	Check Output for Single Error in Position e_i							
	none	e_1	e_2	e_3	e_4	e_5	e_6	e_7
$S_2 = e_7 \oplus e_6 \oplus e_5 \oplus e_4$	0	0	0	0	1	1	1	1
$S_1 = e_7 \oplus e_6 \oplus e_3 \oplus e_2$	0	0	1	1	0	0	1	1
$S_0 = e_7 \oplus e_5 \oplus e_3 \oplus e_1$	0	1	0	1	0	1	0	1

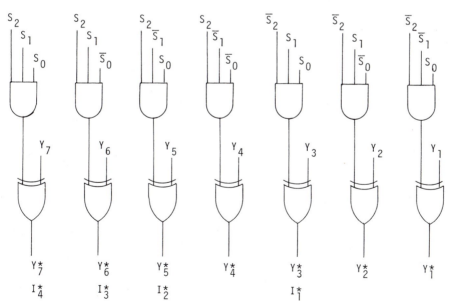

Fig. 7.9 Correction Circuit for 7-Bit Hamming Code

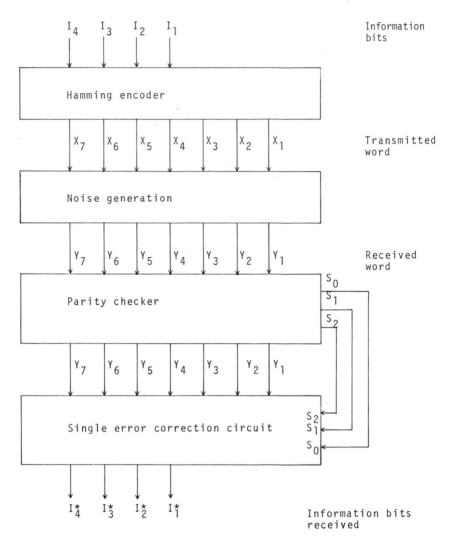

Fig. 7.10 Flow Chart for Hamming 7-Bit Single-Error-Correcting Code

7.4 CODE CONVERSION AND DISPLAY CIRCUITS

In Section 7.3, several binary codes were presented representing the decimal digits. Because of the variety of codes available, there will be times when it is necessary to convert from one code to another, for example, between ordinary BCD 8-4-2-1 code and 2-4-2-1 weighted code. The conversion process can be considered as an exercise in combinational logic design with don't cares for the six unused code words.

The following example will be used to illustrate the conversion process.

Example 7.2 Design a combinational logic unit which will convert the 2-4-2-1 weighted decimal code into normal 8-4-2-1 BCD code. Assume only valid code words are to be converted.

Solution. The first step in the design process is to convert the problem statement into a truth table amenable to solution by the combinational logic design procedures learned so far.

Table 7.7 shows one such truth table with the valid 2-4-2-1 code words as inputs and the corresponding valid 8-4-2-1 code as outputs. The outputs corre-

TABLE 7.7
Code Conversion Table for 2-4-2-1 to 8-4-2-1 Codes

Decimal Number	Input 2-4-2-1				Output 8-4-2-1			
N	W	X	Y	Z	A	B	C	D
0	0	0	0	0	0	0	0	0
1	0	0	0	1	0	0	0	1
2	0	0	1	0	0	0	1	0
3	0	0	1	1	0	0	1	1
4	0	1	0	0	0	1	0	0
5	1	0	1	1	0	1	0	1
6	1	1	0	0	0	1	1	0
7	1	1	0	1	0	1	1	1
8	1	1	1	0	1	0	0	0
9	1	1	1	1	1	0	0	1
-	0	1	0	1	d	d	d	d
-	0	1	1	0	d	d	d	d
-	0	1	1	1	d	d	d	d
-	1	0	0	0	d	d	d	d
-	1	0	0	1	d	d	d	d
-	1	0	1	0	d	d	d	d

-Invalid code words

sponding to the six invalid code words 0101, 0110, 0111, 1000, 1001, and 1010 are treated as don't-care terms.

Figure 7.11 shows the K-maps resulting from this table, with the circled terms corresponding to minimum sum-of-product solutions without attempting to share terms among the outputs.

From Fig. 7.11 we obtain the equations:

$$A(W, X, Y, Z) = XY, \qquad C(W, X, Y, Z) = W\bar{Y} + \bar{W}Y = W \oplus Y,$$
$$B(W, X, Y, Z) = X\bar{Y} + W\bar{X}, \qquad D(W, X, Y, Z) = Z.$$

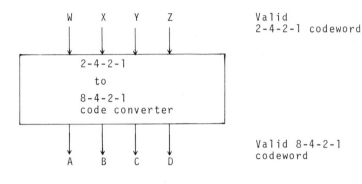

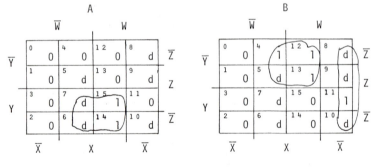

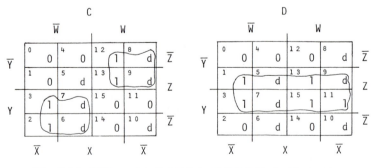

Fig. 7.11 K-Maps for 2-4-2-1 to 8-4-2-1 Code Converter

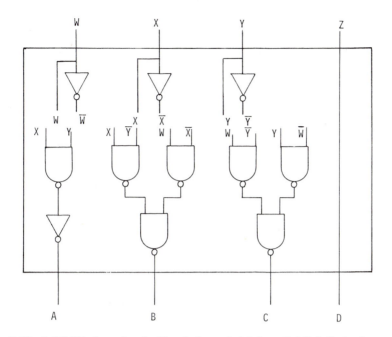

Fig. 7.12 A NAND-Gate Logic Circuit for a 2-4-2-1 to 8-4-2-1 Code Converter

A NAND gate circuit for these functions is shown in Fig. 7.12.

The same procedure can be used to find a code converter for 8-4-2-1 to 2-4-2-1.

Before leaving the subject of code converters, let us consider the problem of designing a combinational logic circuit to drive a standard seven-segment indicator light similar to that shown in Fig. 7.13. The normal decimal code for these indicators is shown in Fig. 7.14.

If we want to display the values of some numerical calculation directly on these indicators, a code converter is necessary to convert from the binary-coded-decimal system used into seven-segment indicator code.

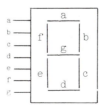

Fig. 7.13 A Typical 7-Segment Indicator Light

N	Pattern	Inputs a	b	c	d	e	f	g
0	⌐¦ └_┘	1	1	1	1	1	1	0
1	¦	0	1	1	0	0	0	0
2	⌐¦ ¦⌐	1	1	0	1	1	0	1
3	─¦ ─¦	1	1	1	1	0	0	1
4	└_┘¦	0	1	1	0	0	1	1
5	¦⌐ ─┘	1	0	1	1	0	1	1
6	¦ └_┘	0	0	1	1	1	1	1
7	─¦ ¦	1	1	1	0	0	0	0
8	⌐¦ └_┘	1	1	1	1	1	1	1
9	⌐¦ ─¦	1	1	1	0	0	1	1

Fig. 7.14 Indicator Patterns for 7-Segment Decoder

Example 7.3 Design a 7-segment indicator driver for the 2-4-2-1 weighted binary code

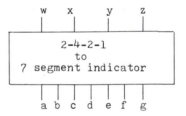

Solution: Using the information in Fig. 7.14, we can develop a truth table for the indicator. One such table is shown in Table 7.8 on page 261.

From Table 7.8 the *K*-maps shown in Fig. 7.15 can be drawn and the logic equations simplified. The resulting output equations from these *K*-maps are:

$$a(W, X, Y, Z) = (\bar{X} + Y + Z)(W + Y + \bar{Z}) = Y + WZ + \bar{X}\bar{Z};$$
$$b(W, X, Y, Z) = (\bar{W} + X)(\bar{W} + Y + Z) = \bar{W} + XY + XZ;$$
$$c(W, X, Y, Z) = W + \bar{Y} + Z;$$
$$d(W, X, Y, Z) = (Y + \bar{Z})(\bar{X} + \bar{Z})(W + \bar{X}) = W\bar{Z} + \bar{X}Y + \bar{X}\bar{Z};$$
$$e(W, X, Y, Z) = \bar{Z}(\bar{X} + W) = W\bar{Z} + \bar{X}\bar{Z};$$
$$f(W, X, Y, Z) = (\bar{Z} + Y)(W + \bar{Y}) = \bar{Y}\bar{Z} + WY;$$
$$g(W, X, Y, Z) = (\bar{Z} + Y)(X + Y) = Y + X\bar{Z}.$$

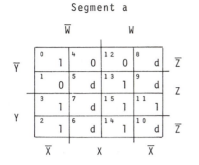

Segment a

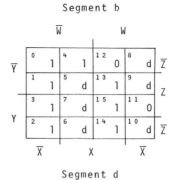

Segment b

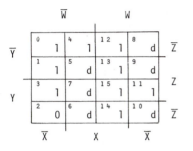

Segment c

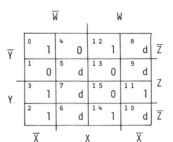

Segment d

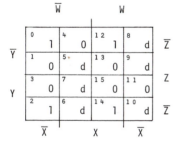

Segment e

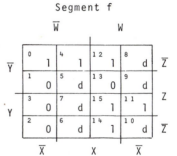

Segment f

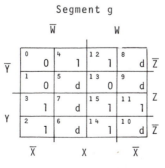

Segment g

Fig. 7.15 *K*-Maps for 2-4-2-1 to 7-Segment Indicator Converter

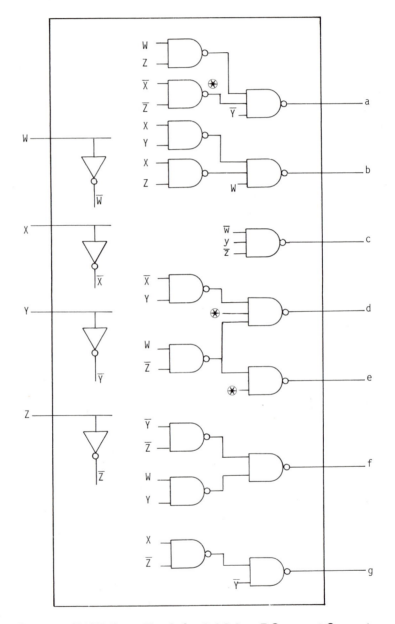

Fig. 7.16 NAND-Gate Circuit for 2-4-2-1 to 7-Segment Converter

TABLE 7.8
Code Conversion Table for 2-4-2-1 to 7-Segment Indicator

Decimal Digit	Input 2-4-2-1				Output Indicator Segments						
N	W	X	Y	Z	a	b	c	d	e	f	g
0	0	0	0	0	1	1	1	1	1	1	0
1	0	0	0	1	0	1	1	0	0	0	0
2	0	0	1	0	1	1	0	1	1	0	1
3	0	0	1	1	1	1	1	1	0	0	1
4	0	1	0	0	0	1	1	0	0	1	1
5	1	0	1	1	1	0	1	1	0	1	1
6	1	1	0	0	0	0	1	1	1	1	1
7	1	1	0	1	1	1	1	0	0	0	0
8	1	1	1	0	1	1	1	1	1	1	1
9	1	1	1	1	1	1	1	0	0	1	1
-	0	1	0	1	d	d	d	d	d	d	d
-	0	1	1	0	d	d	d	d	d	d	d
-	0	1	1	1	d	d	d	d	d	d	d
-	1	0	0	0	d	d	d	d	d	d	d
-	1	0	0	1	d	d	d	d	d	d	d
-	1	0	1	0	d	d	d	d	d	d	d

A shared NAND gate circuit for this converter is shown in Fig. 7.16.

The same design techniques can be used to design converters for the other binary-coded-decimal codes. However, for the more commonly used codes, converters can be obtained as prepackaged IC units.

7.5 DECIMAL ADDING CIRCUITS

The logic symbol for a decimal full adder is shown in Fig. 7.17. This symbol requires the inputs and outputs to be 4-bit binary numbers. However, for most applications, the carry-in, Z, and carry-out, C, will be either 0 or 1 and hence really only require a single input or output line. This modification to the basic full adder greatly reduces the hardware complexity of adder cells. The modified cell symbol is shown in Fig. 7.18. This cell assumes that the inputs A and B are coded using a 4-bit binary code and that they take on the values $[0 \le A, B \le 9]_{10}$ and that C_{in} and C_{out} are either a 0 or 1, indicating the presence or absence of a carry from a preceding or to a succeeding stage. The resulting sum output is again a 4-bit binary-coded-number $(0 \le S \le 9)_{10}$.

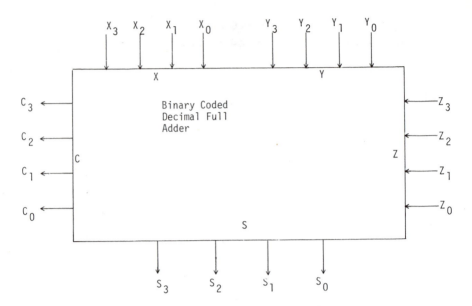

Fig. 7.17 Decimal Full-Adder Logic Symbol

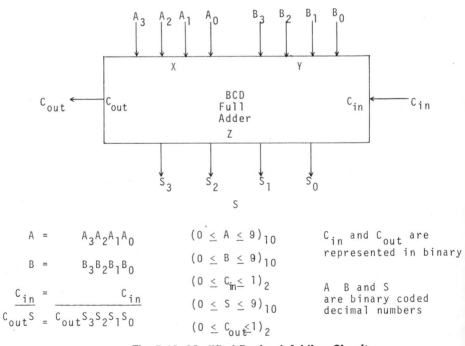

$$A = A_3 A_2 A_1 A_0$$

$$(0 \le A \le 9)_{10}$$

$$B = B_3 B_2 B_1 B_0$$

$$(0 \le B \le 9)_{10}$$

$$C_{in} = \overline{C_{in}}$$

$$(0 \le C_{in} \le 1)_2$$

$$C_{out} S = C_{out} S_3 S_2 S_1 S_0$$

$$(0 \le S \le 9)_{10}$$

$$(0 \le C_{out} \le 1)_2$$

C_{in} and C_{out} are represented in binary

A B and S are binary coded decimal numbers

Fig. 7.18 Modified-Decimal Adding Circuit

A general procedure for designing decimal adders would be to first convert the BCD numbers into binary, add the resulting binary numbers, and then reconvert the binary sum back into the appropriate BCD code. Figure 7.19 shows the general form for such a circuit configuration.

This approach, however, can be very wasteful of circuit components. An alternative is to simply bypass the initial conversion circuit and then add the BCD numbers in binary. The required post-correction and conversion circuitry is only slightly more complex. This alternate model is shown in Fig. 7.20.

This alternate model is not always applicable. There are decimal codes such as the 5-4-2-1 weighted code which yield ambiguous results at the output of the binary adder. However, for the more commonly used decimal codes, the 8-4-2-1, excess-3, excess-6, and 2-4-2-1 codes, this alternate approach can be used.

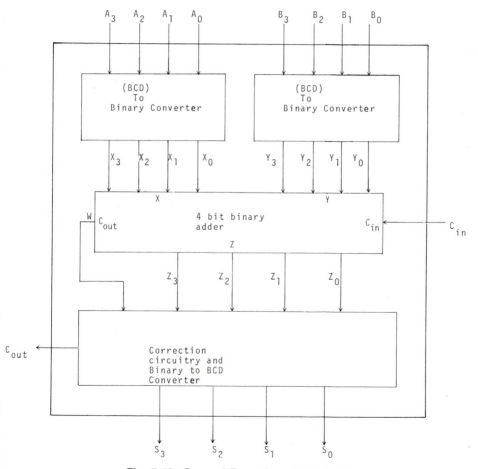

Fig. 7.19 General Form for a BCD Adder

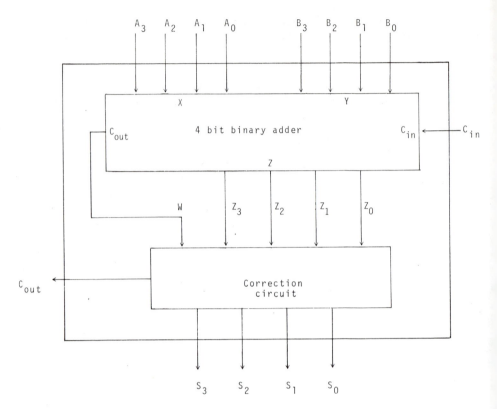

Fig. 7.20 Alternate Form for BCD Adder

Example 7.4 Design a decimal adder for the normal 8-4-2-1 BCD code using the model shown in Fig. 7.20.

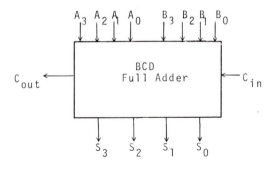

Solution. Table 7.9 shows the output of the 4-bit binary adder ($WZ_3Z_2Z_1Z_0$) obtained by adding all possible input number combinations as if they were 4-bit binary numbers.

TABLE 7.9

BCD Adder Outputs

N_{10}		Intermediate Adder Output $(A + B + C_{in})$ binary					Final Output $(A + B + C_{in})$ BCD				
	W	Z_3	Z_2	Z_1	Z_0		C out	S_3	S_2	S_1	S_0
0	0	0	0	0	0		0	0	0	0	0
1	0	0	0	0	1		0	0	0	0	1
2	0	0	0	1	0	No	0	0	0	1	0
3	0	0	0	1	1	correction	0	0	0	1	1
4	0	0	1	0	0	needed	0	0	1	0	0
5	0	0	1	0	1	$C_{out} = 0$	0	0	1	0	1
6	0	0	1	1	0		0	0	1	1	0
7	0	0	1	1	1		0	0	1	1	1
8	0	1	0	0	0		0	1	0	0	0
9	0	1	0	0	1		0	1	0	0	1
10	0	1	0	1	0		1	0	0	0	0
11	0	1	0	1	1	Correction	1	0	0	0	1
12	0	1	1	0	0	needed	1	0	0	1	0
13	0	1	1	0	1	add 6	1	0	0	1	1
14	0	1	1	1	0	to	1	0	1	0	0
15	0	1	1	1	1	$(A+B+C_{in})_2$	1	0	1	0	1
16	1	0	0	0	0	and set	1	0	1	1	0
17	1	0	0	0	1	$C_{out} = 1$	1	0	1	1	1
18	1	0	0	1	0		1	1	0	0	0
19	1	0	0	1	1		1	1	0	0	1

The desired answer $(C_{out}S_3S_2S_1S_0)$ agrees with the intermediate result until the number 10 is reached. The intermediate binary sums, between 10 and 19 decimal, are off by 6. Thus a correction factor of 6 must be added to the sum to convert the binary answer into the required BCD answer. If we assume only valid inputs are applied to the adder, then the binary sums corresponding to $(20 \text{ to } 31)_{10}$ can be treated as don't-care states. The K-map in Fig. 7.21 can be used to determine when the correction factor is required.

$$X = W + Z_3 Z_2 + Z_3 Z_1$$

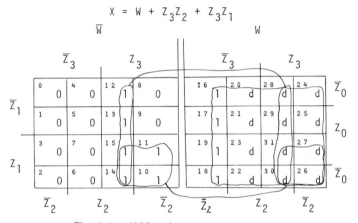

Fig. 7.21 *K*-Map for Correction Factor *X*

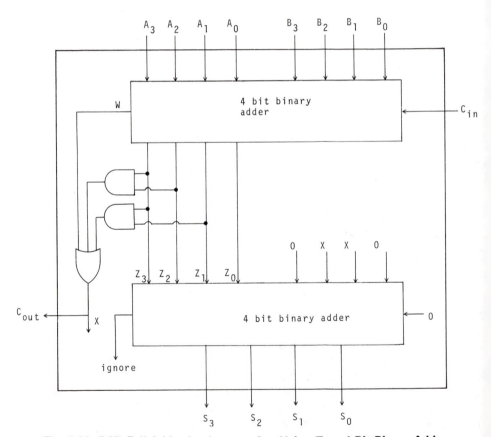

Fig. 7.22 BCD Full-Adder Implementation Using Two 4-Bit Binary Adders

If we let X be a Boolean variable which equals 1 if a correction is required, then from the K-map we find that $X = W + Z_3Z_2 + Z_3Z_1$. We also note from Table 7.9 that C_{out} is also equal to X.

The resulting BCD full adder configuration is shown in Fig. 7.22. A second 4-bit binary adder is used to provide the correction factor when needed.

An alternate correction circuit using full adders and logic gates is shown in Fig. 7.23. This alternate circuit is able to save on gates by noting that the output Z_0 needs no correction. Since the carry out of the correction adder is not needed, an XOR gate can be used to correct the most significant sum bit S_3.

The choice of which circuit to select will depend on the availability of components FA, HA, etc., and the extra wiring cost involved in connecting the discrete logic-gate correction circuit together. Where wiring costs are high, the use of the prepackaged 4-bit adder may be more economical. However, if an MSI circuit for the BCD full adder is to be designed, then by all means the configuration shown in Fig. 7.23 is to be preferred.

7.6 BCD RIPPLE ADDERS

In this section, let us combine the decimal logic circuits studied in the previous sections of this chapter into a useful multidigit decimal adder-subtracter.

Example 7.5 Design an N-digit dual-mode ripple adder-subtracter for 4-bit binary-coded-decimal numbers.

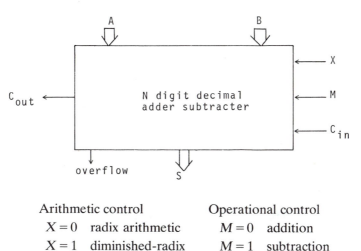

Arithmetic control	Operational control
$X = 0$ radix arithmetic	$M = 0$ addition
$X = 1$ diminished-radix arithmetic	$M = 1$ subtraction

Before attempting to solve this problem, let us quickly review the rules for addition and subtraction using 10's complement (radix) and 9's complement (diminished-radix) arithmetic.

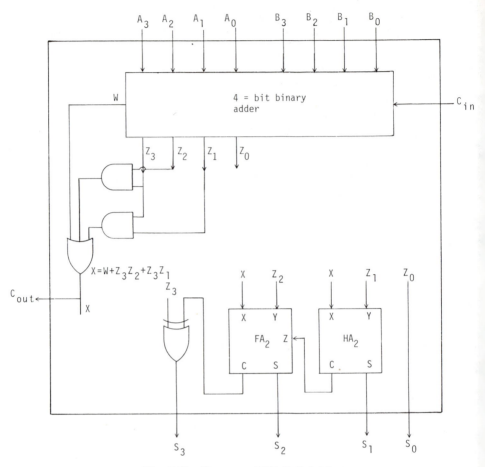

Fig. 7.23 Alternate BCD Full Adder

RULES.

1. 10's complement addition:
   Add all digits including the sign digit. Ignore any carry from the sign position.

2. 10's complement subtraction:
   Take the 10's complement (9's complement and add 1) of the subtrahend and add.

3. 9's complement addition:
   Add all digits including the sign digit. Add the carry out of the sign position (end-around-carry) to the least significant position.

4. 9's complement subtraction:
   Take the 9's complement of the subtrahend and add.

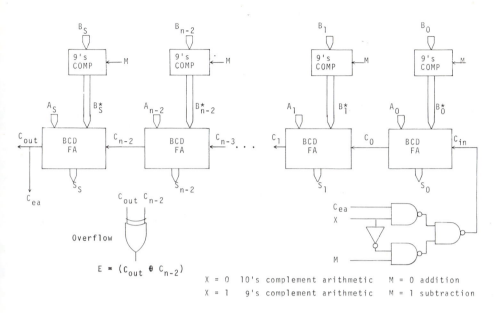

Fig. 7.24 *N*-Bit Ripple Decimal Adder

5. Rules for overflow:
   a) When adding like signed numbers, the resulting sum must have the same sign, otherwise an overflow has occurred.
   b) When adding signed numbers, the carry into the sign position must equal the carry out of the sign position, otherwise an overflow has occurred.

For more detailed rules the reader is referred back to Chapter 2 where the rules summarized above were derived.

Solution. Figure 7.24 shows a cascade interconnection of decimal adders. The numbers are of the form

$$A = A_s A_{n-2} \cdots A_2 A_1 A_0,$$
$$B = B_s B_{n-2} \cdots B_2 B_1 B_0,$$

with

$$A_i = a_{3i} a_{2i} a_{1i} a_{0i}$$

being coded in binary.

For each unit the 9's complement of B_i is added to A_i, when subtraction is required. The end-around-carry, C_{ea}, is added into the least significant position whenever 9's complement arithmetic is requested ($X = 1$). If 10's complement arithmetic is requested ($X = 0$), then a one is added into the least significant position to complete the 10's complement, whenever subtraction is required ($M = 1$).

Thus the carry into the least significant position becomes:

$$C_{in} = XC_{ea} + \bar{X}M. \tag{7.1}$$

The only remaining unit to be designed is the overflow logic unit.

If the carries from each adder cell are available, then overflow rule (b) should be used since it involves less hardware. $E = (C_{out} \oplus C_{n-2})$. The two adder inputs can be checked for agreement by using coincidence gates.

If the carry lines are not available then rule (a) can be used with the resulting logic equation

$$E = \bigvee_{i=0}^{3} (a_{is}b_{is}^{*}\bar{S}_{is} + \bar{a}_{is}\bar{b}_{is}^{*}S_{is}). \tag{7.2}$$

7.7 PROBLEMS

7.1 Construct a decimal code using the weights $+8$, -4, $+2$, and -1.

7.2 Construct a self-complementing decimal code using the weights $+3$, $+3$, $+2$, $+1$.

7.3 Design an alternate controlled 9's complement unit for the BCD (8-4-2-1) code using a 4-bit binary adding cell and 4 XOR gates.

7.4 Design a 9's complement unit for the codes shown in Table 7.2.
   a) excess-6   b) 5-4-2-1
   c) 4-3-2-1   d) Gray

7.5 The Hamming code words shown below were received. What information was sent assuming the data format:
   I_4 I_3 I_2 P_3 I_1 P_2 P_1.
   a) 0 0 0 0 1 0 0
   b) 0 1 0 1 1 0 0
   c) 1 1 0 1 1 1 0
   d) 0 1 0 1 0 1 0
   e) 1 1 1 1 1 1 1

7.6 Construct a table showing the correct Hamming-code words for the following decimal codes
   a) excess 3,   b) 5-4-2-1,   c) Gray.

7.7 Using only a 4-bit binary adder, design decimal code converters for:
   a) 8-4-2-1 to excess-3,   b) Excess-3 to 8-4-2-1,
   c) excess-6 to excess-3.

7.8 Design, using conventional logic gates, decimal-code converters for
   a) Gray to 8-4-2-1,
   b) 8-4-2-1 to Gray,   } Table 7.2
   c) 5-4-2-1 to 8-4-2-1,
   d) 8-4-2-1 to 2 out of 5,   } Table 7.4
   e) 8-4-2-1 to biquinary.

7.9 Design a 7-segment indicator-driver for the
   a) 8-4-2-1 code,
   b) excess-3 code,
   c) Gray code,
   d) excess-6 code.

7.10 Design a decimal full-adder cell for the unit shown in Figure 7.17, assuming the normal 8-4-2-1 BCD code.

7.11 Design a decimal full adder cell assuming
   a) excess-3 code,   b) excess-6 code,
   c) Gray code,   d) 5-4-2-1 code,
   e) 2-4-2-1 code.

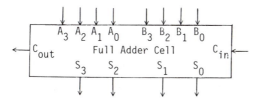

7.12 Show how to interconnect a series of four decimal-adding cells to form a 4-digit ripple adder.
   a) Assume radix arithmetic (10's complement).
   b) Assume diminished-radix arithmetic (9's complement).
   c) Assume sign-and-magnitude arithmetic.
   Be sure to include an overflow indicator.

7.13 The following cell can be used as part of an array for the direct conversion of binary integers into BCD integers using the 8-4-2-1 code. Treating x and y as 4-bit binary numbers

$$y = x, \quad \text{if} \quad 0 \leq x \leq 4;$$
$$y = x + 3, \quad \text{if} \quad 5 \leq x \leq 9.$$

x will never be greater than 9.

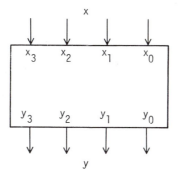

a) Design this basic cell using NAND-WIRED-AND logic. The array below can be used to convert any 6-bit binary number into a 2-digit decimal number in 8-4-2-1 code.

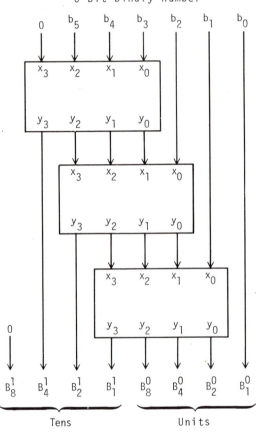

6 bit binary number

2-digit decimal number

b) Using the array above convert the following binary numbers into BCD 8-4-2-1 format. Show the inputs and outputs on all cells in the array.
a) 101 101   b) 110 001
c) 011 011   d) 111 111

c) Design an array which can convert any 12-bit binary number directly into a 4-digit decimal number code in 8-4-2-1 code. Show the conversion of the 12-bit number, $N = 010\ 111\ 001\ 100$, on your array.

7.14 Design a basic decimal multiplication cell.

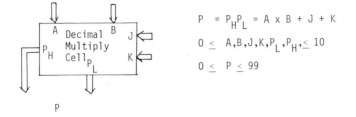

$$P = P_H P_L = A \times B + J + K$$

$$0 \leq A,B,J,K,P_L,P_H, \leq 10$$

$$0 \leq P \leq 99$$

Use a modular design. *Hint:* Use a binary array and the results of Problem 7.13. Assume an 8-4-2-1 code is used. Calculate the worst-case delay time for your circuit.

7.15 Using the basic multiply cell designed in Problem 7.14, construct a 3×3 decimal multiplication array. Calculate the worst-case multiply time for your unit.

7.8 BIBLIOGRAPHY

Boyce, J. C., *Digital Logic and Switching Circuits.* Englewood Cliffs, N.J.: Prentice-Hall, 1975.

Flores, I., *The Logic of Computer Arithmetic.* Englewood Cliffs, N.J.: Prentice-Hall, 1963.

Hamming, R. W., "Error detecting and error correcting codes," *BSTJ*, Vol. 29, No. 2, pp. 147–160, Apr. 1950.

Lin, S., *An Introduction to Error-Correcting Codes.* Englewood Cliffs, N.J.: Prentice-Hall, 1970.

Schmid, H., *Decimal Computation.* New York: Wiley, 1974.

Stein, M. L., and W. D. Munro, *Introduction to Machine Arithmetic.* Reading, Mass.: Addison-Wesley, 1971.

Chapter 8

Introduction to Sequential Circuit Design

8.1 INTRODUCTION

Up till now we have considered only combinational logic circuits. In this chapter, we will introduce the concept of sequential circuits. We will begin by defining the basic circuit characteristics in Section 8.2. A short discussion of logic hazards will be presented in Section 8.3.

In Section 8.4 we will formally define a sequential circuit and also introduce the present-state/next-state table and transition diagrams as design aids.

The basic set-reset (SR) flip-flop will be introduced in Sections 8.5, 8.6, and 8.7. Clocked storage elements will be discussed in Section 8.8. A formal procedure for analyzing SR flip-flops will be given in Section 8.9. Sections 8.10 and 8.11 will be devoted to formal synthesis procedures for SR flip-flops.

Sections 8.12 and 8.13 will be devoted to a discussion of one-shot and digital-clock circuits. Finally, Section 8.14 will discuss a simple method for making bounceless switches.

8.2 BASIC CIRCUIT CHARACTERISTICS

Let us examine a few characteristics of electronic devices and see how they affect the overall device performance. To help relate the terminology to a logic device, a positive logic inverter will be used to illustrate the various concepts. A positive supply voltage and a positive logic assignment will also be assumed. The reader can easily modify the definitions and illustrations to handle the cases where a negative supply voltage or a negative logic assignment is required.

Before proceeding further with this general discussion on electronic device characteristics, a few terms need to be defined and explained. Since every electronic logic device has a dual logic meaning, depending upon whether a positive or negative logic assignment is used (positive NAND, negative NOR), some confusion can arise when terms such as rise time and fall time are used to specify the type of voltage waveshapes associated with the device. The difficulty is due to the unfortunate double meanings given these terms by various authors and manufacturers. Some use the term rise time to mean the time required for the signal voltage to rise from 10% above its lowest logic value to 10% below its highest logic value, irrespective of the logical assignment of zero and one. Others use the same term to measure the time required for the logic level to change from within 10% of a logical zero to within 10% of a logical one regardless of the voltage assignment. The confusion is also compounded by the use of such terms as turn-on and turn-off time, with their equivalent dual meanings.

One way of eliminating any possible confusion is to define the terms on the basis of the actual high (H) and low (L) voltage levels. This practice appears to be gaining approval and will be adopted in this book. Refer to Fig. 8.1 for graphical interpretations of the output transition time and output propagation-delay time.

Definition 8.1 The length of time required for the output voltage of an electronic logic device to transition from 10% above its lowest voltage level to 10% below its highest voltage level will be called the *output low to high transition time* and will be denoted by t_{LH}.

Definition 8.2 During an output voltage transition from low to high, the time required for the output voltage to cross the switching threshold V_s, measured from the time the triggering input signal crossed the same switching threshold, will be called the *output low-to-high propagation delay* and will be denoted by the symbol t_{PLH}.

Definition 8.3 The length of time required for the output voltage of an electronic logic device to transition from 10% below its highest logic voltage to 10% above its lowest logic voltage will be called the *output high-to-low transition time* and will be denoted by t_{HL}.

Definition 8.4 During a transition from high to low, the time required for the output voltage to cross the switching threshold V_s, measured from the time the triggering input signal crossed the same switching threshold, will be called the output *high-to-low propagation delay*, and will be denoted by the symbol t_{PHL}.

There are numerous circuit parameters which can cause variations in both the gate transition times and the gate propagation delays. An increase in the power supply voltage V_{cc} will generally cause the transition times and gate propagation delay times to rise and also cause the switching threshold to increase. Variations in ambient temperature and the type of output loading will cause wide variations in gate propagation delays, with fourfold increases not uncommon. Finally, the

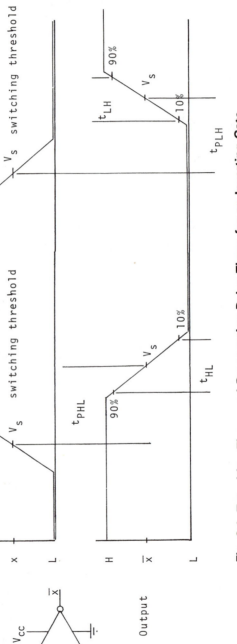

Fig. 8.1 Transition Times and Propagation Delay Times for an Inverting Gate

wave shape of the input signal applied to the gates will affect the output signals as well as the circuit propagation delays.

To prevent erroneous measurements and excessive gate delays, restrictions are usually placed on the input-signal transition times.

Within a given family of logic devices, the product of the average gate-power dissipation per stage P_D and the average gate propagation delay t_{PD}, $= (t_{PHL} + t_{PLH})/2$, tends to be a constant $P_D \cdot t_{PD} \simeq c$. Thus within a given family of logic devices, a decrease in the average propagation delay implies an increase in the average power dissipation. It is the logic designer's job to select the appropriate family of devices which best suit a particular application problem. Designs for telephone-switching equipment should favor both low cost and the lowest power dissipation per stage possible. Designs for computer arithmetic subsystems would generally favor high speed over both cost and power dissipation, due to the relatively few elements involved. Solid-state computer memories are instances where definite compromises must be made since all three goals—low cost, high speed, and low power—are sought.

In many design problems, logic gates with varying propagation times and power dissipations are intermixed, since a few well-placed high-speed circuit elements can greatly increase overall circuit performance.

The operating temperature range of a logic circuit informs the designer of the type of temperature environment which the device will tolerate. In general, the wider the operating range, the higher the cost. The accepted nominal operating temperature is 25°C, with an operating range between 0°C and 70°C. The overall circuit performance is usually degraded as the operating temperature varies appreciably above or below this value.

In order to obtain realistic values for critical circuit parameters, the input signal waveforms should approximate the signals that will normally be encountered in the actual circuits application. At best, laboratory measured values or manufacturers' supplied average time specifications should serve as a design guideline and not be accepted as gospel, since circuit parameters unfortunately tend to vary widely with temperature, load conditions, and input-signal wave shape. This makes it difficult to accurately predict final design delays. In many applications, the worst-case delay times are used and the decreased delay time of the actual circuits are considered a bonus. The two times which are important to a multilevel logic circuit are the gate propagation delay times, t_{PHL} and t_{PLH}. These delay values add at each stage, while the low-to-high or high-to-low transition time enters only once at the output of the final stage.

Due to different physical phenomenon (saturation and cut off), the low-to-high and high-to-low transition and the propagation delays are generally not equal, and in some cases may differ by as much as a 10-to-1 ratio.

Because an inverting type electronic gate tends to be either conducting or not conducting, the transition times are sometimes referred to as the turn on and turn off times for the logic device.

Since a great many logic circuits are designed to use two levels of logic, both types of propagation delays will be encountered. For this reason, a commonly used design parameter, the gate-pair propagation delay denoted by t_{GPD}, is often specified, with $t_{\text{GPD}} = t_{\text{PHL}} + t_{\text{PLH}}$.

Dividing the pair propagation delay by 2 yields the average propagation delay denoted by t_{PD}. This average delay is frequently used instead of the worst-case delay time, when working with multilevel logic circuits. Using $t_{\text{GPD}}/2$ as an average gate propagation delay, instead of worst-case delay time, tends to give closer correlation to the estimated design delay times and experimentally measured delay times. In this book, we will use t_{PD} as an average delay propagation time of an inverting gate and refer to it as a unit of gate delay. We will also generally neglect the gate transition times when drawing timing charts for our circuits. The reasonableness of neglecting this time can be questioned; however, the ease of calculation and ability to quickly draw timing charts more than offset the slight error which may be introduced. Since many logic circuits are designed using cascaded sections of two levels of NAND's or NOR's, the signals passing through these two-level units encounter both types of propagation delay.

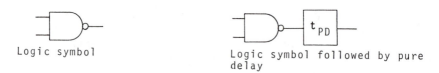

Logic symbol   Logic symbol followed by pure
                                                delay

Fig. 8.2 Pure Delay Model of a Logic Device

Using the average gate propagation delay t_{PD} as a pure delay, we can model our logic gates as a device which instantaneously follows input changes, followed by an element which delays the output change by t_{PD}. This circuit model is shown in Fig. 8.2.

Rather than add all the extra delay boxes to our logic circuit, we will understand that the gates have this average propagation delay associated with them.

Figure 8.3 shows a NAND gate circuit and its associated timing chart assuming a pure delay model.

A better model, suitable for computer use, would require the delay box to be a function of the gate circuit parameters. Thus the delays would vary as a function of temperature, power supply voltage, and output loading.

An area of concern sometimes overlooked by beginning design engineers who are used to assembling and testing logic devices under almost ideal laboratory conditions is the ability of the logic device to withstand the adverse effects of electrical noise.

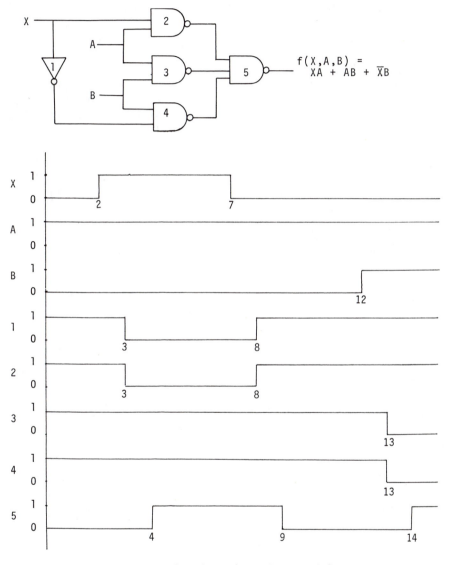

$$f(X,A,B) = XA + AB + \overline{X}B$$

t - time in units of gate delay

Fig. 8.3 Timing Chart for the Two-Level Logic Circuit

Typically, the differences between the nominal output voltages and the nominal switching threshold are referred to as the high and low *DC noise-immunity margins* V_{NL} and V_{NH} with

$$V_{NL} = V_s - V_{OL} \quad \text{and} \quad V_{NH} = V_{OH} - V_s.$$

In Fig. 8.4, these high- and low-noise immunity values are illustrated. What they essentially mean is that if a noise source was situated between a gate output with a nominal output voltage of V_{OH}, then an additive noise pulse of $V_{OH} - V_s$ volts would be needed before the output of the following gate would be changed

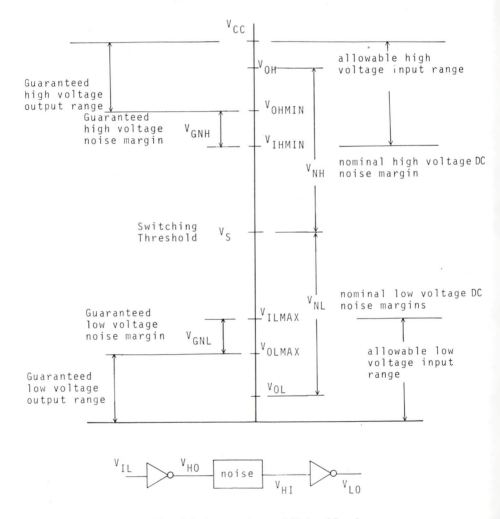

Fig. 8.4 Comparison of Noise Margins

erroneously from low to high. However, the noise figure stated above is almost useless to the circuit designer, since logic circuits seldom operate at their nominal values due to variations in ambient temperature, supply voltage, and output loading. A more accurate noise measure is obtained by placing limits on the input and output voltage values permitted to represent valid logic signals. The highest low-level input signal voltage V_{ILMAX} which yields this worst-case high-level output voltage V_{OHMIN} is used instead of the switching threshold voltage V_s when calculating the low-level noise-immunity figure. The highest guaranteed low-level output voltage, V_{OLMAX}, is used to replace the nominal output voltage V_{OL}. This noise figure is commonly referred to as the *guaranteed low-level noise margin* V_{GNL}, with $V_{GNL} = V_{ILMAX} - V_{OLMAX}$. An input signal below V_{ILMAX} will result in an output voltage above the guaranteed minimum high-level output voltage, V_{OHMIN}.

When calculating the guaranteed high-level noise margin, V_{GNH}, the lowest permissible high-level input voltage V_{IHMIN}, which yields the maximum allowable low-level output voltage V_{OLMAX}, is used along with the guaranteed minimum high output V_{OHMIN}. Thus $V_{GNH} = V_{OHMIN} - V_{IHMIN}$ is the *guaranteed high-level noise margin*. Figure 8.4 illustrates the differences between the two noise margins.

The *output driving capability* or fan-out of a logic device refers to the number of equivalent logic gates which can be attached to the output of the device and still operate satisfactorily within its design specifications.

8.3 HAZARDS IN COMBINATIONAL LOGIC CIRCUITS

The gate propagation delay, inherent in all logic devices, can only be detected at the output of a combinational logic circuit following changes to its input signals.

The spurious output signals resulting from input changes will not adversely affect the operation of purely combinational circuits, since the outputs of these circuits are only a function of the steady-state input signals. However, we will see shortly that if the output signals of a combinational circuit are used as inputs to a sequential circuit, then the possibility of erroneous operation due to circuit transients will exist.

The extraneous output signals which result are called logical hazards. These hazards can be classified as being either static or dynamic depending upon the initial and final value of the output. If the initial and final output of a logic device are the same, then the extraneous output signals, which result from input signal changes, will be called *static hazards*. If the initial and final outputs of the logic device differ, then the extraneous output signals which result from input signal changes will be called *dynamic hazards*.

Note that the output signal will change an even number of times during a static hazard and an odd number of times during a dynamic hazard. An example of a static hazard is shown in Fig. 8.5. The circuit illustrated is a simple electronic

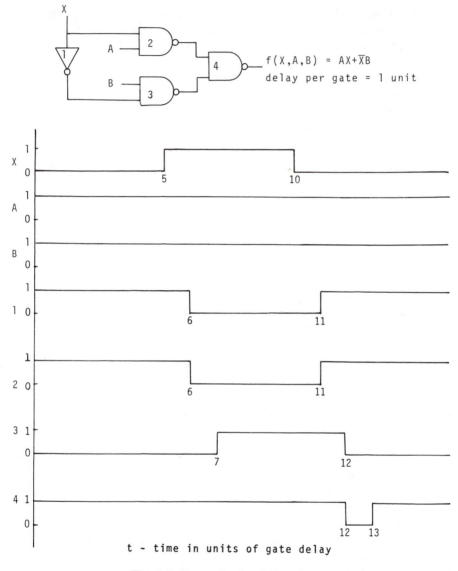

Fig. 8.5 Example of a Static Hazard

equivalent of a single-pole double-throw switch. If both inputs *A* and *B* are equal to one and the control input *X* is changed from a one to a zero, then the output signal should not change. However, due to the inherent propagation delay of the logic gates, a momentary zero output signal will occur.

Figure 8.6 shows the effect of changing all three inputs into a cascaded connection of two exclusive OR gates constructed with NAND gates. The

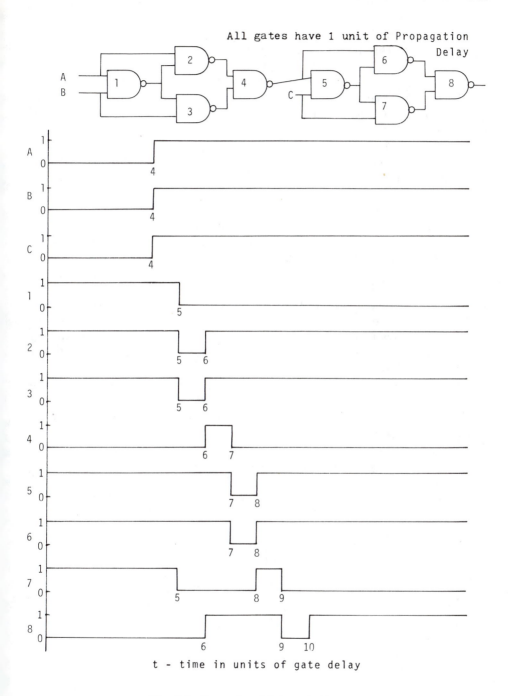

Fig. 8.6 Example of Dynamic Hazard

simultaneous change of A and B from zero to one yields a static hazard at the output of gate 4. This hazard, in combination with the change in input C to a one, yields the dynamic hazards at the output of gates 7 and 8.

If only a single input to a logic circuit is allowed to change at a given time, and the effect of this change is allowed to propagate through the circuit before any additional input changes, then it is possible to implement a Boolean function which is free of both static and dynamic hazards.

One simple, but expensive, way to eliminate logical hazards for single input-changing circuits is to realize the function using all of its prime implicants in sum-of-product form. The resulting two-level logic circuit will be free of static and dynamic hazards regardless of variations in propagation delays among the gates used to implement the circuit.

It is easily shown that, when using the sum-of-product form with all of the prime implicants, a static one hazard cannot occur, i.e., the output will not momentarily drop to zero. If the function $f(x_1, \ldots, x_n)$ equals 1 for some set of inputs $i_1, i_2, \ldots, i_j, \ldots, i_n$ and also equals 1 for the set of inputs $i_1, i_2, \ldots, \bar{i}_j, \ldots, i_n$, then either the product term $x_1^{i_1} x_2^{i_2} \cdots x_{j-1}^{i_{j-1}} x_{j+1}^{i_{j+1}} \cdots x_n^{i_n}$ is a prime implicant or it is covered by a prime implicant. In either case, the function will be held constant by this covering product term during the change in the input variable x_j. A similar argument will show that static zero hazards cannot occur, i.e., the output will not momentarily rise to one. It is left as a simple exercise for the reader to verify that no dynamic hazards can occur.

Thus it appears that the existence of hazards depends not so much on the function being realized but rather on the physical circuit configuration and restrictions placed on the manner in which input variables are allowed to change. By adding the product term AB (thus including all of the prime implicants) to the circuit in Fig. 8.5, we can obtain a hazard-free circuit if we restrict input changes to a single variable at a time. The hazard-free circuit is shown in Fig. 8.7.

Unfortunately, when more than one input signal is allowed to change simultaneously, it is not always possible to prevent hazards from occurring. Functions

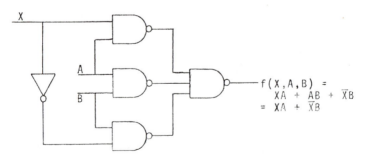

$$f(X, A, B) = XA + AB + \bar{X}B$$
$$= XA + \bar{X}B$$

Fig. 8.7 Hazard-Free Switch

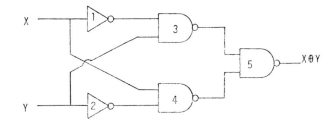

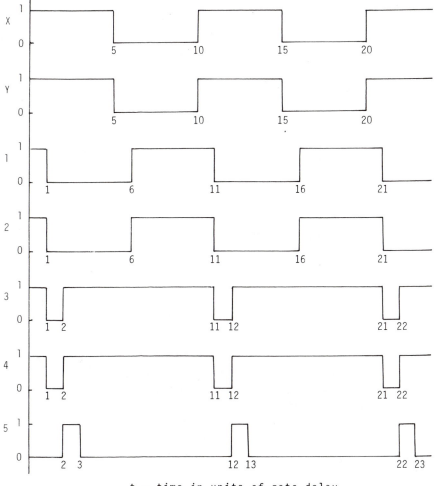

t - time in units of gate delay

Fig. 8.8 Timing Chart for an XOR Circuit

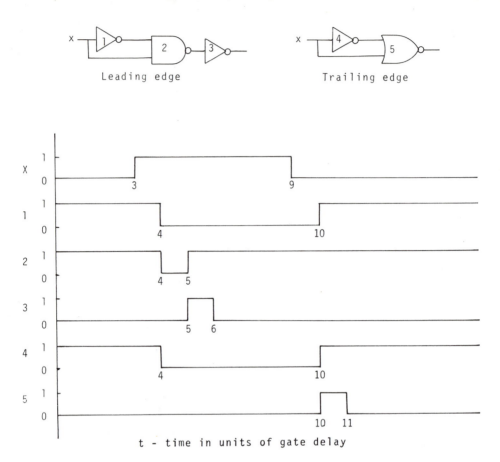

Fig. 8.9 Leading- and Trailing-Edge Detector Circuits and Their Timing Charts

which have nonadjacent product terms are inherently hazardous when subjected to simultaneous input changes. An example of such a function is $f(x, y) = x\bar{y} + \bar{x}y = x \oplus y$. If we assume each gate, in the circuit shown in Fig. 8.8, has a pure propagation delay of unit width and *apply a square wave as an input to both x and y*, then the output should be a constant zero.

The small unit-wide pulse in the output wave shape is a static hazard and is due to the finite delay-propagation time of the gates. If we had used a square wave on x and the inverse of this square wave on y, we would again have seen a small hazard in the output wave shape.

Because of the varying electronic characteristics found in different types of logic gates, it is usually necessary to examine each circuit realization for the presence of logical hazards. Due to the varying effects of temperature and output loading on the gate propagation delays, the transient characteristics of the same circuit configuration can vary greatly.

It should again be stressed that the effects of these logic hazards on the output of a purely combinational logic circuit are usually not serious and generally they are ignored. It is only when the output of the combinational circuit is being used as an input to a sequential circuit which responds to input signal changes that the problems of hazard-free circuit design become important.

Finally, it should be pointed out that there are circuits which take advantage of the existence of hazards. For example, the two circuits shown in Fig. 8.9 are commonly used to detect the leading and trailing edge of a randomly changing input signal.

The duration of the output pulse produced by these two circuits is dependent upon the delay of the input inverter. In many cases an odd number of inverters is used to yield an output pulse of several gate propagation delays in width.

8.4 INTRODUCTION TO SEQUENTIAL CIRCUITS

As long as the steady-state output depended only upon the present inputs to the logic circuit we call these circuits combinational. We will call all other circuits sequential.

Let us now consider what might happen if we were to allow feedback to occur from some output terminal to the input. In Fig. 8.10 a two input NAND gate is shown with its output returned to one of its two inputs.

If a pure delay of τ seconds is assumed, the output of the NAND gate B, will satisfy the time-dependent logic equation

$$B(t + \tau) = \overline{A(t) \cdot B(t)} = \bar{A}(t) + \bar{B}(t).$$

Thus whenever A transitions from a one to a zero, the output B will be forced to assume the value one after a time delay of τ seconds. A later transition of the input signal A from a zero to a one will now cause the circuit to oscillate with a period of 2τ seconds. This oscillation results from the delayed response time of the NAND gate. In effect, this circuit acts as a gated oscillator with input A as the gating signal.

In actual practice a feedback path around a single NAND gate may cause the resulting output not to oscillate but instead to assume an output value somewhere between a logical zero and a logical one. This discrepancy is inherent in our model, which assumes a pure delay response with instantaneous transitions between logic levels. Real circuits require a short but finite transition time. However, in general, the addition of two extra inverters to increase the delay in the feedback loop is usually sufficient to offset this problem. Figure 8.11 shows how a quad 2-input NAND gate integrated circuit can be interconnected to yield a simple clocking circuit with a buffered output.

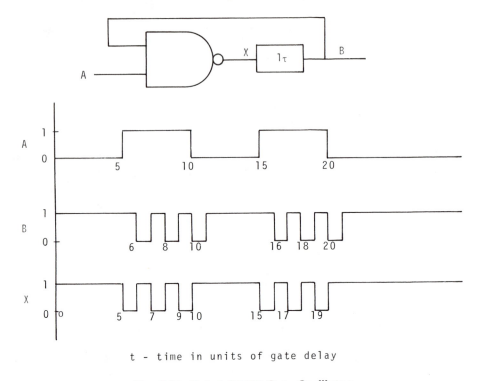

t - time in units of gate delay

Fig. 8.10 Gated NAND-Gate Oscillator

However, due to variations in temperature, output loading, and chip electronics, the square-wave output frequency can easily vary 25% or more. For this reason, this type of clocking unit should only be employed where timing is not a critical problem. The output frequency F_0 for this circuit is approximately $F_0 = 1/6\tau$, where τ is the average propagation delay in seconds, for the NAND gate utilized.

Examination of Fig. 8.11 also reveals that the width of the final output pulse will vary and will be dependent upon when the gate signal G returns to zero.

Let us now try to analyze the circuit shown in Fig. 8.10 from a slightly different viewpoint. In order to aid in our analysis, we will restrict the gate propagation delays to being integer multiples of τ seconds. We will also require all input-signal changes to occur only at the beginning of these discrete time intervals.

This restriction allows us to freeze the action of the circuit into discrete time intervals of τ units. Once a discrete circuit model has been obtained, it is generally quite easy to relax these restrictions. By choosing τ, very short, continuous circuit operations can be modeled quite closely.

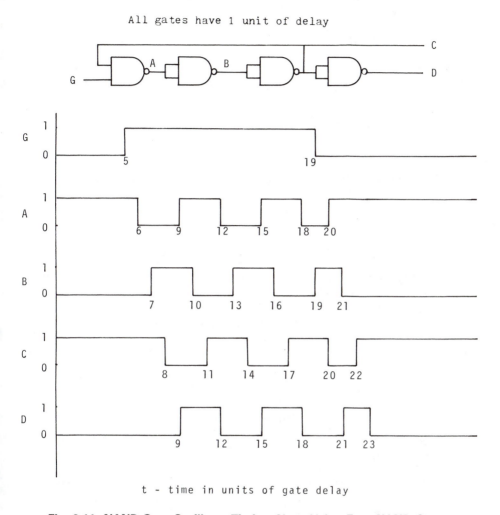

Fig. 8.11 NAND-Gate Oscillator Timing Chart Using Four NAND Gates

In order to understand the operation of this circuit, we will introduce the concept of the *state* of the circuit. The state of the circuit depends on the values of all the outputs of the gates in the circuit. Thus if we assign a binary variable to each gate output in Fig. 8.10, we can describe the operation of this circuit by predicting the next state of the circuit in terms of the present state (gate outputs) and the present inputs. The state diagram shown in Fig. 8.12 shows the resulting state transitions which can occur.

A *state diagram* consists of one labeled node [circle] for each state and a set of labeled transition arrows leaving each state and terminating on the next state,

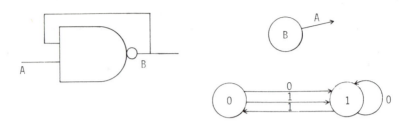

Fig. 8.12 State Transition Diagram for a Gated Oscillator

which results from the present values of the inputs and the gate outputs. There is one arrow leaving each node for each possible set of inputs. The state diagram, shown in Fig. 8.12, has two nodes with two arrows leaving each node.

An examination of this state diagram reveals that regardless of the value of the present output B, if A is set to zero the circuit will eventually reach the state 1 and remain there until the input signal A is changed back to a one.

It is also clear from Fig. 8.12 that, once the input A is changed from a zero to a one, the circuit will oscillate between states 0 and 1 until the input is again returned to a logical zero.

It is apparent that under the input condition $A = 0$, state 1 is a *stable* state, while state 0 is a *transitional* state. For the input $A = 1$, the pair of states 0 and 1 form a pair of *oscillating* states.

As the number of gates and inputs increase, the number of states and transition arrows can grow at a fantastic rate. In the binary case with n gates and k external inputs, the number of nodes needed would be equal to 2^n, with each node having 2^k state-transition arrows.

Let us consider the slightly more complicated three-gate circuit shown in Fig. 8.13. For this circuit, assuming each gate has a unit gate delay, the time-dependent output equations for each gate can be written as:

$$Z(t + \tau) = X(t) + Y(t),$$
$$X(t + \tau) = A(t) \cdot B(t),$$
$$Y(t + \tau) = B(t) \cdot C(t).$$

Combining these equations yields:

$$Z(t + 2\tau) = X(t + \tau) + Y(t + \tau),$$
$$= A(t)B(t) + B(t)C(t),$$
$$= B(t)[A(t) + C(t)].$$

The resulting timing chart for this circuit is shown in Fig. 8.13.

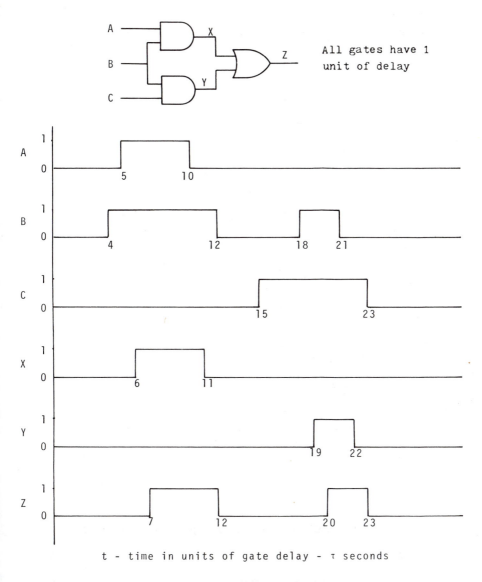

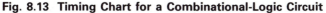

t - time in units of gate delay - τ seconds

Fig. 8.13 Timing Chart for a Combinational-Logic Circuit

Let us now replace input C with a feedback loop from gate Z. This change will cause the time-dependent output equation to become

$$Z(t + 2\tau) = B(t)[A(t) + Z(t)].$$

The resulting timing chart for this modified circuit is shown in Fig. 8.14. In order to draw this timing chart, it is necessary to know the initial condition

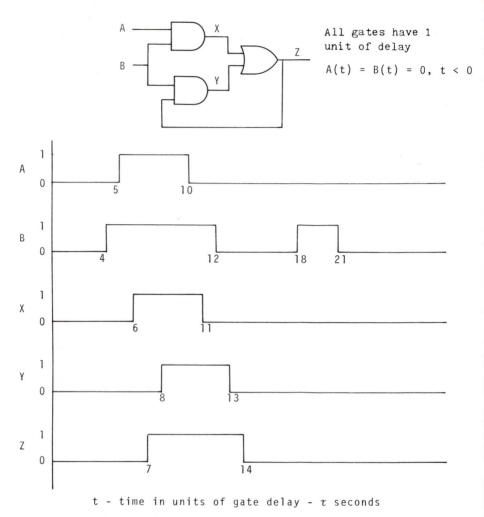

Fig. 8.14 Timing Chart for a 3-Gate Logic Circuit with Feedback

(output) of each gate in the circuit. With this information, it is an easy matter to fill in the output of each gate during the time interval t to $(t + \tau)$, since we know the output of each gate during the interval $(t - \tau)$ to t.

In Fig. 8.14, we are told that the external inputs are both zero for all time < 0, and hence we can deduce that $Z(t)$ also equals zero for $t < 0$.

Let us draw a state-transition diagram for this circuit, labeling the transition arrows between the states in binary with the input variable order AB and the states by the variables XYZ.

A cursory examination of this state diagram, shown in Fig. 8.15, reveals that there is too much information present for easy visual interpretation. Of particular interest to the designer are the state transitions which occur under fixed-input conditions. This information can be obtained either directly from the circuit equations or from the state transition diagram shown in Fig. 8.15.

The present-state/next-state (PS–NS) table shown in Fig. 8.16 lists, at the intersection of the present input column and present state row, the next state which will occur after one interval of time τ.

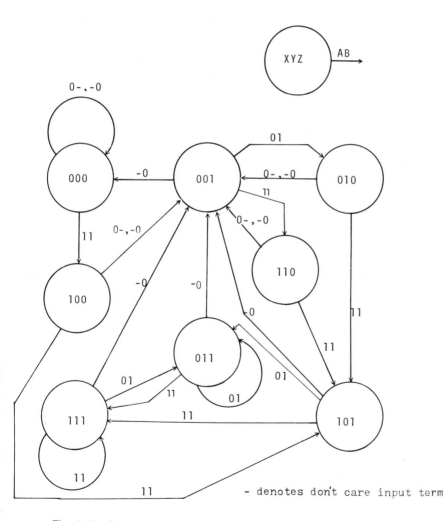

Fig. 8.15 State Transition Diagram for Circuit Shown in Fig. 8.14

Present State	Next State X(t+τ)Y(t+τ)Z(t+τ)			
	Present Input A(t)B(t)			
X(t)Y(t)Z(t)	00	01	10	11
000	(000)	(000)	(000)	100
001	000	010	000	110
010	001	001	001	101
011	001	(011)	001	111
100	001	001	001	101
101	001	011	001	111
110	001	001	001	101
111	001	011	001	(111)

Fig. 8.16 Present-State/Next-State Table for the Circuit Shown in Fig. 8.14

In this case, the intersection points represent $X(t + \tau)Y(t + \tau)Z(t + \tau)$ where

$$X(t + \tau) = A(t)B(t)$$
$$Y(t + \tau) = B(t)Z(t)$$

and

$$Z(t + \tau) = X(t) + Y(t).$$

States which do not change for a fixed set of inputs are called *stable states* and are indicated on the PS–NS table by circles.

Examination of the present-state/next-state table shown in Fig. 8.16 reveals that there are five stable state conditions possible, as indicated by the circles.

It should also be apparent from an examination of the logic circuit, the state transition diagram, and the logic equations, that the outputs X and Y do not affect the final outcome of the circuit. With this in mind, we could redraw the logic circuit lumping the gate delay at the output of an equivalent logic box which realizes the same logic equation. The gates inside the logic box are then assumed to have zero-delay, i.e., ideal gates. One such equivalent circuit is shown in Fig. 8.17.

Careful examination of Figs. 8.14 and 8.18 reveals that the two output waveforms for Z are the same. However, if the actual gate delay values had been

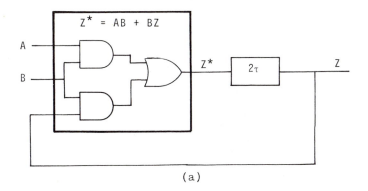

(a)

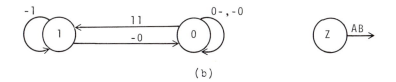

(b)

Present	Next State Z(t+2τ)			
State	Present Input A(t)B(t)			
Z(t)	00	01	10	11
0	⓪	⓪	⓪	1
1	0	①	0	①

(c)

Fig. 8.17 Lumped-Logic Model. (a) Logic Diagram; (b) State Transition Diagram; (c) Present-State/Next-State Table

unequal, then the equivalent circuit shown in Fig. 8.18 would have been required to use the worst-case delay, and minor discrepancies in the actual waveform would have been apparent.

One obvious reason for using a lumped analysis approach is to keep the size of our tables and diagrams manageable. A ten-gate circuit with five inputs would require 1024 nodes each with 32 transition arrows to describe completely. By judiciously using lumped logic analysis, many of these nodes can be eliminated.

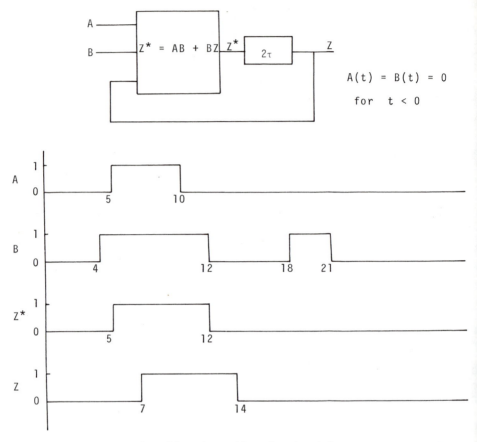

t - time in units of gate delay

Fig. 8.18 Timing Chart for the Lumped-Logic Configuration

8.5 BINARY-STORAGE ELEMENTS

In this section we will determine how the various types of binary-storage elements work. In particular, we will consider the operation of binary latches, and set-reset flip-flops. The purpose of this section is to provide the reader with a little insight into the reasons for various restrictions placed on the input signals of commercially available flip-flops.

With the possible exception of the set-reset flip-flops, the reader will (and should) use the commercially available storage elements rather than attempting to design and fabricate his or her own units.

Probably the most primitive form of memory element is a circuit which would remember if a binary signal ever reached a particular logic value. For example, suppose we wanted to monitor the output of a digital fire-alarm system. Due to the possibility of a fire destroying the detection circuitry, a memory element is needed. The output of this memory element must equal zero until the input alarm signal first goes to a one. The output signal must then remain a one regardless of the fluctuations on the alarm input signal. Labeling the output signal Z and the alarm or set input signal S, we can write a time-dependent output equation for the operation of the memory element just described.

$$Z(t + \tau) = Z(t) + S(t) \qquad (8.1)$$

with

$$Z(0) = 0.$$

A possible circuit configuration for Eq. 8.1 is shown in Fig. 8.19 using NOR gates. From this circuit we can obtain the time dependent output equation

$$Z(t + \tau_1 + \tau_2) = S(t) + Z(t),$$

assuming distributed gate delays of τ_1 and τ_2 sec.

In order to assure that the initial output $Z(t) = 0$ for $t = 0$, it will be necessary to first open the feedback connection and place a logic zero on both inputs to the NOR gate. If the ground signal corresponds to a logical zero, then it would only be necessary to ground the output terminal and alarm input prior to connecting the actual alarm signal to the device.

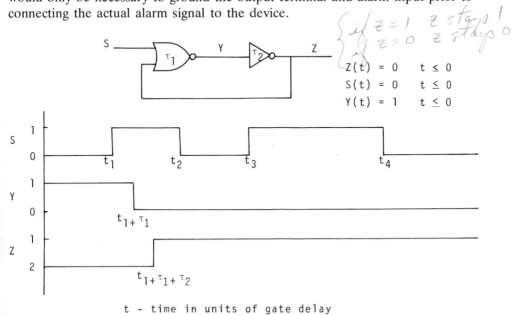

$$Z(t) = 0 \qquad t \le 0$$
$$S(t) = 0 \qquad t \le 0$$
$$Y(t) = 1 \qquad t \le 0$$

t - time in units of gate delay

Fig. 8.19 NOR-Gate Digital Latching Circuit and its Associated Timing Chart

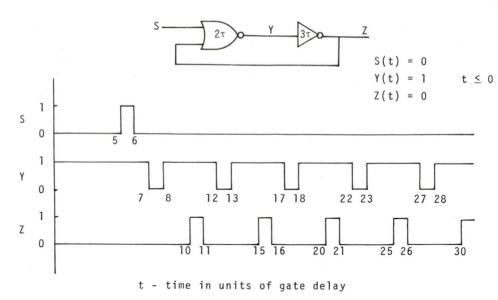

Fig. 8.20 Output Response of a NOR Gate Latching Circuit to a Short Duration Pulse

One drawback to this latching circuit is the possibility of a short noise pulse of duration less than the sum of the gate delays τ_1 and τ_2 triggering an oscillatory output at point Z. This type of response is illustrated in Fig. 8.20. Assuming gate delays on the order of 10 nanoseconds or less, a simple low-pass filter can be placed on the input signal S if high-frequency noise pulses are a problem.

Although in practice this type of oscillatory response is quite rare due to the natural low-pass filter effect inherent in the circuit wiring, the reader should be aware of and anticipate the possibility.

8.6 NOR IMPLEMENTED SET-RESET (SR) FLIP-FLOPS

For those applications where it is desirable or convenient to have a resetting capability, only a simple modification to the circuit shown in Fig. 8.19 is required. The resulting circuit is shown in Fig. 8.21. By replacing the inverter with a second NOR gate and attaching a reset line R, it is clear that as long as R is a logical one, then the output Z will be forced to a logical zero. This resulting circuit configuration is commonly called a *set-reset storage element* or an SR *flip-flop*.

It is also clear from this figure that

$$Y(t + \tau_1) = \overline{Z(t) + S(t)}$$

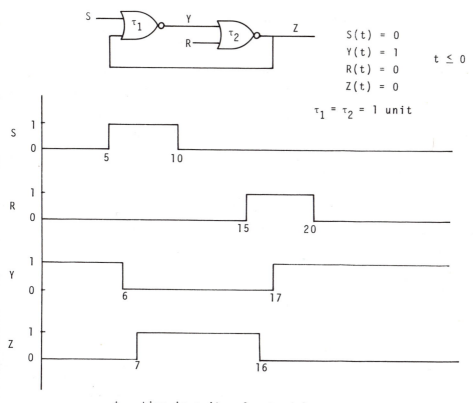

Fig. 8.21 A NOR-Implemented SR Flip-Flop and Its Associated Distributed-Delay Timing Chart

and that

$$Z(t + \tau_2) = \overline{Y(t) + R(t)}.$$

Combining these two equations yields, for the distributed logic model,

$$Z(t + \tau_1 + \tau_2) = \overline{Y(t + \tau_1) + R(t + \tau_1)}$$
$$= \overline{\overline{Z(t) + S(t)} + R(t + \tau_1)}.$$
$$Z(t + \tau_1 + \tau_2) = \bar{R}(t + \tau_1)[Z(t) + S(t)]. \qquad (8.2)$$

Assuming equal gate delays, the resulting equation becomes

$$Z(t + 2\tau) = \bar{R}(t + \tau)[Z(t) + S(t)].$$

The lumped logic model shown in Fig. 8.22 yields the time dependent equation

$$Z(t + 2\tau) = \bar{R}(t)[Z(t) + S(t)]. \qquad (8.3)$$

This latter lumped-circuit model, while exhibiting the same steady-state behavior, does not provide the designer with the necessary transient data resulting from input-signal changes.

In this book, we will use both circuit models with the distributed delay model being applicable to hand analysis of circuits with only a few gates. For large gate

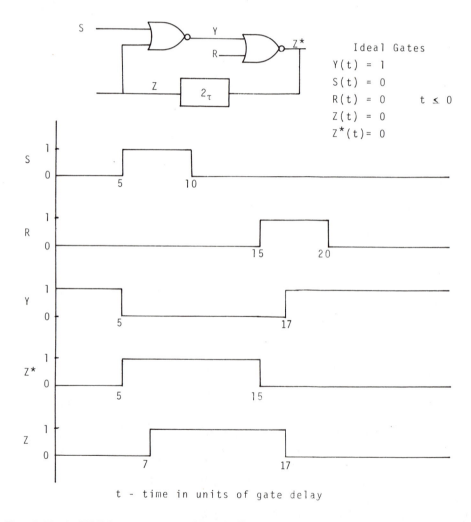

Fig. 8.22 A NOR-Implemented SR Flip-Flop and Its Associated Lumped-Delay Timing Chart

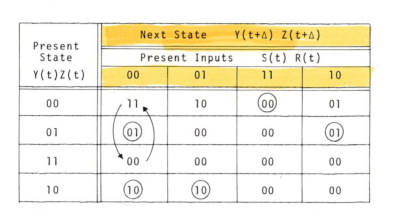

Present State $Y(t)Z(t)$	Next State $Y(t+\Delta)$ $Z(t+\Delta)$			
	Present Inputs $\quad$ $S(t)$ $R(t)$			
	00	01	11	10
00	11	10	00	01
01	01	00	00	01
11	00	00	00	00
10	10	10	00	00

Fig. 8.23 Present-State/Next-State Table for a NOR-Implemented Set-Reset Flip-Flop Assuming a Lumped Delay of Δ

configurations, a computer program can be easily written to provide an accurate timing chart using either distributed or lumped models.

Figure 8.23 shows the present-state/next-state (PN–NS) table for the SR flip-flop assuming equal distributed delays for each gate.

Close inspection of the PS–NS table reveals that it is possible for the circuit to oscillate between states 00 and 11 as long as the inputs are both held at zero. The oscillation is easily stopped by raising either of the two input lines to a logical one. We also see that the oscillation can be started from either stable state 01 or 10 is by setting both S and R to logical ones thereby driving the system into the state 00.

With the system in state 00, if both S and R are simultaneously set to zero, the system will proceed into oscillation. However, if these two inputs are not dropped to zero simultaneously, the system will proceed to state 10 if S reaches zero first or to state 01 if R reaches zero first. This situation is commonly referred to as a *critical race condition*. The final state is determined not by the steady-state values of the inputs but instead by the transient behavior of the circuit resulting from the double input change. Figure 8.24 shows the expected timing chart for an SR flip-flop using a distributed gate delay model and assuming equal delay for each gate.

Because of the large discrepancies between our pure delay model and the actual circuit delay characteristics for short duration pulses, only rarely will an SR flip-flop actually break into oscillation even when both S and R are driven by a

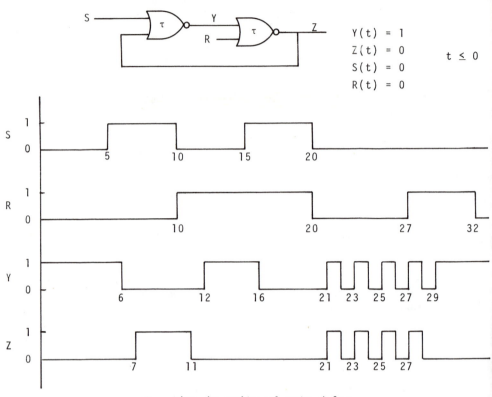

Fig. 8.24   Timing Chart for a NOR-Implemented SR Flip-Flop Using a Distributed-Delay Gate Model

square-wave input. In general, a given circuit will default to one or the other of the two stable states, 01 or 10.

However, since either stable state is possible, a wise practice when using this simple two-gate set-reset flip-flop is to never allow both S and R to equal one at the same time or to be sure and never return both S and R to zero simultaneously if the condition should arise. Also short pulses $(<2\tau)$ on S and R should not be used.

Examination of the timing chart shown in Fig. 8.22 reveals that except for the transient period following changes to either input, the outputs at points Y and Z are complementary. During the transient period both outputs are momentarily equal to zero. It is also clear from the circuit symmetry that the S (set) input sets output Z to a one and at the same time acts as a reset input for point Y. The R (reset input), while setting output Z to zero, also acts as a set input for point Y.

If we add the restriction that $S(t) \cdot R(t) = 0$ we can relabel the Y output as $\bar{Z}$. The resulting circuit configuration is usually replaced by the logic symbol shown in

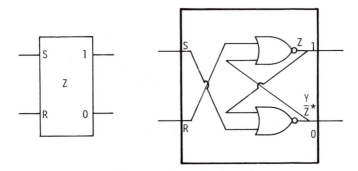

Fig. 8.25 NOR-Gate SR Flip-Flop Logic Symbol

Fig. 8.25, with the implicit understanding that S and R will never be simultaneously set equal to one.

The one output corresponds to the set output which equals one when $S = 1$ and $R = 0$. The zero or reset output equals one when $R = 1$ and $S = 0$.

The set of time-dependent output equations used to describe this circuit assuming a lumped-delay analysis are:

$$Z(t + \Delta) = \bar{R}(t)[S(t) + Z(t)] \tag{8.4}$$

and

$$Y(t + \Delta) = \bar{S}(t)[R(t) + Y(t)]$$
$$\bar{Z}^*(t + \Delta) = \bar{S}(t)[R(t) + \bar{Z}^*(t)] \tag{8.5}$$

with the restriction for normal use that

$$S(t) \cdot R(t) = 0. \tag{8.6}$$

Δ is the time required for the outputs to reach steady state. The asterisk has been appended to the output variable $\bar{Z}$ in order to remind the reader that the Y output can only be treated as the complement of Z as long as the input restriction $S(t) \cdot R(t)$ equal to zero holds, and the circuit remains in a steady-state condition.

8.7 NAND IMPLEMENTED SET-RESET (S̄R̄) LATCH

If we complement Eqs. 8.4 to 8.6 we obtain the equations

$$\bar{Z}(t + \Delta) = R(t) + \bar{S}(t) \cdot \bar{Z}(t) \tag{8.7}$$
$$Z^*(t + \Delta) = S(t) + \bar{R}(t) \cdot Z^*(t) \tag{8.8}$$
$$\bar{S}(t) + \bar{R}(t) = 1 \tag{8.9}$$

and the complementary circuit realization shown in Fig. 8.26.

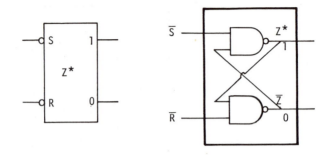

Fig. 8.26 NAND-Gate $\overline{\text{Set}}$-$\overline{\text{Reset}}$ Latch

A sample timing chart for this circuit is shown in Fig. 8.27. A comparison of Figs. 8.21 and 8.27 reveal again the complementary nature of the two flip-flop storage elements. The rest state for the NAND-gate implementation requires that both inputs be at a logical one, while the rest state for the NOR gate implementation requires that both inputs be equal to zero.

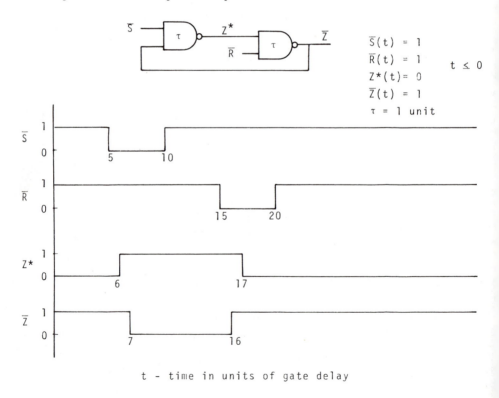

t - time in units of gate delay

Fig. 8.27 Timing Chart for $\overline{\text{S}}\overline{\text{R}}$ NAND-Implemented Flip-Flop

Present State $Z^*(t)\bar{Z}(t)$	Next State $\quad Z^*(t+\Delta)\ \bar{Z}(t+\Delta)$			
	Present Inputs $\quad \bar{S}(t)\ \bar{R}(t)$			
	00	01	11	10
00	11	11	11	11
01	11	11	(01)	(01)
11	(11)	10	00	01
10	11	(10)	(10)	11

Fig. 8.28 Present-State/Next-State Table for NAND-Implemented $\overline{\text{Set}}$-$\overline{\text{Reset}}$ Flip-Flop Assuming a Lumped Delay of Δ

It is also clear from the present-state/next-state table (Fig. 8.28) that a critical-race condition exists if both inputs are set to zero and then simultaneously set to logical ones.

8.8 CLOCKED-STORAGE ELEMENTS

In the preceding section we briefly introduced two asynchronous circuit models for the set-reset storage element or flip-flop. It should be pointed out that for reliable operation of such a set-reset flip-flop, the user must prevent stray pulses on the order of a few gate delays from appearing on either the set or reset input. These random pulses are commonly generated by logical hazards in the combinational network attached to the S and R inputs.

The simplest method of preventing these hazards from causing malfunctions is to provide a separate logic signal which is ANDed with both the set and reset input. This enabling signal allows the user to gate the S and R information to the device. This extra signal is usually called a clock line. The purpose of the clock signal is to allow the user to return and hold the SR flip-flop in its rest state, while changes occur on the S and R inputs. Once the set and reset lines have settled into their new steady-state values, the clock line can again be set to a logical one, thereby allowing the altered inputs to determine the next state. In Fig. 8.29, clock lines have been added to both the NOR and NAND SR flip-flops.

The corresponding output equations for these logic circuits can be easily obtained from this figure. Using a lumped circuit model analysis, the output

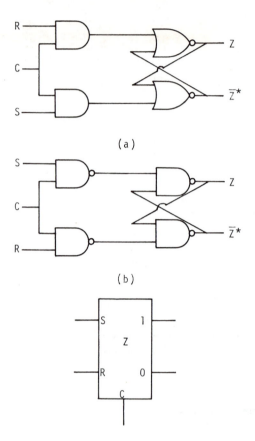

Fig. 8.29 Clocked SR Flip-Flop Circuits. (a) Clocked NOR-Gate Circuit; (b) Clocked NAND-Gate Circuit; (c) Logic Symbol for a Clocked SR Flip-Flop

equations for the NOR gate configurations are

$$Z(t + \Delta) = [\bar{R}(t) + \bar{C}(t)][S(t)C(t) + Z(t)] \tag{8.10}$$

$$\bar{Z}^*(t + \Delta) = [\bar{S}(t) + \bar{C}(t)][R(t)C(t) + \bar{Z}^*(t)]. \tag{8.11}$$

when $C(t) = 0$

$$Z(t + \Delta) = Z(t), \qquad \bar{Z}^*(t + \Delta) = \bar{Z}^*(t)$$

and when $C(t) = 1$

$$Z(t + \Delta) = \bar{R}(t)[S(t) + Z(t)],$$

$$\bar{Z}^*(t + \Delta) = \bar{S}(t)[R(t) + \bar{Z}^*(t)].$$

Using a lumped-circuit model analysis, the output equations for the NAND configuration are

$$Z(t + \Delta) = C(t)S(t) + [\bar{R}(t) + \bar{C}(t)]Z(t) \qquad (8.12)$$

and

$$\bar{Z}^*(t + \Delta) = C(t)R(t) + [\bar{S}(t) + \bar{C}(t)]\bar{Z}^*(t). \qquad (8.13)$$

when $C(t) = 0$

$$Z(t + \Delta) = Z(t), \qquad \bar{Z}^*(t + \Delta) = \bar{Z}^*(t)$$

and when $C(t) = 1$

$$Z(t + \Delta) = S(t) + \bar{R}(t)Z(t),$$
$$\bar{Z}^*(t + \Delta) = R(t) + \bar{S}(t)\bar{Z}^*(t).$$

Finally the restriction that $R(t) \cdot S(t) \cdot C(t) = 0$ should be imposed on the general use of either circuit. For, if $S(t) \cdot R(t) = 1$ and the clock signal is set to zero, resulting critical-race condition will make it difficult to predict with certainty what the rest state for the flip-flop will be.

In general, when dealing with clocked flip-flops, it is convenient to suppress the explicit reference to the clock in the output equations and instead to simply refer to the next state as occurring after a complete clocking cycle.

In the case of the SR flip-flop, a complete clocking cycle would start with the clock line at a logical zero and end the next time the clock returns to zero following a transition to a logical one. This clocking cycle is illustrated in Fig. 8.30. One advantage which this definition allows is the use of variable lengths as well as asymmetric clocking signals.

However, in order to effectively utilize a clock, a few general guidelines should be followed.

First, the S and R inputs to the clocked flip-flop should be allowed to change during the time interval when the clock line is equal to zero.

Second, the S and R inputs should not be allowed to change during the time interval when the clock line is equal to one.

Third, the clock line must remain at a logical one long enough to assure that the outputs assume their new steady-state values.

Fourth, the condition $S \cdot R = 1$ should not be allowed as a steady-state condition when the clock signal is equal to a one.

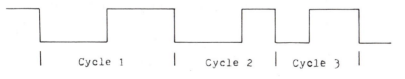

Cycle 1   Cycle 2   Cycle 3

Fig. 8.30 SR Flip-Flop Clocking Cycle

S(t)	R(t)	Z(t+Δ)	$\overline{Z}^*(t+Δ)$
0	0	Z(t)	$\overline{Z}^*(t)$
0	1	0	1
1	0	1	0
1	1	?	?

Fig. 8.31 Clocked SR Flip-Flop Logic Description

If we follow the guidelines above, it is possible to define a universal SR flip-flop whose steady-state behavior will be independent of its logic-circuit configuration. The resulting composite-logic description is shown in Fig. 8.31.

8.9 ANALYSIS OF SR FLIP-FLOP CIRCUITS

Having established a lumped-delay-circuit model for a clocked SR flip-flop in the preceding section, let us see how we can utilize this information to analyze logic circuits which contain SR flip-flops. A straightforward analysis procedure is given below.

ALGORITHM 8.1

SR analysis procedure

STEP OPERATION

1.   Determine the combinational-logic equations for the S and R inputs assuming ideal gates.

2.   Determine if $S \cdot R = 0$.
Note: If $S \cdot R \neq 0$, this procedure should be terminated, since the circuit may contain critical races. (Because of these critical races, the output equations may have to be determined experimentally or graphically using a distributed gate delay model.)

3.   Apply either the NAND or NOR circuit model equation to determine the next state equations.

$$Z(t + \Delta) = S(t) + \bar{R}(t)Z(t) \qquad \text{NAND}$$
$$Z(t + \Delta) = \bar{R}(t)[S(t) + Z(t)] \qquad \text{NOR}$$

Note that when $S \cdot R = 0$, these two equations are equivalent.

Let us apply this analysis procedure to a few selected examples.

Example 8.1 Determine the next-state equation and present-state/next-state table for the following clocked SR flip-flop.

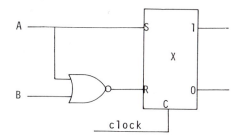

Using our analysis procedure, we have:

1. $S(t) = A(t)$, $R(t) = \overline{A(t) + B(t)} = \overline{A(t)} \cdot \overline{B(t)}$
2. $S(t) \cdot R(t) = A(t) \cdot \overline{A(t)}\,\overline{B(t)} = 0$. Hence we can proceed.
3. $X(t + \Delta) = S(t) + \bar{R}(t)X(t)$
$$= A(t) + [A(t) + B(t)]X(t)$$
$$= A(t) + B(t)X(t)$$

Having determined the next-state equation, its present-state/next-state table is easily determined.

Present Input		Present State	Next State
A(t)	B(t)	X(t)	X(t+Δ)
0	0	0	0
0	0	1	0
0	1	0	0
0	1	1	1
1	0	0	1
1	0	1	1
1	1	0	1
1	1	1	1

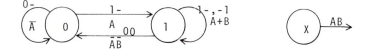

The state transition diagram for this circuit is easily obtained from the PS–NS Table. It should be noted that the format for this table has been changed from that in Fig. 8.16. This new format in many cases is easier to construct and use. The dashes on the state diagram indicate don't-care input conditions.

Example 8.2 Analyze the following SR flip-flop circuit.

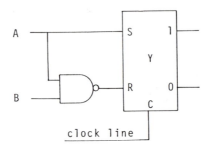

Applying our analysis procedure, we have:

1. $S(t) = A(t), R(t) = \overline{A(t) \cdot B(t)}$
2. $S(t) \cdot R(t) = A(t) \cdot [\bar{A}(t) + \bar{B}(t)] = A(t)\bar{B}(t) \neq 0$

Hence, this circuit cannot be analyzed further without detailed timing information about the input signals A and B. If A is equal to one and B is equal to zero, the indeterminate input pair $S = 1$ and $R = 1$ will cause a critical-race condition when the clock line is returned to zero.

Example 8.3 Analyze the following SR flip-flop circuit; i.e., write out the set of next-state equations.

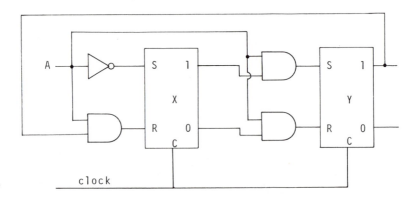

Using our analysis procedure, we obtain

1. $S_x(t) = \bar{A}(t)$ $S_y(t) = A(t)X(t)$

   $R_x(t) = A(t)Y(t)$ $R_y(t) = A(t)\bar{X}(t)$

2. $S_x(t)R_x(t) = \bar{A}(t) \cdot A(t)Y(t) = 0$

   $S_y(t)R_y(t) = A(t)X(t)A(t)\bar{X}(t) = 0$

3. $X(t + \Delta) = S_x(t) + \bar{R}_x(t)X(t)$

   $\qquad\qquad = \bar{A}(t) + [\bar{A}(t) + \bar{Y}(t)]X(t)$

   $\qquad\qquad = \bar{A}(t) + X(t)\bar{Y}(t)$

   $Y(t + \Delta) = S_y(t) + \bar{R}_y(t)Y(t)$

   $\qquad\qquad = A(t)X(t) + [\bar{A}(t) + X(t)]Y(t)$

   $\qquad\qquad = A(t)X(t) + \bar{A}(t)Y(t) + X(t)Y(t)$

   $\qquad\qquad = A(t)X(t) + \bar{A}(t)Y(t)$

8.10 SYNTHESIS OF SEQUENTIAL CIRCUITS USING SR FLIP-FLOPS

In the preceding section we considered the problems associated with analyzing a binary logic circuit which included clocked SR flip-flops. In this section we will consider the inverse problem. Given a logical description for a sequential circuit, we would like to establish a few simple design procedures to aid in the circuit implementation. As a starting point we must establish a general circuit model for realizing an arbitrary sequential circuit. One such model using clocked SR flip-flops is shown in Fig. 8.32.

This model assumes a number of external binary input signals $I_1 \cdots I_k$, and a number of internal state variables $X_1 \cdots X_n$ which are the output terminals of n clocked-SR flip-flops, operating from a common clock.

Combinational logic networks are used to provide the necessary excitation signals for the S and R inputs of each flip-flop. These logic equations are functions of both the present inputs and the present-state variables.

$$S_{x_j}(t) = F_{S_{x_j}}(I_1(t), I_2(t), \ldots, I_k(t); X_1(t), X_2(t), \ldots, X_n(t))$$
$$R_{x_j}(t) = F_{R_{x_j}}(I_1(t), I_2(t), \ldots, I_k(t); X_1(t), X_2(t), \ldots, X_n(t))$$
$$1 \leq j \leq n$$

The circuit also allows for m binary output signals $Z_i(t)$, where $1 \leq i \leq m$, which are also functions of the present inputs and present state of the circuit.

$$Z_i(t) = F_{Z_i}(I_1(t), I_2(t), \ldots, I_k(t); X_1(t), X_2(t), \ldots, X_n(t))$$

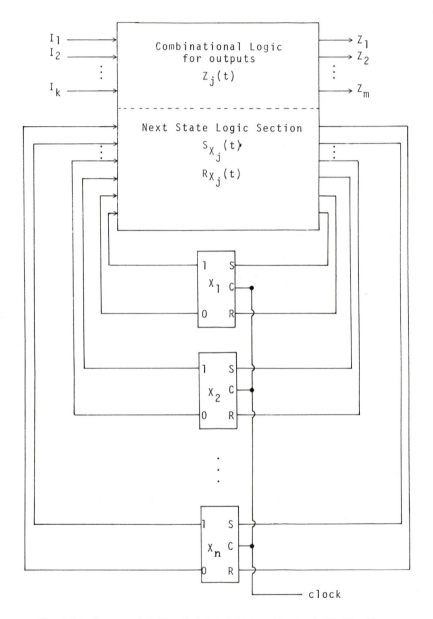

Fig. 8.32 Sequential Circuit Model Using Clocked SR Flip-Flops

We will require that the input signals, I_j's, be changed only while the clock signal is a zero. We will also require all the S_{x_i} and R_{x_i} inputs reach steady-state values prior to setting the clock signal to a logical one. Finally we apply the restriction that $S_{x_j}(t) \cdot R_{x_j}(t) \cdot C(t) = 0$ for all flip-flops in order that the next-state equation

$$X_j(t + \Delta) = S_{x_j}(t) + \bar{R}_{x_j}(t)X_j(t), \qquad 1 \le j \le n,$$

will hold. Once the clock line is returned to a logical zero, the new excitation inputs S_{x_j} and R_{x_j} can be calculated for each flip-flop along with the set of binary outputs.

This model basically assumes a lumped logic delay approach. Because of the clocking configuration inherent in the SR flip-flop design, the flip-flop output, i.e., the internal state variables, will be hazard free. However, the output lines, Z_i's and the excitation lines, S_{x_j}'s and R_{x_j}'s may, and probably will, contain logical hazards following changes to either input variables or internal state variables.

The assumptions just discussed allow us to treat a sequential circuit as a network whose states change only at discrete intervals of time, controlled by the clock. Keeping this idea in mind, we will proceed in our synthesis of sequential circuits using clocked SR flip-flops by converting the problem statement into a form amenable to solution using the circuit model shown in Fig. 8.32.

8.11 EXCITATION TABLE FOR SR FLIP-FLOPS

Before formalizing a synthesis procedure, let us consider the question of how to determine the proper set and reset excitation equations while satisfying the restriction that $S \cdot R = 0$. A little thought reveals that there are only four cases which must be considered.

CASE 1. Present state is a zero, $X(t) = 0$, and next state is to be a zero, $X(t + \Delta) = 0$.

From the present-state/next-state table shown in Fig. 8.33, there are only two input combinations which keep $X(t + \Delta)$ at zero and $S \cdot R = 0$. These are $S(t) = 0$, $R(t) = 0$ and $S(t) = 0$, $R(t) = 1$. From this we can conclude that if we want to retain $X(t + \Delta)$ at zero, we must set input S to a zero. The R input can be treated as a don't-care input since its value will not affect the next state.

CASE 2. The present state is a zero, $X(t) = 0$, and the next state is to be a one, $X(t + \Delta) = 1$.

Examination of the PS–NS Table in Fig. 8.33 reveals that the S input must be a one and the reset input must be a zero in order to accomplish the transition.

CASE 3. The present state is a one, $X(t) = 1$, and the next state is to be zero, $X(t + \Delta) = 0$.

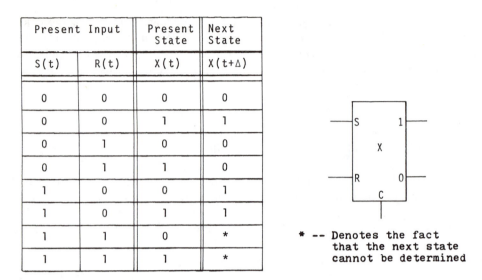

Present Input		Present State	Next State
S(t)	R(t)	X(t)	X(t+Δ)
0	0	0	0
0	0	1	1
0	1	0	0
0	1	1	0
1	0	0	1
1	0	1	1
1	1	0	*
1	1	1	*

* -- Denotes the fact that the next state cannot be determined

Fig. 8.33 Present-State/Next-State Table for a Clocked SR Flip-Flop

Examination of the PS–NS table in Fig. 8.33 reveals that the S input must be set to a zero and the reset input set to a one in order to accomplish this transition.

CASE 4. The present state is a one, $X(t) = 1$, and the next state is also a one, $X(t + \Delta) = 1$.

The PS–NS table in Fig. 8.33 reveals two input combinations which will keep the next state at a logical one while keeping $S \cdot R = 0$. The input combinations are $S = 1$, $R = 0$, and $S = 0$, $R = 0$. Thus, in order to retain $X(t + \Delta)$ at a

X(t)	X(t+Δ)	$S_X(t)$	$R_X(t)$
0	0	0	d
0	1	1	0
1	0	0	1
1	1	d	0

Fig. 8.34 SR Flip-Flop Excitation Table

value of one, the reset input must be set to a logical zero. The set input can then be treated as a don't-care input as its value will not affect the next state output.

The results of this discussion are summarized in the form of the SR flip-flop excitation table shown in Fig. 8.34.

ALGORITHM 8.2

Sequential Circuit Synthesis Procedure Using Clocked SR Flip-Flops

STEP   OPERATION

1.   Using the known next-state equations, construct a present-state/next-state table for the circuit to be realized.

2.   Add a pair of excitation columns $S_{Q_i}(t)$ and $R_{Q_i}(t)$ for each internal state variable Q_i. Fill in the entries in these columns using the row pairs $[Q_i(t), Q_i(t + \Delta)]$ as a table lookup entry point for the SR flip-flop excitation table.

$Q_i(t)$	$Q_i(t+\Delta)$	$S_{Q_i}(t)$	$R_{Q_i}(t)$
0	0	0	d
0	1	1	0
1	0	0	1
1	1	d	0

3.   Write out the logic equations for each excitation column $S_{Q_i}(t)$ and $R_{Q_i}(t)$.

4.   Implement these excitation and output equations subject to any parts constraint imposed. Minimize the logic equations as much as time and/or resources permit.

5.   Check your work against careless errors by analyzing each flip-flop using the general next-state equation

$$Q_i(t + \Delta) = S_{Q_i}(t) + \bar{R}_{Q_i}(t)Q_i(t).$$

Also be sure that $S_{Q_i}(t) \cdot R_{Q_i}(t) = 0$.

Let us illustrate how we can utilize Algorithm 8.2 when designing sequential circuits.

Example 8.4 Design a clocked sequential circuit using an SR flip-flop which satisfies the next-state equation

$$X(t + \Delta) = A(t)X(t) + B(t)$$

Solution. Using the next-state equation, we can construct a present-state/next-stable table.

Present Input		Present State	Next State	Excitation Values	
$A(t)$	$B(t)$	$X(t)$	$X(t+\Delta)$	$S_x(t)$	$R_x(t)$
0	0	0	0	0	d
0	0	1	0	0	1
0	1	0	1	1	0
0	1	1	1	d	0
1	0	0	0	0	d
1	0	1	1	d	0
1	1	0	1	1	0
1	1	1	1	d	0

The next-state column is filled in by suppressing the time dependence and treating $X(t + \Delta)$ as the output function of a combinational logic circuit.

The excitation columns $S_x(t)$ and $R_x(t)$ are completed a line at a time by using the present-state $X(t)$ and next-state $X(t + \Delta)$ values in each row as table lookup entries in the excitation table shown in Fig. 8.34. Thus for row one we have $X(t) = 0$ and $X(t + \Delta) = 0$, yielding $S(t) = 0$ and $R(t) = d$ from the excitation table. Proceeding down the table, the second row has $X(t) = 1$ and $X(t + \Delta) = 0$, yielding $S(t) = 0$ and $R(t) = 1$. Finally for the last row we have $X(t) = 1$ and $X(t + \Delta) = 1$, yielding $S(t) = d$ and $R(t) = 0$ from the excitation table.

Once we have filled in the excitation columns for our circuit, we suppress the time dependence and treat S_x and R_x as combinational outputs of a logic circuit with input variables of A, B, and X.

In this example, let us use K-maps to aid us in obtaining a circuit realization.

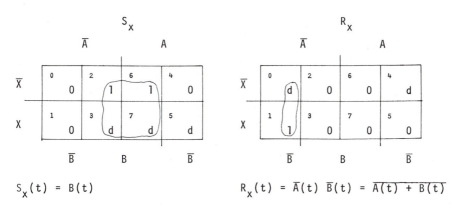

$$S_x(t) = B(t)$$

$$R_x(t) = \bar{A}(t)\ \bar{B}(t) = \overline{A(t) + B(t)}$$

The circuit realization as shown below

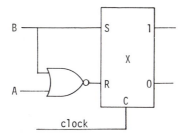

Example 8.5 Design a clocked sequential circuit using SR flip-flops which satisfies the pair of next-state equations

$$X(t + \Delta) = A(t)X(t) + \bar{A}(t)Y(t)$$
$$Y(t + \Delta) = A(t)\bar{X}(t) + \bar{Y}(t)$$

Solution. Using the next-state equations, we can construct a present-state/next-state table.

With the aid of the excitation table in Fig. 8.34, the S_x and R_x excitation values can be obtained using $X(t)$ and $X(t + \Delta)$ value for each row as the table entry points. The S_y and R_y values are obtained in the same manner with $Y(t)$ and $Y(t + \Delta)$ values as the table entry points. (See Fig. 8.35.) From these maps we obtain the excitation equations

$$S_x(t) = \bar{A}(t)Y(t), \qquad R_x(t) = \bar{A}(t)\bar{Y}(t),$$
$$S_y(t) = \bar{Y}(t) \quad \text{and} \quad R_y(t) = \bar{A}(t)Y(t) + X(t)Y(t).$$

One circuit realization for these equations is shown in Fig. 8.36.

Present Input	Present State		Next State		Excitation Values			
A(t)	X(t)	Y(t)	X(t+∆)	Y(t+∆)	$S_x(t)$	$R_x(t)$	$S_y(t)$	$R_y(t)$
0	0	0	0	1	0	d	1	0
0	0	1	1	0	1	0	0	1
0	1	0	0	1	0	1	1	0
0	1	1	1	0	d	0	0	1
1	0	0	0	1	0	d	1	0
1	0	1	0	1	0	d	d	0
1	1	0	1	1	d	0	1	0
1	1	1	1	0	d	0	0	1

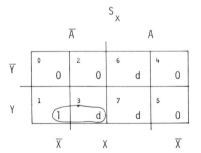

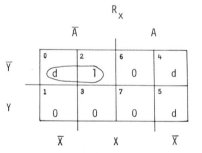

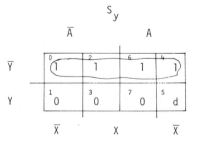

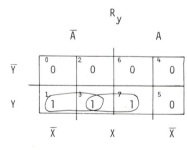

Figure 8.35

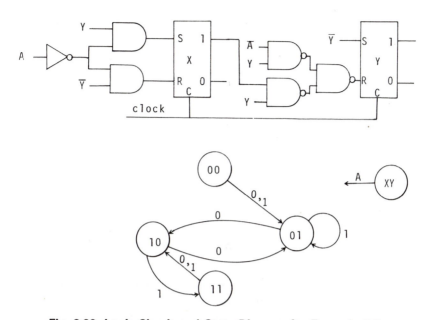

Fig. 8.36 Logic Circuit and State Diagram for Example 8.5

8.12 SPECIALIZED DIGITAL-LOGIC
CIRCUITS—MONOSTABLE MULTIVIBRATORS

This section and the next will be devoted to a brief discussion of a few of the rather specialized logic circuits used in the implementation of sequential logic circuits. Introduced in these sections will be digital clocks, one-shots, and gated oscillators. The operation of these devices will be explained using simple logic-circuit models. More detailed explanations on the design and operation of these units can be found in most digital electronic textbooks.

A monostable multivibrator or one-shot is an electronic circuit which produces a single pulse output when properly triggered. An idealized unit along with its associated timing chart is shown in Fig. 8.37.

When the command line is changed from a logical zero to a logical one, the unit is said to have been triggered. The output of the monostable multivibrator will be forced to transition to a logical one. However, unlike a bistable element such as a flip-flop, the output will return to zero after a predetermined time delay of Δ seconds. Because the unit yields only a single output pulse for each trigger signal, the monostable multivibrator is commonly referred to as a *one-shot* logic unit.

The amount of delay before the output returns to zero is generally a variable which can be controlled by the addition of a simple resistor-capacitor (R-C)

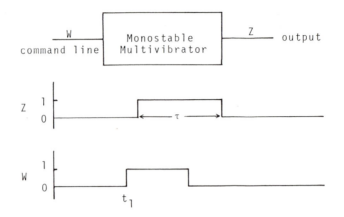

Fig. 8.37 Idealized Timing Chart for a Monostable Multivibrator

network. Monostable multivibrators with a variable pulse duration are also referred to as *logic delay units* or simply as delay units.

A simple one-shot circuit can be easily constructed using a NOR gate, an inverter, and an R-C coupling network. One such circuit configuration is shown in Fig. 8.38.

This circuit is seen as a modification of the basic NOR implemented latch discussed in Section 8.4.

Assuming ideal logic gates (0 delay) and a positive logic convention with a logical one represented by a positive voltage, V_L, and a logical zero corresponding to zero volts, then initially, input W and output Z will be logical zeroes (≈ 0 volts).

Points Y and X will then correspond to logical ones ($+ V_L$ volts) and the voltage across the capacitor C will be 0 volts, i.e., the capacitor will be discharged. When the input W is changed to a logical one, ($+ V_L$ volts), at some time t_s, point X, the output of the NOR gate will be forced to become a logical zero (≈ 0 volts). This sudden voltage change at point X will cause the voltage at point Y to drop to zero (logical zero), which in turn will cause point Z to transition to a logical one.

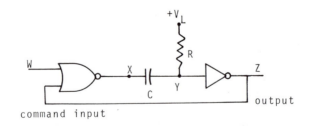

Fig. 8.38 Simple One-Shot Logic Circuit

Once point Z becomes a one ($+ V_L$ volts), the input W can be removed without affecting the value of X. However, the voltage at point Y will not remain fixed at zero, but instead will slowly return to a logical one, as the capacitor C is charged. The voltage at point Y will be

$$V_Y = V_L(1 - e^{-(t-t_s)/RC}), \quad t > t_s.$$

After a short delay, point Y will reach the switch threshold voltage of the inverter (V_T). At this point, $t = t_d$, Z will again return to zero. The delay, $\tau = t_d - t_s$, for this circuit will be the time required for the voltage to reach the switching threshold. Thus we have

$$V_Y = V_T = V_L(1 - e^{-\tau/RC}).$$

Solving for τ we have

$$V_L - V_T = V_L e^{-\tau/RC}$$

$$\ln\left(\frac{V_L}{V_L - V_T}\right) = \frac{\tau}{RC}$$

$$\tau = RC \ln\left(\frac{V_L}{V_L - V_T}\right).$$

Thus if $V_L = 3.0$ volts, $V_T = 1.2$ volts, $R = 400$ ohms and $C = 250$ μfd, then τ would equal $0.1 \ln\left(\frac{3.0}{1.8}\right) = 51$ milliseconds.

Once Z returns to zero, the voltage at point Y will continue to rise exponentially toward V_L. After t_f seconds, the voltage will equal

$$V_f = V_L(1 - e^{-(t_f-t_s)/RC}).$$

At this time, the voltage at point W is returned to zero. This change disables the one-shot and causes the voltage at point X to again return to a logical one ($+ V_L$ volts). This sudden change will cause the voltage at point Y to rise to $V_f + V_L$ volts. This voltage will now decay toward V_L volts with

$$V_Y = (V_f + V_L) - V_f(1 - e^{-(t-t_f)/RC})$$
$$= V_L + V_f e^{-(t-t_f)/RC}, \quad t > t_f.$$

An idealized timing chart for this circuit is shown in Fig. 8.39, with the voltage at point Y equal to

$$V_L \qquad\qquad \text{for} \qquad 0 \le t < t_s$$
$$V_L(1 - e^{-(t-t_s)/RC}) \quad \text{for} \qquad t_s \le t < t_f$$
$$V_L + V_f e^{-(t-t_s)/RC} \quad \text{for} \qquad t_f \le t$$

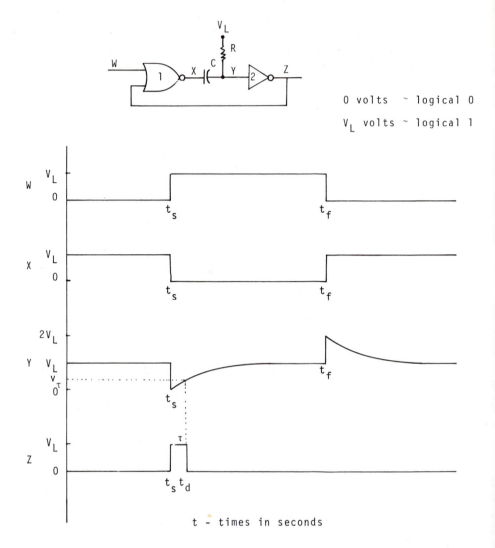

0 volts ∼ logical 0

V_L volts ∼ logical 1

Fig. 8.39 Idealized Timing Charts for a NOR-Implemented One-Shot Logic Circuit

The voltage at point Z will be

$$
\begin{array}{lll}
0 & \text{for} & 0 \le t < t_s \\
V_L & \text{for} & t_s \le t < t_d \\
0 & \text{for} & t_d < t
\end{array}
$$

where $\tau = t_d - t_s = RC \ln\left(\dfrac{V_L}{V_L - V_Y}\right).$

If we take our output from point Z we have, in effect, a one-shot pulse generator. Each time we trigger the unit, a single pulse of duration τ occurs.

The delay time τ is variable and can be controlled by a potentiometer. In general, there will be a minimum delay time τ_{min} and some maximum delay time τ_{max} which the device can provide. In a steady-state condition, the output of the device, Z, will be at a logical zero. In order to activate the device, the command line, W, must be at a logical zero for an enable time on the order of three times the value of RC in order to ensure that the capacitor can be discharged. Once the capacitor has been discharged, a level change on the command line to a logical one will cause the output of the circuit to rise to a logical one and to remain at a logical one for the specified delay time of τ nanoseconds, after which the output of the device will again return to a logical zero. Once the device has been fired, the input line will have no effect on the circuit until the device has returned to a logical zero. The device cannot be reliably activated a second time until the minimum recovery time needed to discharge the capacitor has elapsed. Note that the recovery time will not begin until the command line is returned to a logical zero. Triggering the unit prematurely will result in pulses with a shorter delay value. Additional internal circuitry can be added to reduce the recovery time by quickly discharging the capacitor.

We will assume that all monostable multivibrators used in this book have fast-recovery circuitry. In order to prevent stray noise pulses from accidentally triggering the one-shot pulse unit, an AND gate is sometimes added to the input line. Then both a command signal and an enable signal will be required for the monostable multivibrator to operate. The logic symbols used in this book are shown in Fig. 8.40.

It should be emphasized that the one-shot circuits discussed in this section are commercially available. The commercial units are also vastly superior in terms of stability, operating environment, and low price. The range of delay values available for these units is from a few nanoseconds to several minutes.

As a sequential control element, monostable multivibrators are relatively easy to use and are particularly suited to tasks which require a sequence of operations

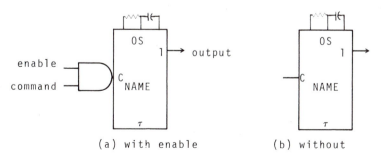

(a) with enable   (b) without

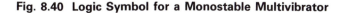

Fig. 8.40 Logic Symbol for a Monostable Multivibrator

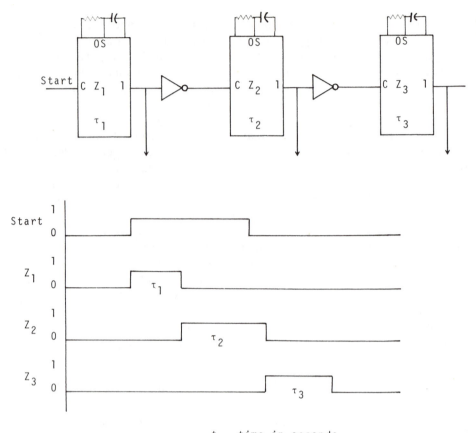

Fig. 8.41 Simple Pulse Generator Using Monostable Multivibrators (One Shots)

each requiring a different amount of time to complete. The circuit shown in Fig. 8.41 shows a simple series connection of one-shots, which provides a sequence of three timing pulses. The only restriction placed on the operation of this type of serial unit is that the pulse duration of each unit be longer than the time required to discharge the capacitors in the timing circuit of the following unit.

8.13 DIGITAL CLOCKS

A simple digital clocking unit is shown in Fig. 8.42, requiring only a single resistor and capacitor. A timing chart for this clock is given in Fig. 8.43.

Assuming idealized gates (0 gate delay), a logical one voltage of V_L volts, a logical zero voltage of 0 volts, and a switching threshold voltage of V_T volts, this

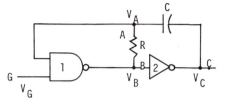

Note: Depending on the actual NAND gates used, a load resistor may be needed.

Fig. 8.42 A Simple Digital-Clock Circuit

circuit can be readily analyzed. With the gate signal G at a logical zero, the output of gate 1 will be held at a voltage value corresponding to a logical one, V_L volts, and gate 2 will be held at a voltage value corresponding to a logical zero, 0 volts. Under this condition the capacitor, C, will be charged to V_L volts.

The voltage at point A will remain at V_L volts until the gate signal G is changed to a logical one (V_L volts), at some time t_1. This is noted by region 1 on the timing chart shown in Fig. 8.43. The voltage V_A in region 1 is thus $V_A = V_L,\ 0 \le t < t_1$.

When V_G is set to V_L volts, the output of gate 1 will drop to zero volts in turn causing the output of gate 2 to change to V_L volts. Since the voltage across the capacitor cannot change instantaneously, the voltage at point A will rise to $2V_L$ volts. The voltage at point A will now exponentially decrease toward zero with

$$V_A = 2V_L e^{-(t-t_1)/RC} \qquad t_1 \le t < t_2 \qquad (8.14)$$

At some time t_2, the voltage at point A will decay to that of the switching threshold of gate 1. The time required, $\tau_1 = t_2 - t_1$, can be found by setting $V_A = V_T$ and solving for τ_1. Thus we have

$$V_T = 2V_L e^{-(\tau_1/RC)} \qquad (8.15)$$

$$\ln\left(\frac{V_T}{2V_L}\right) = -\frac{\tau_1}{RC}$$

$$\tau_1 = RC \ln\left(\frac{2V_L}{V_T}\right). \qquad (8.16)$$

If we let $V_L = 3.0$ volts, $V_T = 1.2$ volts, $R = 400$ ohms, and $C = 250\ \mu$fd,

$$\tau_1 = 0.1 \ln(5) = 0.161 \text{ seconds.}$$

This region of operation is marked by a 2 in Fig. 8.43.

When the voltage crosses the switching threshold, V_T, the output of gate 1 will rise to V_L volts, in turn causing the output of gate 2 to fall to zero volts. This change in voltage at the output of gate 2 will be reflected at point A by a drop in voltage from V_T volts to $(V_T - V_L)$ volts.

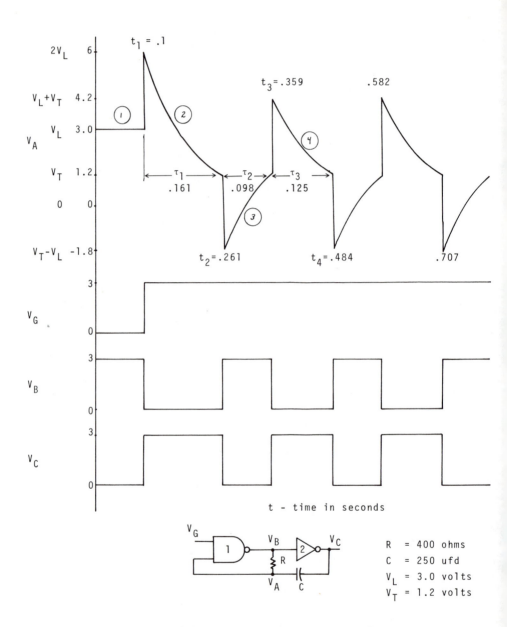

Fig. 8.43 Timing Chart for a Digital Clock

The voltage at point A will now begin to rise toward V_L volts, a rise of $(2V_L - V_T)$ volts with

$$V_A = (V_T - V_L) + (2V_L - V_T)(1 - e^{-[(t-t_2)/RC]})$$
$$= V_L - (2V_L - V_T)e^{-[(t-t_2)/RC]}, \qquad t_2 \leq t < t_3. \tag{8.17}$$

Before the voltage reaches V_L volts it will reach the switching threshold of V_T at some time, t_3. The time required to reach the threshold $\tau_2 = t_3 - t_2$ can be found by setting $V_A = V_T$ in Eq. 8.17 and solving for τ_2. Thus we have

$$V_T = V_L - (2V_L - V_T)e^{-(\tau_2/RC)}$$
$$\tau_2 = RC \ln\left(\frac{2V_L - V_T}{V_L - V_T}\right) \tag{8.18}$$

If we let $V_L = 3.0$ volts, $V_T = 1.2$ volts, $R = 400$ ohms, and $C = 250$ μfd,

$$\tau_2 = 0.1 \ln(2.67) = 0.098 \text{ seconds.}$$

The region between t_2 and t_3 is marked by a 3 in Fig. 8.43.

When the voltage at point A again crosses the switching threshold, the output of gate 1 will drop to zero in turn causing the output of gate 2 to rise to V_L volts. This sudden voltage change will cause the voltage at point A to rise immediately to $(V_T + V_L)$ volts. The voltage at point A will then begin to decay toward zero with the voltage at point A expressed by

$$V_A = (V_T + V_L)e^{-[(t-t_3)/RC]}, \qquad t_3 \leq t < t_4. \tag{8.19}$$

At some time, t_4, the voltage will decay to the switching threshold V_T. The time required $\tau_3 = t_4 - t_3$ can be found by setting $V_A = V_T$ in Eq. 8.19 and solving for τ_3.

Thus we have

$$V_T = (V_L + V_T)e^{-(\tau_3/RC)}$$
$$\tau_3 = RC \ln\left(\frac{V_L + V_T}{V_T}\right) \tag{8.20}$$

If we let $V_L = 3.0$ volts, $V_T = 1.2$ volts, $R = 400$ ohms, and $C = 250$ μfd,

$$\tau_3 = 0.1 \ln(3.5) = 0.125 \text{ seconds.}$$

The voltage curve in this region is marked on Fig. 8.43 by a 4.

At this point the waveshapes for regions 3 and 4 will repeat until the voltage at V_G is again reduced to zero. The steady-state output frequency, F, for this

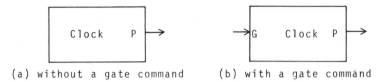

(a) without a gate command   (b) with a gate command

Fig. 8.44 Basic Logic Symbols for a Clock Unit. (a) Without a Gate Command; (b) With a Gate Command

clock is given by

$$F = \frac{1}{\tau_2 + \tau_3},$$

$$F = \frac{1}{RC \ln\left(\dfrac{2V_L - V_T}{V_L - V_T} \cdot \dfrac{V_L + V_T}{V_T}\right)}.$$

Unless $V_T = V_L/2$, the output of the clock will be asymmetric.

For low-frequency operations this equation and analysis model are very good. However, at high frequencies, the effect of a finite gate delay will have to be taken into account. The analysis of such a high-frequency clock model will be left as an exercise. With proper selection of R and C, this clock can operate over a range of fractions of hertz to megahertz.

Because of the inherent frequency instability of the circuits shown in Fig. 8.42, commercial-clock circuits are recommended for use in most critical-control-circuit applications. These clocking units will contain a considerable amount of additional electronics needed to maintain an accurate frequency under a wide variety of operating conditions. The logic symbols we will use to designate a clock circuit are shown in Fig. 8.44.

8.14 SWITCH-INTERRUPT CIRCUITS

In order to start, stop, and provide information to many digital systems, mechanical switches are utilized. Unfortunately, most mechanical switches have a tendency to bounce, i.e., when the switch is transferred from one setting to another, the switch contacts open and close several times before settling into the new switch position. Even though the duration of the switch bounce is small, in most cases only a few milliseconds, these momentary signal excursions can cause erratic operation of a high-speed digital-logic circuit. For example, a binary counter which has as its clock signal a single-pole single-throw switch, will count several dozen times each time the switch is used.

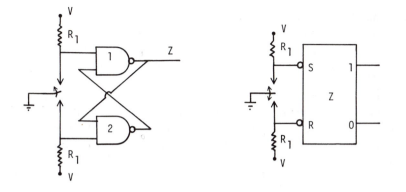

Fig. 8.45 Basic Bounce-Free Switch Circuit

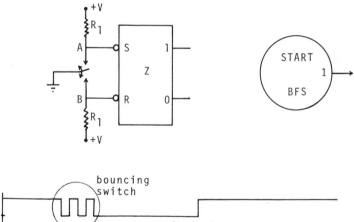

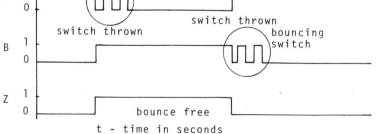

Fig. 8.46 Typical Operation of a Bounce-Free Switch

Fortunately, it is easy to eliminate the unwanted pulses generated by a bouncing switch. Using a pair of NAND gates, a pair of resistors, and any break-before-make single-pole double-throw switch, a relatively inexpensive bounce-free switch-input signal can be obtained. A typical bounce protection circuit is shown in Fig. 8.45.

In this figure we are assuming that the logic units are operating with 0 corresponding to 0 volts and a 1 corresponding to some positive voltage V_L. The NAND-gate circuit is easily recognized as the standard $\overline{S}\overline{R}$ flip-flop studied in Section 8.7. Because of the break-before-make restriction on the switches to be used with this circuit, it will be impossible to apply logical zeroes to both of the inputs simultaneously. With the switch in the lower position, gate 2 will be a one and gate 1 will be a zero. When the switch is changed to the upper position, the output of gate 1 will immediately transition to a one. The bouncing action of the switch will not affect the output Z until the switch is returned to the lower position. Typical operation of this circuit is shown in Fig. 8.46.

It will be assumed from here on that all command switches used to input information to a digital logic system will be bounce-protected.

For simplicity, the switch inputs will be represented by a circle or box surrounding the switch name. The letters BFS will be used to indicate that the switch is bounce-free.

8.15 PROBLEMS

8.1 Apply the input signals for X, A, and B given in Fig. 8.5 to the circuit shown in Fig. 8.3. Compare the resulting timing chart with the timing chart given in Fig. 8.5.

8.2 a) Construct a minimum sum-of-product realization for the logic function given on the K-map below. Assume only the true values of x, y, and z are available. Use NAND gates and assume 1 unit of gate delay per gate.

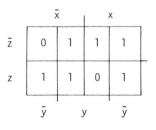

b) Construct a sum-of-product realization for this same logic function using all of the prime implicants.

c) Construct timing charts for both circuits using the signals below.

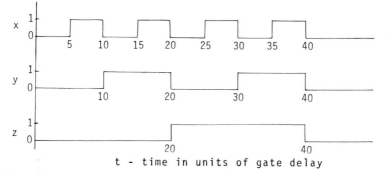

d) Comment on the output signals you obtained.

8.3 Construct timing charts for the XOR gate shown below using the signals given. Assume the gates have 1 unit of delay.

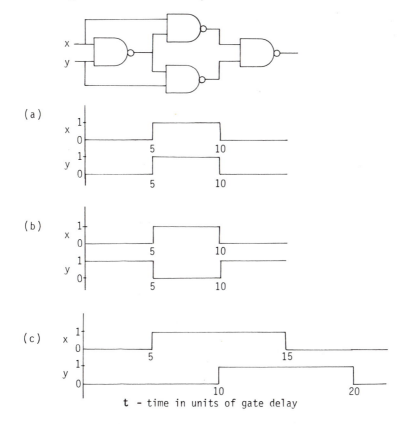

Comment on your observed output patterns.

8.4 Give a simple argument explaining why a circuit constructed using all of its prime implicants will not contain any dynamic hazards assuming only one input variable is allowed to change at a time.

8.5 Analyze the circuits shown below.
   a) Construct a present-state/next-state table, circling the stable states.
   b) Draw a state-transition diagram for the circuit.
   c) Write out a set of time-dependent output equations for your circuits assuming a lumped delay model. (Hint: Use the PS–NS table.)

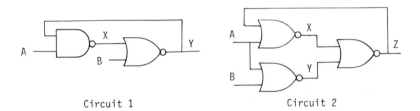

Circuit 1   Circuit 2

8.6 Draw a timing chart for the $\overline{S}\overline{R}$ flip-flop below.

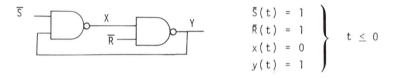

$$\overline{S}(t) = 1$$
$$\overline{R}(t) = 1$$
$$x(t) = 0$$
$$y(t) = 1$$
$$\left.\right\} \quad t \leq 0$$

Assume one unit of gate delay for each gate.

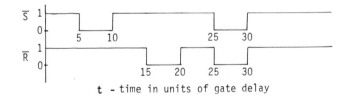

t - time in units of gate delay

8.7 Draw a timing chart for the circuit shown below.

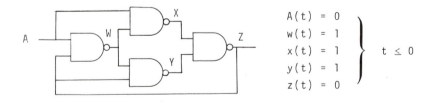

$$A(t) = 0$$
$$w(t) = 1$$
$$x(t) = 1$$
$$y(t) = 1$$
$$z(t) = 0$$
$$\left.\right\} \quad t \leq 0$$

Assume one unit of gate delay for each gate. Comment on your observations.

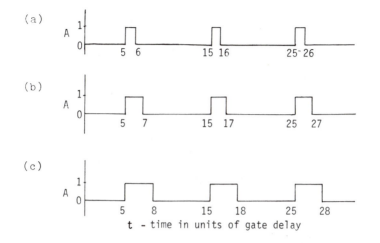

(a)

(b)

(c)

t - time in units of gate delay

8.8 One problem inherent in the basic NOR-gate flip-flop is the fact that when S and R are both set to 1, the outputs are both equal to zero.

a) Modify the basic circuit such that when S and R both equal 1,

$$Z = 1 \quad \text{and} \quad \bar{Z}^* = 0 \text{ (set dominant).}$$

b) Modify the basic circuit such that when S and R both equal 1,

$$Z = 0 \quad \text{and} \quad \bar{Z}^* = 1 \text{ (reset dominant).}$$

Challenge: Can you do it by adding only one gate?

8.9 Analyze the following clocked-SR flip-flop circuits.

a) Write out the next state equation.

b) Construct a present-state/next-state table.

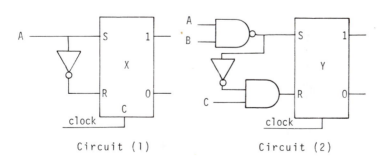

Circuit (1)   Circuit (2)

8.10 Write out the set of next-state equations for the clocked-SR flip-flop circuit shown below. Also construct a present-state/next-state table for the unit.

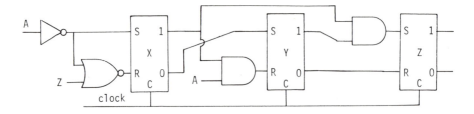

8.11 Using Algorithm 8.2, synthesize the SR flip-flop circuits given by the next-state equations below.

a) $x(t + \Delta) = A(t) \oplus B(t)x(t)$

b) $x(t + \Delta) = A(t)B(t) + \bar{A}(t)\bar{x}(t)$

c) $x(t + \Delta) = A(t)y(t)$

   $y(t + \Delta) = \bar{x}(t) + A(t)$

d) $x(t + \Delta) = A(t)$

   $y(t + \Delta) = x(t)$

   $z(t + \Delta) = y(t)$

e) $x(t + \Delta) = x(t) \oplus [y(t)z(t)A(t)]$

   $y(t + \Delta) = y(t) \oplus [z(t)A(t)]$

   $z(t + \Delta) = z(t) \oplus [A(t)]$

      Draw a state diagram for each of your circuits.

8.12 Analyze the NAND-gate one-shot circuit shown below.

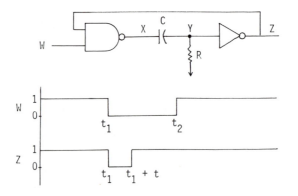

a) Explain how the circuit works.

b) Assuming ideal components, calculate the delay t in terms of V_L, V_T, R, and C.

c) Assuming $R = 500$ ohms,

$$C = 600 \ \mu\text{fd},$$
$$V_T = 1.2 \text{ volts, and}$$
$$V_L = 3.0 \text{ volts}$$

calculate τ.

d) Sketch carefully the voltage waveshape for the values in (c) assuming w is set to zero at $t = 1$ second and that w returns to one at $t = 2$ seconds.

8.13 Assuming $V_T = 1.2$ volts and $V_L = 3.0$ volts
   a) Determine values for R and C for the monostable circuit in Fig. 8.38 which will yield a delay τ of 0.1 seconds.
   b) What value must C be if $R = 250$ ohms, and $\tau = 0.1$ seconds?
   c) What value must R be if $C = 330 \ \mu$fd and $\tau = 0.1$ seconds?

8.14 Assuming $V_T = 1.2$ volts and $V_L = 3.0$ volts
   a) Determine values for R and C for the clock shown in Fig. 8.42 which will yield a clock frequency of 10,000 hertz.
   b) If R is held at 1500 ohms, determine the value of C required in (a) as well as the delay τ_1, τ_2, and τ_3 as indicated in Fig. 8.43.

8.15 Show how a NOR-implemented SR flip-flop can be used to form a bounceless switch.

8.16 BIBLIOGRAPHY

Booth, T. L., *Digital Networks and Computer Systems*. New York: Wiley, 1971.

Boyce, J. C., *Digital Logic and Switching Circuits*. Englewood Cliffs, N.J.: Prentice-Hall, 1975.

Carr, W. N., and J. P. Mize, *MOS/LSI Design and Application*. New York: McGraw-Hill, 1972.

Gill, A., *Introduction to the Theory of Finite State Machines*. New York: McGraw-Hill, 1962.

Hennie, F. C., *Finite-State Models for Logical Machines*. New York: Wiley, 1968.

Hill, F. J., and G. R. Peterson, *Introduction to Switching Theory and Logical Design*. 2nd ed., New York: Wiley, 1974.

Larsen, D. G., and P. R. Rony, *The Bugbook I*. Derby, Conn.: E and L Instruments, 1974.

Larsen, D. G., and P. R. Rony, *The Bugbook II*. Derby, Conn.: E and L Instruments, 1974.

Maley, G. A., *Manual of Logic Circuits*. Englewood Cliffs, N.J.: Prentice-Hall, 1970.

Moore, E. F., *Sequential Machines, Selected Papers*. Reading, Mass.: Addison-Wesley, 1964.

Unger, S. H., *Asynchronous Sequential Switching Circuits*. New York: Wiley-Interscience, 1969.

Chapter 9

Practical
Flip-Flop
Circuits

9.1 INTRODUCTION

In this chapter we will consider the design of reliable flip-flop circuits. We will initially discuss the design of delay toggle and JK flip-flops using the SR flip-flop developed in Chapter 8. We will see that, due to the severe restrictions placed on the clock signal, the conventional SR flip-flop circuit is basically unreliable. However, we will show how the clocking requirements can be relaxed by modifying the basic design. The master-slave and edge-triggered flip-flops developed in Sections 9.6 and 9.7 will be used throughout the remainder of this book.

Special emphasis will be paid to the JK flip-flop in Section 9.8 where formal analysis and synthesis procedures are outlined. The master-slave JK flip-flop and the edge-triggered delay flip-flop will be used a great deal in this book. Finally, in Section 9.10, the use of preset and preclear inputs and their implementation will be discussed.

9.2 THE SR DELAY FLIP-FLOP

A few circuit configurations occur frequently enough to warrant special attention. In this section we will introduce these devices using the SR flip-flop as a building block. We will attempt to point out the special design problems which confront the user of these SR circuit configurations. This material will thereby serve as motivation material for the master-slave and edge-triggered circuit configurations to be presented in the next section.

PI	PS	NS	EI	
D(t)	X(t)	X(t+Δ)	$S_x(t)$	$R_x(t)$
0	0 ✓	0	0	d
0	1	0	0	1
1	0 ✓	1	1	0
1	1	1	d	0

Figure 9.1

Often it is desirable to have a sequential circuit which simply retains the input data value between clock pulses, i.e., a device with the next state equation of

$$X(t + \Delta) = D(t).$$

Such a circuit can be synthesized using a single SR flip-flop as a building block, as shown in Fig. 9.1.

The resulting SR circuit implementation along with an expanded NAND gate circuit for the device is shown in Fig. 9.2. This unit is usually denoted by its own logic symbol as shown in Fig. 9.3. It should be pointed out that this unit tends to follow the input signal, D, as long as the clock line is a one. However, once the clock line is returned to zero, the output remembers (latches on to) the last input value until the clock line is again returned to a logical one.

Since the input can only be passed to the output when the clock is a logical one, the clock line is sometimes referred to as a GATE and this delay flip-flop referred to as a GATED LATCH.

Although the NAND-gate circuit shown in Fig. 9.2 is used, a slight variation of this delay circuit is more frequently used. This modified circuit shown in Fig. 9.4 takes advantage of the special properties of NAND-gate logic to eliminate one gate and still obtain the steady-state values at points A and B of $\overline{DG}$ and

$$\overline{G \cdot \overline{DG}} = \overline{\overline{DG}}.$$

These values are the same as would be obtained using the conventional configuration.

The distributed time-dependent output equation for X can be easily derived assuming all gates have 1 unit of pure delay.

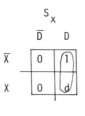

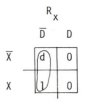

$$S_X(t) = D(t)$$

$$R_X(t) = \overline{D}(t) = \overline{S}_X(t)$$

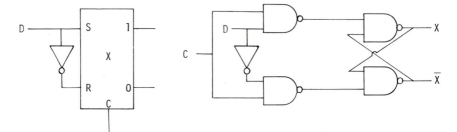

Fig. 9.2 NAND-Gate Delay Flip-Flop

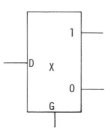

Fig. 9.3 Symbol for a Gated Delay Latch

We have from Fig. 9.4

$$X(t + \tau) = \overline{Y(t)A(t)}$$

$$Y(t + \tau) = \overline{X(t)B(t)}$$

$$X(t + 2\tau) = \overline{Y(t + \tau)A(t + \tau)}$$

$$A(t + \tau) = \overline{D(t)G(t)}$$

$$X(t + 2\tau) = X(t)B(t) + D(t)G(t)$$

$$B(t) = \overline{G(t - \tau)\overline{D(t - 2\tau)G(t - 2\tau)}}$$

Therefore

$$X(t + 4\tau) = X(t + 2\tau)[\bar{G}(t + \tau) + D(t)G(t)] + D(t + 2\tau)G(t + 2\tau). \qquad (9.1)$$

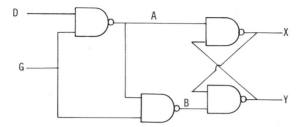

Fig. 9.4 Modified NAND-Gate Delay Latch

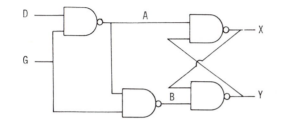

All gates have a
pure delay of
1 unit

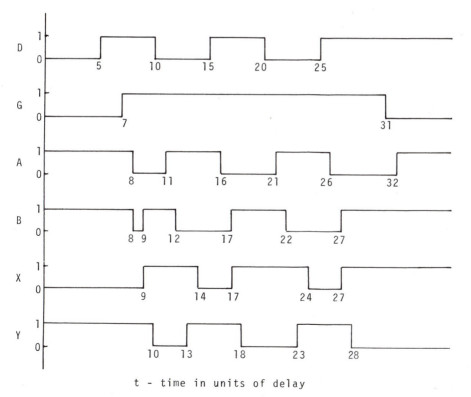

t - time in units of delay

Fig. 9.5 Sample Timing Chart for a Gated Latch

The steady-state condition with $G = 0$ yields $X(t + \Delta) = X(t)$ and the steady-state condition with $G = 1$ and D held constant yields $X(t + \Delta) = D(t)$. Δ indicates the time required for a complete clocking cycle.

Figure 9.5 shows a typical timing chart for a delay latch. From this figure and also from Eq. 9.1 it is easily seen that the gate signal must be held for a period of at least four gate-delay units if proper operation is to be assured.

In Section 9.7 we will show how we can construct delay flip-flops which will sample the D input thereby allowing the D input to change when the clock or gate input is a logical 1 without affecting the output of the flip-flop. Sequential circuit implementation using a delay flip-flop is quite simple once the next state equations are known. There is only one excitation equation $D(t)$ for each flip-flop output $X(t)$ and it is always equal to $X(t + \Delta)$. There are no don't-care terms in the excitation table for a delay flip-flop. See Fig. 9.6.

D(t)	X(t + Δ)	$\overline{X}$(t + Δ)
0	0	1
1	1	0

(a)

$$X(t + \Delta) = D(t)$$

(b)

X(t)	X(t + Δ)	D(t)
0	0	0
0	1	1
1	0	0
1	1	1

(c)

Fig. 9.6 Delay Flip-Flop Information. (a) Logic Description; (b) Next-State Equation; (c) Excitation Table

Example 9.1 Design, using delay flip-flops, a sequential logic circuit which satisfies the set of next-state equations

$$X(t + \Delta) = A(t)X(t) + Y(t)$$

$$Y(t + \Delta) = \bar{A}(t)Y(t) + \bar{X}(t)$$

Solution. Since $Q(t + \Delta) = D(t)$ for a delay flip-flop we have

$$D_x(t) = X(t + \Delta) = A(t)X(t) + Y(t)$$

and

$$D_y(t) = Y(t + \Delta) = \bar{A}(t)Y(t) + \bar{X}(t)$$

which yields the circuit configuration shown in Fig. 9.7.

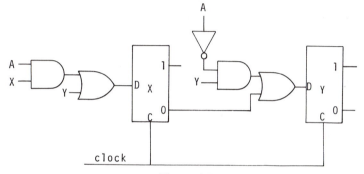

Figure 9.7

9.3 THE TOGGLE OR T FLIP-FLOP

Another commonly used flip-flop, particularly in counting circuits, is the toggle flip-flop shown in Fig. 9.8 which satisfies the next-state equation

$$X(t + \Delta) = \bar{X}(t).$$

On each clock pulse the output of the flip-flop is complemented. The SR implementation of this circuit is straightforward; however, *it is also unreliable.*

Close examination of the timing chart (Fig. 9.9) for the NAND-gate realization of this circuit reveals that the clock pulse must be precisely controlled in order to prevent multiple output changes (oscillations) when the clock line is a one. Again it should be pointed out that, although our circuit model exhibits an oscillatory output, a real circuit probably would return to one of two stable states $(X = 0, Y = 1)$ or $(X = 1, Y = 0)$ following the return of the clock or toggle input to zero. This discrepancy in our model is due to the ideal pure-delay assumption which is not valid for short timing intervals.

One of our basic assumptions in using our SR circuit model was that the S and R input never changes while the clock is high. From our timing chart it is clear that this condition will not hold for all toggle input signals which equal one for longer than 2τ units of gate delay. Unless the toggle input is held at a width of

X(t)	X(t+Δ)	S_x	R_x
0	1	1	0
1	0	0	1

$$S_x(t) = \bar{X}(t)$$
$$R_x(t) = X(t)$$

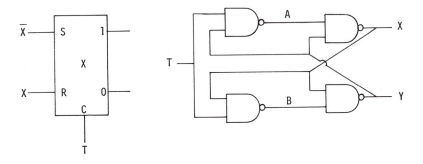

Fig. 9.8 NAND-Gate Realization for a Toggle Flip-Flop Not Reliable

precisely 2τ, the circuit will malfunction. This malfunction, in the form of an oscillatory output response, is shown clearly in Fig. 9.9 for a toggle pulse of 3τ in width. The reader should check that a pulse of 1τ in width will also cause the circuit to malfunction.

In Sections 9.6 and 9.7 we will see how we can eliminate the critical timing requirements from this circuit and thereby implement reliable toggle flip-flops.

9.4 THE JK FLIP-FLOP

The JK flip-flop, one of the most versatile, can be described by the logic table given in Fig. 9.10.

If we compare this table with that of an SR flip-flop (Fig. 8.31), we see that the first 3 rows are the same and that when $J(t) = K(t) = 1$, the device acts as a toggle flip-flop. The time-dependent next-state output equation for the JK flip-flop can be easily obtained from this truth table.

$$X(t + \Delta) = \bar{J}(t)\bar{K}(t)X(t) + J(t)\bar{K}(t) + J(t)K(t)\bar{X}(t)$$
$$= J(t)\bar{X}(t) + \bar{K}(t)X(t). \tag{9.2}$$

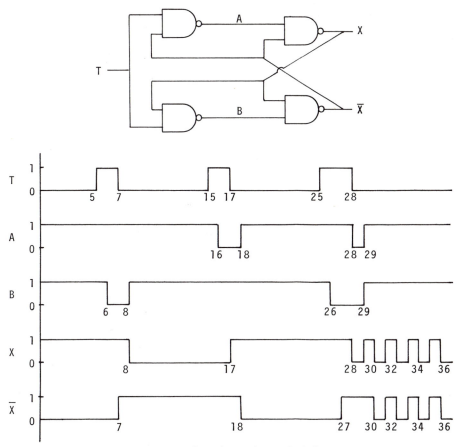

t – time in units of delay

Fig. 9.9 Timing Chart for a Toggle SR Flip-Flop

J(t)	K(t)	X(t+Δ)	X̄(t+Δ)
0	0	X(t)	X̄(t)
0	1	0	1
1	0	1	0
1	1	X̄(t)	X(t)

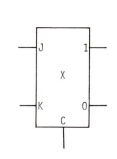

Fig. 9.10 JK Flip-Flop Specification

Present Input		Present State	Next State	Excitation Inputs	
J(t)	K(t)	X(t)	X(t+Δ)	$S_X(t)$	$R_X(t)$
0	0	0	0	0	d
0	0	1	1	d	0
0	1	0	0	0	d
0	1	1	0	0	1
1	0	0	1	1	0
1	0	1	1	d	0
1	1	0	1	1	0
1	1	1	0	0	1

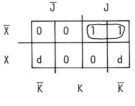

S_X

$$S_X(t) = J(t)\ \overline{X}(t)$$

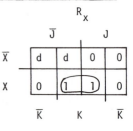

R_X

$$R_X(t) = K(t)\ X(t)$$

Figure 9.11

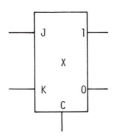

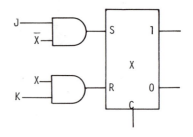

Figure 9.12

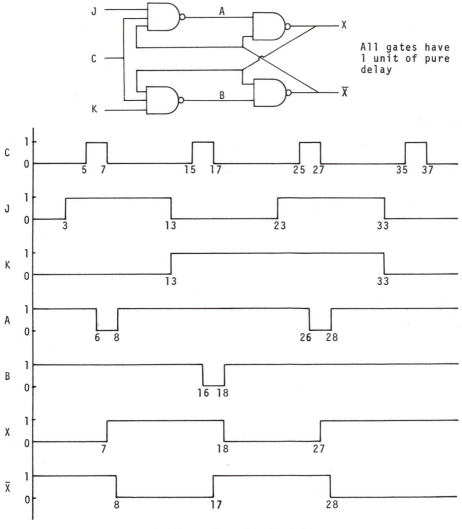

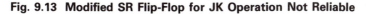

t - time in units of gate delay

Fig. 9.13 Modified SR Flip-Flop for JK Operation Not Reliable

By symmetry we also obtain

$$\bar{X}(t + \Delta) = \bar{J}(t)\bar{X}(t) + K(t)X(t) \tag{9.3}$$

As we have already done with the toggle and delay flip-flops, let us realize this next state equation (Fig. 9.11). These excitation equations lead to the SR circuit realization in Fig. 9.12. An expanded NAND-gate realization for this circuit is shown in Fig. 9.13 along with a typical timing chart. In this circuit the input AND gates have been merged with the input NAND gates of the basic SR flip-flop circuit to conserve parts.

Comparing this circuit with that for the toggle flip-flop shown in Fig. 9.9 reveals that this modified SR-JK flip-flop will have the same timing problem whenever $J = K = 1$ and thus *this circuit configuration is also unreliable.*

9.5 IMPROVED FLIP-FLOP DESIGN TECHNIQUES

In the previous section we encountered a number of timing problems when attempting to design reliable delay, toggle, and JK flip-flops using the basic clocked SR flip-flop as a building block. The major difficulty arose from our inability to prevent the S and R inputs from changing while the clock line was a logical one. There are two ways we can eliminate the undesirable effect of the feedback from the output to the input. The first method we will discuss uses the master-slave principle which essentially isolates the input and output from one another by an additional SR delay flip-flop (the slave).

The second method eliminates the effect of the feedback by using a pair of SR flip-flops to retain the old input values while the clock time is high. This latter technique results in the clock line being effective for only a few units of gate delays. Because of this it is usually referred to as the edge-triggered technique.

Let us now examine each of these design procedures in detail to better understand both how and why commercially available flip-flops operate and also how these same principles can be applied to problems in other areas of logic design where feedback isolation is important.

9.6 THE MASTER-SLAVE PRINCIPLE

The key to the master-slave principle is the operation of the basic clocked-SR flip-flop. When the clock line is a zero, the S and R inputs have no effect on the output. Only when the clock line is a one is the output able to change.

In Fig. 9.14 we have interconnected a pair of clocked SR flip-flops in such a way as to allow the second stage (the *slave*) to copy the contents of the first stage (the *master*) on each clocking cycle.

The inverted clock line is used both to inform the user that the data is transferred on the trailing edge (1 to 0) of the clock signal and also because the data is stored in the slave unit. The output terminals for the master-slave flip-flop ($X, \bar{X}$) are taken from the slave while the input terminals of the master are used as the master-slave SR-input terminals.

The secret to successful operation of this device lies in not synchronously clocking both SR flip-flops, but rather in alternately turning them on and off. This out-of-phase antisynchronization is accomplished by placing an inverter on the

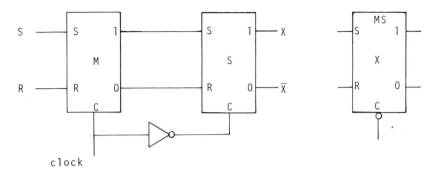

Fig. 9.14 Basic Master-Slave Flip-Flop Configuration and its Logic Symbol

clock line between the master and slave. Thus, when the clock line is a logical one, the master is on and the slave is off. As long as the S and R inputs are held constant and $S \cdot R = 0$ while the clock input is a one, the master will eventually assume the value required by the next-state equation.

Once a steady-state condition is reached, the clock line is returned to a zero thereby turning off the master (freezing its data) and then after one gate delay provided by the clock line inverter, the slave is turned on. Thus after a short delay the slave outputs will copy the data stored in the master. Since the data in the master is fixed, the slave outputs X and $\bar{X}$ can only change once.

Now that the clock line to the master is a zero, we can safely allow the combinational S and R inputs to change in order to properly excite the master flip-flop for the next-state transition. The clock line should be retained at a zero value until the S and R inputs have reached their steady state values. When the clock line is again returned to a logical one, the master flip-flop is first turned on followed by the slave flip-flop being turned off one gate delay later. There is sufficient delay interval in the master flip-flop to prevent the master from responding to input changes before the slave can be turned off. With the slave turned off, the data is again fixed at the outputs of the slave flip-flop.

Since the output is taken from the slave flip-flop and is frozen when the clock transitions from a one to a zero, the logic symbol for a master-slave flip-flop usually shows an inverted clock line as shown in Fig. 9.14. The letters MS at the top of the flip-flop symbol are preferred by the author to indicate that the flip-flop is a master-slave device. This convention is by no means standard. The author uses this as a reminder that the clock line used to sequence these flip-flops is subject to a few minor restrictions.

In general, these devices are driven by a symmetrical square wave signal whose upper frequency limit is dictated by the amount of time required to transfer the data. Typically on the order of 3 to 4 units of gate delays are required for the master and slave. The time needed for the input combinational logic to stabilize must be added to this.

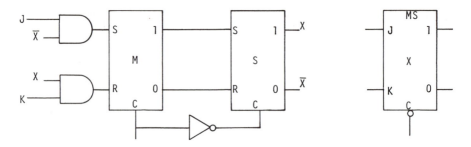

Fig. 9.15 JK Master-Slave Flip-Flop

The synthesis procedures developed in Section 8.11 for the basic SR flip-flop also hold for the master-slave SR flip-flop. The S and R inputs to the master unit should be stabilized prior to sending the clock input to a logical one. Failure to do this can lead to erroneous behavior of the master unit. The types of failure will be illustrated shortly in connection with a study of the master-slave JK flip-flops.

Rather than repeat the derivation of the S and R input-excitation equations needed to convert the basic master-slave SR flip-flop (MS-SR) into a master-slave JK flip-flop (MS-JK), let us use the results determined in Section 9.4.

Thus we have $S_M = J\bar{X}$ and $R_M \doteq KX$. The resulting master-slave implementation is shown in Fig. 9.15, along with the logic symbol we will use to denote a master-slave JK flip-flop. An expanded NAND gate circuit configuration for this unit is shown in Fig. 9.16. In this figure the input AND gates have been merged with the S and R input NAND's of the basic clocked SR NAND flip-flop in order to conserve parts and also reduce the overall logic delay through the two-stage unit.

If we require that J and K be stabilized prior to the clock transitioning to a logical one, then one additional modification can be made to this NAND gate

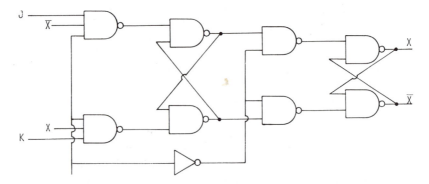

Fig. 9.16 Expanded NAND Gate Circuit for a Master-Slave JK Flip-Flop

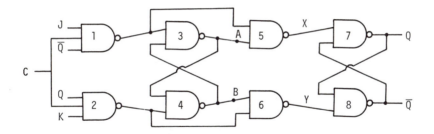

Fig. 9.17 Nonisolated Master-Slave JK Flip-Flop Configuration

circuit. The inverter can be eliminated with the inverted clock inputs on gates 5 and 6 replaced by connections to gates 1 and 2 respectively. The resulting nonisolated master-slave JK flip-flop is shown in Fig. 9.17.

Under the assumption that J and K do not change when the clock input is at a logical one, the basic operation can be explained as follows. At the start of a clock cycle, the clock input is a logical zero holding gates 1 and 2 at logical one. In this mode, the slave (gates 5, 6, 7, and 8) simply copies and holds the information stored in the master. When the clock line is set to a one, *at most* either gate 1 or 2 will be returned to zero depending on the feedback and JK input values.

Ignoring the time dependence for the moment, the steady-state outputs of gates 5 and 6 can be expressed as

$$X = \overline{\overline{J \cdot C \cdot \bar{Q}} \cdot A}$$

and

$$Y = \overline{\overline{K \cdot C \cdot Q} \cdot B},$$

respectively. With $C = 0$, the steady-state values of A and B are Q and $\bar{Q}$ respectively. With $C = 0$, we have

$$X_{\text{steady state}}|_{C=0} = \bar{Q}$$

and

$$Y_{\text{steady state}}|_{C=0} = Q.$$

These values are required to hold the slave flip-flop at Q. When the clock is sent to a logical one, A will assume a new steady-state value of $(J\bar{Q} + \bar{K}Q)$ and B will assume a steady-state value of $(\bar{J}\bar{Q} + KQ)$. Thus with $C = 1$, the new steady-state values of X and Y become

$$X_{\text{steady state}}|_{C=1} = \overline{\overline{J \cdot \bar{Q}} \cdot (J\bar{Q} + \bar{K}Q)}$$
$$= \overline{\overline{K}Q}$$

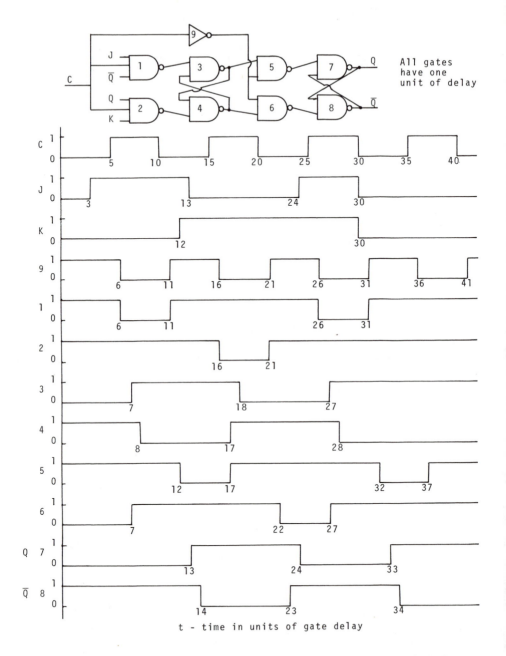

All gates
have one
unit of delay

t - time in units of gate delay

Fig. 9.18 Timing Chart for a Master-Slave Flip-Flop with Inverter Isolation

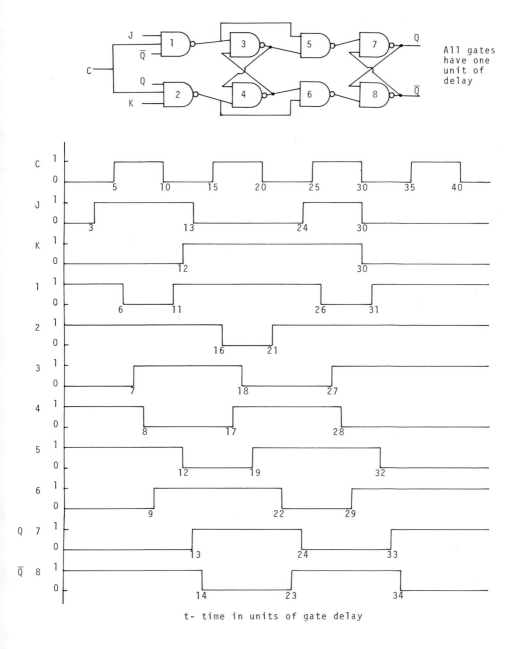

t- time in units of gate delay

Fig. 9.19 Timing Chart for a Master-Slave Flip-Flop without Inverter Isolation

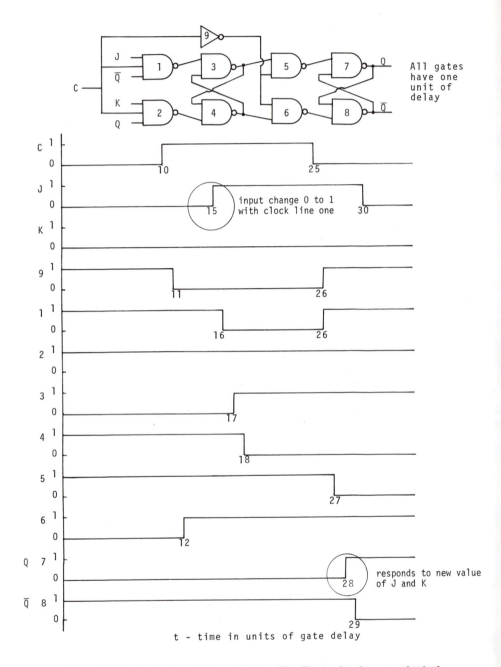

Fig. 9.20 Timing Chart for a Master-Slave Flip-Flop with Inverter Isolation

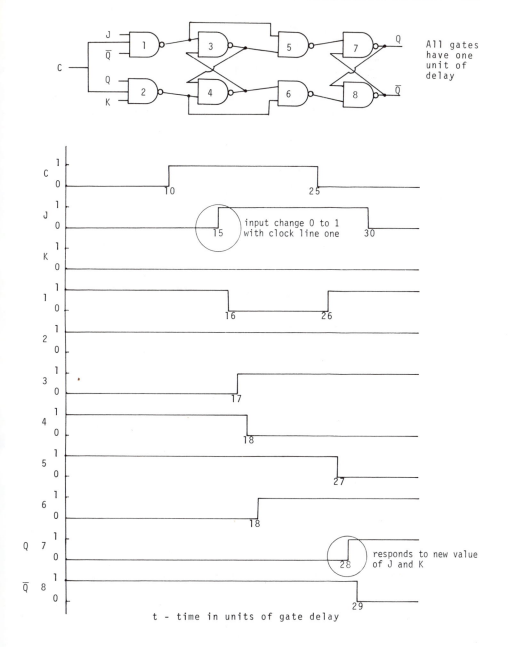

Fig. 9.21 Timing Chart for Master-Slave Flip-Flop without Inverter Isolation

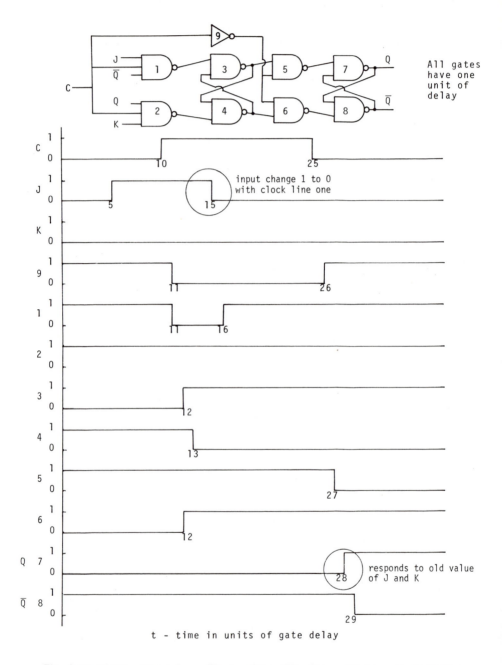

All gates
have one
unit of
delay

input change 1 to 0
with clock line one

responds to old value
of J and K

t - time in units of gate delay

Fig. 9.22 Timing Chart for a Master-Slave Flip-Flop with Inverter Isolation

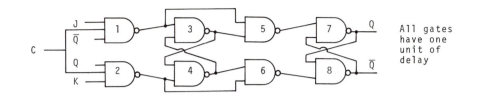

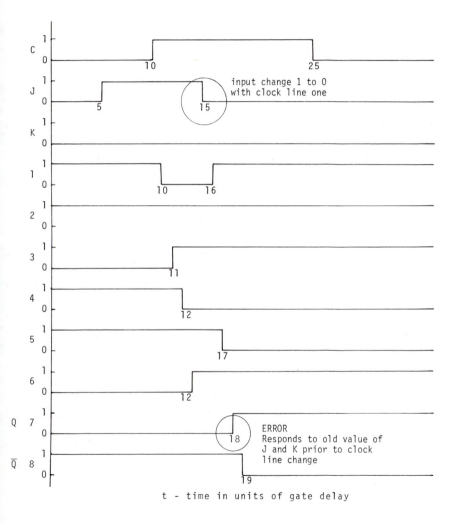

t - time in units of gate delay

Fig. 9.23 Timing Chart for Master-Slave Flip-Flop without Inverter Isolation

and

$$Y_{\text{steady state}}|_{C=1} = \overline{\overline{K \cdot Q} \cdot (\overline{J\bar{Q}} + KQ)}$$
$$= \overline{\overline{J\bar{Q}}}.$$

These values thus hold gate 7 at

$$\overline{\bar{Q} \cdot \overline{\overline{KQ}}} = Q$$

and gate 8 at

$$\overline{Q \cdot \overline{\overline{J\bar{Q}}}} = \bar{Q},$$

thereby preventing the output of gates 7 and 8 from changing while the clock line is a one. When the clock line returns to zero, the slave now copies the new value from the master with

$$Q_{\text{new steady state}} = J\bar{Q}_{\text{old}} + \bar{K}Q_{\text{old}}.$$

Again it must be emphasized, that *these results are only valid provided J and K do not change while the clock line is at a logical one.* Figures 9.18 and 9.19 illustrate the typical behavior of these two master-slave flip-flop configurations. The minimum *data setup time*, t_{setup}, the time required for the data to be transferred into the master, is three units of gate delay. The *input hold time*, t_{hold}, the amount of time the data must be held constant once the clock transitions back to zero, is zero for both of these units. However, three units of delay must be allowed for the transfer of the data from the master into the slave. Thus both of these units have an upper clock-frequency limit of $1/6\Delta$ hertz where Δ is the average gate-propagation delay.

Figures 9.20 and 9.21 reveal that these two flip-flops will correctly respond (i.e., they will respond to the J and K values present at the time the clock line is returned to zero) to input changes from 0 to 1 after the clocking line has been set to a one provided that these inputs are held at one for at least three units of delay prior to the return of the clock line to zero. However, an examination of Figs. 9.22 and 9.23 reveals that an input transition from 1 to 0 after the clock line has been set to one will cause these units to malfunction. The malfunction occurs immediately in the nonisolated MS-JK flip-flop with the unit immediately changing state and on the trailing edge of the clocking line for the isolated MS-JK flip-flop. In this case, the unit responds to the original values on the J and K inputs prior to the clock input being set to a one. These latter two figures again point out the warning to the user not to allow the inputs to a master-slave JK flip-flop to change while the clock line is at one.

9.7 EDGE-TRIGGERED SR FLIP-FLOPS

It is desirable to eliminate the input restriction imposed on the designer using master-slave flip-flops. A useful design would be a JK flip-flop which would allow input changes to take place while the clock input is a one. We will refer to such a

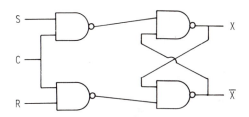

Fig. 9.24 Basic NAND SR Flip-Flop

unit as an edge-triggered flip-flop in order to distinguish it from a master-slave unit.

The logical starting place for the development of an edge-triggered clocked SR flip-flop is the standard clocked SR flip-flop circuit. In this section, we will develop our circuit model using NAND gates. The development of a dual circuit model using NOR gates will be left as an exercise.

The procedure for designing edge-triggered circuits is simple. Let us start with the standard clocked NAND gate SR flip-flop shown in Fig. 9.24 and then systematically convert this circuit into a leading-edge-triggered circuit.

The first step in the transformation is a slight modification of the clocking circuitry by replacing each input NAND gate with a flip-flop.

This change will allow us to remember the initial values of the inputs should they later change to zero while the clock line is at a logical one. The modified circuit configuration is shown in Fig. 9.25. This eliminates the problem noted in the master-slave units.

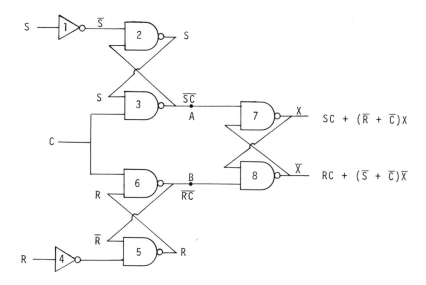

Fig. 9.25 SR Flip-Flop with an Improved Clocking Circuit

It is easily seen that the steady-state behavior of both of these circuits is basically the same, i.e., as long as the S and R input flip-flops or latches are allowed to reach a steady-state condition prior to setting the clock line to a logical one and are not allowed to change while the clock is at one, these two circuits will behave in a similar manner. One basic difference which we will explore is the behavior of the input latches when the S and R inputs are allowed to return to zero while the clock is at one. For unlike the standard SR circuit configuration, the one inputs are remembered until the clock signal is returned to zero. Note also that if at any time while the clock line is a one, the S and R are set to a logical one, points A and/or B will be held at zero until the clock line is returned to zero.

Assuming S is initially a one, both units respond adversely to the change in the value of the reset input from a zero to a one while the clock line is a one, resulting in multiple output changes when the clock line is returned to zero.

A thoughtful examination of Fig. 9.25 reveals that the unwanted output changes can only be caused by the reset input driving point B to a zero. The only way we can prevent NAND gates 3 and 6 from going to zero simultaneously is to maintain a zero input on at least one of the inputs to gates 3 and 6. This can be accomplished by adding an additional feedback line from point A to the input of NAND gate 6 and an additional feedback line from point B to the input of NAND gate 3. Now, if the $S(R)$ input only is initially a one, then the zero output from gate 3(6) will prevent later changes to the $R(S)$ input from adversely affecting the circuit (Fig. 9.26).

Thus as long as either the S or R input is set to a one prior to setting the clock line to a one, the circuit will ignore subsequent changes to either or both inputs while the clock input is held at a one. Now if S is released prior to setting the reset input to a one, the input latch retains the original input value of S prior to clocking and, by means of the feedback line from point A, is able to successfully block the reset input by holding point B at a one.

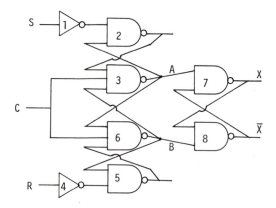

Figure 9.26

This circuit, however, still has two minor flaws. First, if prior to setting the clock line to a one both S and R are equal to zero, then a circuit malfunction will occur if either S or R is subsequently set to a one while the clock line is a logical one. A second but anticipated malfunction will occur if, prior to clocking, both S and R inputs are equal to one. In this latter case, both outputs of the flip-flop will be held at one until the clock is returned to zero, at which time due to the critical race condition, it is difficult to predict the final output state. The pure delay model yields in this case an oscillatory output.

The latter problem is inherent in the logic specification of an SR flip-flop and cannot be avoided without redefining the unit. Therefore avoiding this pitfall is the responsibility of the designer. The use of the excitation table when calculating the excitation equations will help.

The former problem resulting from both inputs being initially at zero can be eliminated by a simple circuit modification. If we replace the set and reset input with the basic next-state equation $S(t) + \bar{R}(t)X(t)$ and $R(t) + \bar{S}(t)\bar{X}(t)$, the feedback terms $\bar{R}(t)X(t)$ and $\bar{S}(t)\bar{X}(t)$ will force gate 4 to zero if $X(t) = 1$ or gate 8 to zero if $X(t) = 0$ when the clock is set to a one. This final circuit configuration shown in Fig. 9.27 can now be considered a reliable edge-triggered SR flip-flop with *the only timing restriction being that the S and R input be maintained constant for 3 gate delay units prior to setting the clock line to a logical one*. Because of the internal feedback paths from gate 4 to gates 3 and 8 and also the feedback paths from gate 8 to gates 4 and 7, the S and R input can be changed simultaneously with the setting of the clock line to a logical one. Hence this unit is indeed an edge-triggered device.

Using the results of the JK flip-flop derivation obtained in Section 9.4, we can easily convert this edge-triggered SR flip-flop into the more versatile JK flip-flop. This is accomplished by replacing the input inverters with NAND gates. Thus S is replaced by $J \cdot \bar{X}$ and R by $K \cdot X$. The resulting edge-triggered JK flip-flop is shown in Figure 9.28.

If we compare this circuit configuration with that for the master-slave JK shown in Figs. 9.16 and 9.17 we can quickly surmise why the latter circuits are more commonly employed in logic circuits. The master-slave configurations require a total of 8 or 9 NAND gates, while the edge-triggered JK flip-flop requires 10 gates. Thus, from a fabrication viewpoint, the master-slave configuration is both cheaper and easier to construct.

However, for those applications where the inputs cannot be held constant while the clock line is a one, then the edge-triggered flip-flop is a good alternative. It should again be pointed out that the addition of gates 2 and 6 on the circuit in Fig. 9.28 were necessary because both the SR and JK flip-flop contain a rest state when both inputs are equal to zero. For sequential circuits which do not allow this rest state to occur, these extra gates are no longer required to hold either gate 3 or gate 7 at a logical one prior to clocking. Thus for sequential circuits where $S \cdot R = 0$ and $S + R = 1$, i.e., where $R = \bar{S}$, we can reduce the edge-triggered circuit complexity to be more competitive with the master-slave circuit configura-

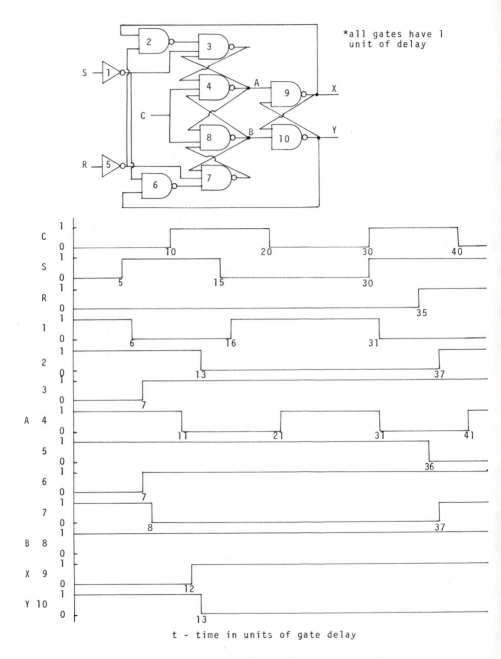

Fig. 9.27 Edge-Triggered SR Flip-Flop and its Timing Chart

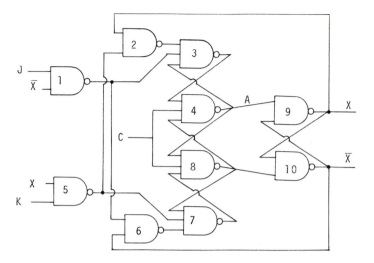

Fig. 9.28 Basic Edge-Triggered JK Flip-Flop

tions. In particular, toggle and delay flip-flops generally favor edge-triggered circuit configurations. Each of these circuits is able to eliminate redundant gating, thereby reducing the required gate count to only 6 NAND gates apiece. The basic circuit realizations and reduced configurations for these special circuits are shown in Fig. 9.29.

The following circuit modifications are used to reduce redundancy in the basic delay flip-flop. Starting with the SR flip-flop in Fig. 9.29(a), an inverter gate can be added to yield a delay flip-flop. However, gates 1, 9, and 4 can be eliminated from the basic circuit of Fig. 9.26(b) by noting that when the clock line is a logical zero, the output of gate 5 equals $\bar{D}$, the required input to gate 2. You should also satisfy yourself that the feedback line from gate 6 to 3 is no longer required.

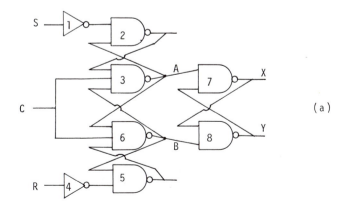

Fig. 9.29 Edge-Triggered Delay-and-Toggle Flip-Flops

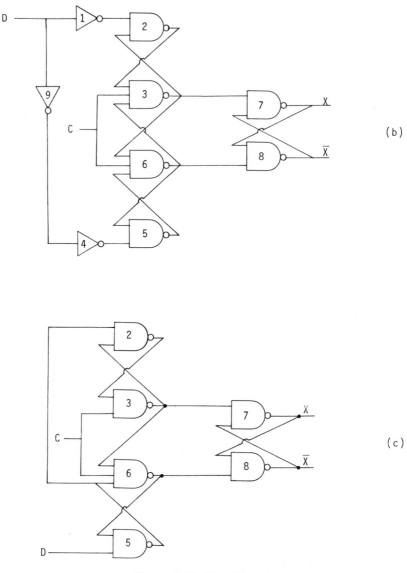

Figure 9.29 (*Continued*)

The basic SR edge-triggered toggle flip-flop shown in Fig. 9.29(d) can be improved by removing the redundant inverters, gates 1 and 4, and using the already available complemented outputs from gates 5 and 6.

Because it is easier to convert a clocked delay flip-flop into an unclocked toggle flip-flop, these toggle flip-flops are generally not readily available. An edge-

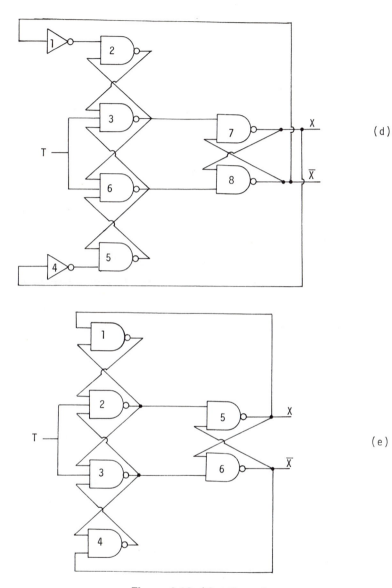

Figure 9.29 (Continued)

triggered delay flip-flop can be converted to a toggle flip-flop by connecting the $\bar{X}$ output into the D input and using the clock input as the toggle input.

For this reason the edge-triggered delay flip-flop (Fig. 9.30) appears to be more versatile. Its low parts count also makes it an attractive and economical building block.

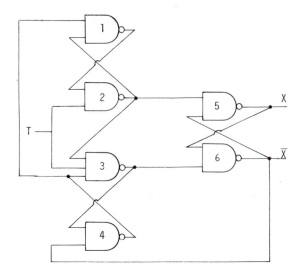

Fig. 9.30 Toggle Flip-Flop Circuit Using an Edge-Triggered Delay Flip-Flop

The conversion of an unclocked toggle flip-flop into a delay flip-flop requires the addition of an XOR gate and an AND gate on the toggle input as shown in the following circuit.

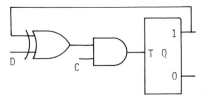

When only a clocked-toggle flip-flop is required, a single AND gate can be added to the T or toggle input and either externally or internally incorporated into the input NAND gates.

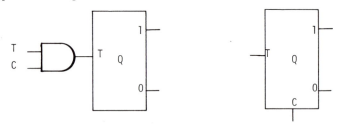

The clocked-toggle flip-flop has the next-state equation of

$$Q(t +\dot{}\Delta) = Q(t) \oplus T(t)$$

which makes both circuit analysis and synthesis quite easy.

Example 9.2 Analyze the sequential circuit below:

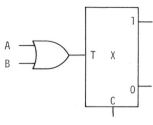

Solution. Using the next-state equation

$$X(t + \Delta) = X(t) \oplus T(t)$$

we have

$$T(t) = A(t) + B(t),$$

therefore

$$X(t + \Delta) = X(t) \oplus (A(t) + B(t))$$
$$= X(t)\bar{A}(t)\bar{B}(t) + \bar{X}(t)A(t) + \bar{X}(t)B(t).$$

Example 9.3 Synthesize a sequential circuit, using clocked-toggle flip-flops, which satisfies the next-state equation

$$X(t + \Delta) = A(t)X(t) + B(t).$$

Solution. Using the general next-state equation

$$X(t + \Delta) = T(t) \oplus X(t)$$

we have

$$T(t) = X(t) \oplus X(t + \Delta)$$
$$= (A(t)X(t) + B(t)) \oplus X(t)$$
$$= B(t)\bar{X}(t) + \bar{A}(t)\bar{B}(t)X(t).$$

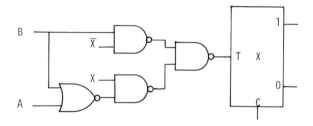

We can conclude from this discussion that both the master-slave principle and the edge-triggered-circuit technique will yield reliable JK delay and toggle flip-flop circuit configurations, with the master-slave principle yielding simpler JK flip-flop circuits and the edge-triggered techniques yielding simpler toggle-and-delay flip-flop configurations. The edge-triggered circuit also tends to be more reliable in a low-frequency environment where there is a lot of electrical noise. This, of course, is due to the low setup-time requirements of these units and to their less severe input requirement.

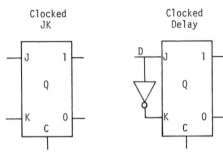
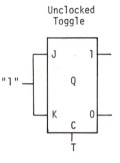

(a)

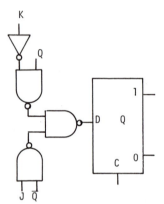
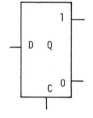
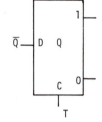

(b)

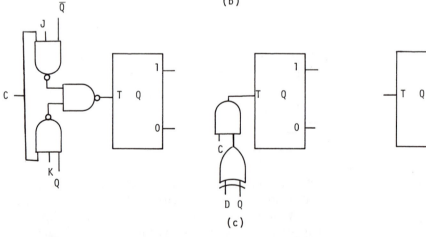

(c)

Fig. 9.31 Circuit Transformations. (a) Using JK Flip-Flops; (b) Using Delay Flip-Flops; (c) Using Toggle Flip-Flops

Finally, it should be noted that the master-slave flip-flop we developed changed state on the 1 to 0 transition or trailing edge, where as the edge-triggered flip-flop changed state on the 0 to 1 transition or the leading edge.

It should also be pointed out that these three basic flip-flops can be easily converted from one to the other by the addition of a small amount of external logic. A table of the required circuit transformation is shown in Fig. 9.31. You should attempt to satisfy yourself as to the validity of these circuits.

9.8 ANALYSIS AND SYNTHESIS OF SEQUENTIAL CIRCUITS USING JK FLIP-FLOPS

Having established reliable circuits for realizing JK flip-flops in the preceding section, let us utilize our knowledge about the operation of these devices to analyze sequential circuits incorporating these devices. A straightforward analysis procedure is given below.

ALGORITHM 9.1

JK Flip-Flop Analysis Procedure

STEP   OPERATION

1.   Determine the combinational logic equations for the J and K inputs.

2.   Obtain the next-state logic equation, assuming a lumped delay model, from the equation:

$$X(t + \Delta) = J(t)\bar{X}(t) + \bar{K}(t)X(t)$$

Example 9.4 Analyze the following sequential circuit.

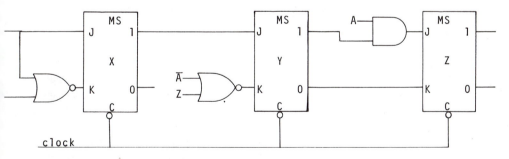

Solution. Using our analysis procedure we have

1.
$$J_x(t) = A(t) \qquad\qquad K_x(t) = \bar{A}(t)Y(t)$$
$$J_y(t) = X(t) \qquad\qquad K_y(t) = \overline{\bar{A}(t) + Z(t)}$$
$$J_z(t) = A(t)\ Y(t) \qquad K_z(t) = \bar{Y}(t),$$

2.
$$Q(t + \Delta) = J_Q(t)\bar{Q}(t) + \bar{K}_Q(t)Q(t)$$
$$X(t + \Delta) = A(t)\bar{X}(t) + A(t)X(t) + \bar{Y}(t)X(t)$$
$$= A(t) + \bar{Y}(t)X(t)$$
$$Y(t + \Delta) = X(t)\bar{Y}(t) + \bar{A}(t)Y(t) + Z(t)Y(t)$$
$$Z(t + \Delta) = A(t)Y(t)\bar{Z}(t) + Y(t)Z(t)$$
$$= A(t)Y(t) + Y(t)Z(t).$$

Let us now turn our attention to the opposite problem, i.e., that of synthesizing a sequential logic circuit using JK flip-flops given a set of next-state logic equations. A general circuit model for realizing arbitrary sequential circuits using clocked JK flip-flops is shown in Fig. 9.32.

This model assumes a number of external binary input signals, $I_1, \ldots, I_k$, and a number of internal-state variables, $X_1, \ldots, X_n$, which correspond to the output terminals of the JK flip-flops. It is also assumed that all of the flip-flops are driven from a common clock line.

Combinational logic circuits are used to provide the necessary excitation signals for the J and K inputs of each flip-flop. These excitation equations are functions of both the present inputs and the present state variables

$$J_{x_i}(t) = F_{J_{x_i}}(I_1(t), I_2(t), \ldots, I_k(t); X_1(t), X_2(t), \ldots, X_n(t))$$
$$K_{x_i}(t) = F_{K_{x_i}}(I_1(t), I_2(t), \ldots, I_k(t); X_1(t), X_2(t), \ldots, X_n(t))$$
$$1 \le i \le n.$$

The circuit model also provides for m binary output signals,

$$Z_i(t), \qquad 1 \le i \le m.$$

These output signals are also functions of both the present input signals and the present-state variables

$$Z_i(t) = F_{Z_i}(I_1(t), I_2(t), \ldots, I_k(t); X_1(t), X_2(t), \ldots, X_n(t)).$$

In order to avoid timing problems, we will require that the external input signals, I_j's be changed only while the clock signal is at a logical zero. We will also require the J_{x_i} and K_{x_i} excitation inputs reach steady-state values prior to setting the clock signal to a logical one.

The reader should be warned that because this circuit model assumes lumped gate delays, we cannot assure hazard-free operation of the circuits. The clocking method utilized by either master-slave or edge-triggered JK flip-flops assures us that the internal state variable will be hazard free. However, the output lines and the input excitation signals Z_i's, J_{x_i}'s, and K_{x_i}'s may and probably will contain logical hazards following changes to either the input variables or the internal state variables.

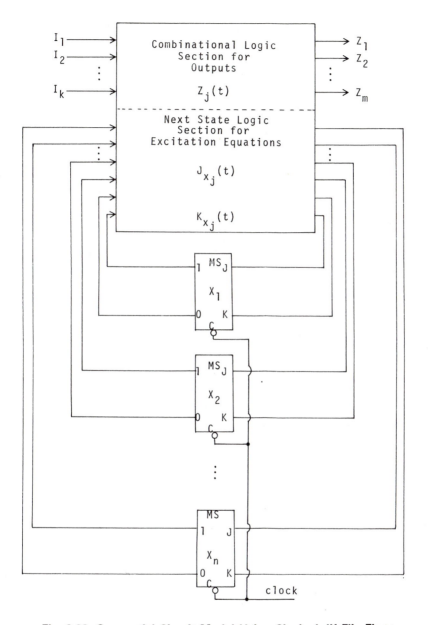

Fig. 9.32 Sequential Circuit Model Using Clocked JK Flip-Flops

This circuit model, along with the assumptions on its utilization, allows us to treat our sequential circuits as networks whose state variables change only at discrete intervals of time controlled by a clocking signal.

9.9 EXCITATION TABLE FOR A JK FLIP-FLOP

Before formalizing a synthesis procedure, let us digress briefly and develop an auxiliary excitation table as was done for the SR flip-flop. Using the expanded logic table (shown in Fig. 9.33) of the JK flip-flop as an aid, let us examine the four possible present-state/next-state situations which can arise. In Fig. 9.34, this present-state/next-state table has been reordered in order to show which input combinations will cause the various state changes.

Simple logic manipulations on data in the input section of the table in Fig. 9.34 yields the desired JK flip-flop excitation table shown in Fig. 9.35.

ALGORITHM 9.2

Synthesis procedure for designing clocked sequential circuits using JK flip-flops from known next-state equations using excitation tables.

STEP   OPERATION

1.   Construct a present-state/next-state table for a circuit to be realized.

2.   Add a pair of excitation columns $J_{Q_i}(t)$ and $K_{Q_i}(t)$ for each internal state variable Q_i. Fill in the entries in these columns using the row pairs $\{Q_i(t), Q_i(t+\Delta)\}$ as table lookup entry points on the JK flip-flop excitation table below.

$Q(t)$	$Q(t+\Delta)$	$J_Q(t)$	$K_Q(t)$
0	0	0	d
0	1	1	d
1	0	d	1
1	1	d	0

3.   Write out the logic equations for each excitation column J_{Q_i}, K_{Q_i} and for each output column Z_i.

4. Implement these excitation and output equations subject to any parts constraint that is imposed. Minimize the logic equation as much as time or resources permit.

5. Check your work against careless errors by analyzing each flip-flop using the general next-state equation

$$Q_i(t + \Delta) = J_{Q_i}(t)\bar{Q}_i(t) + \bar{K}_{Q_i}(t)Q_i(t)$$

and comparing the result against the original equations.

Present Input		Present State	Next State
$J_X(t)$	$K_X(t)$	$X(t)$	$X(t+\Delta)$
0	0	0	0
0	0	1	1
0	1	0	0
0	1	1	0
1	0	0	1
1	0	1	1
1	1	0	1
1	1	1	0

Fig. 9.33 Expanded Present-State/Next-State for a JK Flip-Flop

Present State	Next State	Present Input	
$X(t)$	$X(t+\Delta)$	$J_X(t)$	$K_X(t)$
0	0	0	0
		0	1
0	1	1	0
		1	1
1	0	0	1
		1	1
1	1	0	0
		1	0

Fig. 9.34 Reordered Present-State/Next-State Table for a JK Flip-Flop

X(t)	X(t+Δ)	$J_x(t)$	$K_x(t)$
0	0	0	d
0	1	1	d
1	0	d	1
1	1	d	0

d-don't
care

Fig. 9.35 JK Flip-Flop Excitation Table

Example 9.5 Design a clocked sequential circuit using JK flip-flops which will satisfy the pair of next-state equations and output function:

$$X(t + \Delta) = A(t)X(t) + \bar{Y}(t)$$
$$Y(t + \Delta) = X(t)Y(t) + \bar{A}(t)\bar{Y}(t)$$
$$Z(t) = X(t)Y(t)$$

Solution. Using the next-state equations, we can construct the present-state/next-state table shown in Table 9.1.

The next-state column $X(t + \Delta)$ is filled in by suppressing the time dependence and treating $X(t + \Delta)$ as the output function of a combinational-logic circuit.

The excitation columns $J_x(t)$ and $K_x(t)$ are completed, a line at a time, by using the present-state $X(t)$ and the next-state $X(t + \Delta)$ values in each row as table-lookup entries for the excitation table.

The next-state column $Y(t + \Delta)$ is filled in by suppressing the time dependence and treating $Y(t + \Delta)$ as the output function of a combinational-logic circuit.

The excitation columns $J_y(t)$ and $K_y(t)$ are completed, a line at a time, by using the present-state $Y(t)$ and the next-state $Y(t + \Delta)$ values in each row as table-lookup entries for the excitation table. The excitation columns are then filled in using the excitation table for the JK flip-flop.

We can now map these excitation columns (Fig. 9.36) in order to obtain a reasonably good circuit realization. Thus we obtain $J_x(t) = \overline{Y(t)}$, $K_x(t) = \bar{A}(t)Y(t)$, $J_y(t) = \bar{A}(t)$, and $K_y(t) = \bar{X}(t)$. The reduced output equation is $Z(t) = X(t)Y(t)$. The resulting circuit configuration is shown in Fig. 9.37. As a check on our work

TABLE 9.1

Present Input	Present State		Next State		Present Output	Excitation Values			
$A(t)$	$X(t)$	$Y(t)$	$X(t+\Delta)$	$Y(t+\Delta)$	$Z(t)$	$J_x(t)$	$K_x(t)$	$J_y(t)$	$K_y(t)$
0	0	0	1	1	0	1	d	1	d
0	0	1	0	0	0	0	d	d	1
0	1	0	1	1	0	d	0	1	d
0	1	1	0	1	1	d	1	d	0
1	0	0	1	0	0	1	d	0	d
1	0	1	0	0	0	0	d	d	1
1	1	0	1	0	0	d	0	0	d
1	1	1	1	1	1	d	0	d	0

against careless errors, let us analyze the resulting circuit configuration using the general JK flip-flop next-state equation.

$$Q(t + \Delta) = J_Q(t)\bar{Q}(t) + \bar{K}_Q(t)Q(t)$$

$\therefore$
$$X(t + \Delta) = \bar{Y}(t)\bar{X}(t) + [\bar{Y}(t) + A(t)]X(t)$$
$$= \bar{Y}(t) + A(t)X(t)$$

and

$$Y(t + \Delta) = \bar{A}(t)\bar{Y}(t) + X(t)Y(t)$$

The synthesis procedure above is long and subject to the introduction of careless errors at each step. It would be nice if we could somehow avoid the task of generating the present-state/next-state table. We can accomplish this by expanding the next-state logic equation about the variable $Q(t)$ to obtain

$$Q(t+\Delta) = F(f(t), g(t), Q(t)) = f(t)\bar{Q}(t) + g(t)Q(t)$$

where $f(t)$ and $g(t)$ are no longer functions of $Q(t)$. If we compare this expanded function with the general next-state equation for a JK flip-flop,

$$Q(t + \Delta) = J_Q(t)\bar{Q}(t) + \bar{K}_Q(t)Q(t),$$

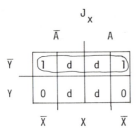

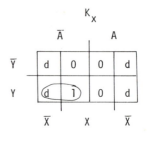

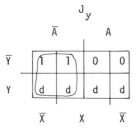

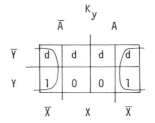

Z

	$\overline{A}$		A	
$\overline{Y}$	0	0	0	0
Y	0	1	1	0
	$\overline{X}$	X	$\overline{X}$	

Figure 9.36

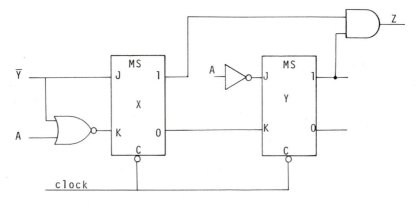

Figure 9.37

it is clear that

$$f(t) = J_O(t)$$

and that

$$\bar{g}(t) = K_O(t).$$

Hence, by simply setting $Q(t) = 0$ in the next-state equation, we can obtain directly the excitation equation for the J_Q input:

$$Q(t + \Delta) = F(f(t), g(t), 0) = f(t) \cdot 1 + g(t) \cdot 0 = f(t)$$
$$Q(t + \Delta) = J_Q(t)\bar{Q}(t) + \bar{K}_Q(t)Q(t) = J_Q(t) \cdot 1 + \bar{K}_Q(t) \cdot 0 = J_Q(t)$$
$$\therefore \qquad f(t) = J_Q(t).$$

By simply setting $Q(t) = 1$ in the next-state equation, we can also obtain directly the excitation equation for the K_Q input.

$$Q(t + \Delta) = F(f(t), g(t), 1) = f(t) \cdot 0 + g(t) \cdot 1 = g(t)$$
$$Q(t + \Delta) = J_Q(t)\bar{Q}(t) + \bar{K}_Q(t)Q(t) = J_Q(t) \cdot 0 + \bar{K}_Q(t) \cdot 1 = \bar{K}_Q(t)$$
$$\therefore \qquad g(t) = \bar{K}_Q(t) \rightarrow \bar{g}(t) = K_Q(t).$$

From the preceding discussion it is clear that it is not necessary to use the cumbersome excitation tables as a synthesis aid when designing sequential circuits utilizing JK flip-flops. The discussion also leads to the following design procedure.

ALGORITHM 9.3

Synthesis procedure for designing clocked sequential circuits using JK flip-flops from known next-state equations using expansions.

STEP   OPERATION

1.   Write the next-state equations for each flip-flop using the expanded form,

$$Q_i(t + \Delta) = F(f_i(t), g_i(t), Q_i(t)) = f_i(t)\bar{Q}_i(t) + g_i(t)Q_i(t).$$

2.   Set $Q_i(t) = 0$ then

$$J_{O_i}(t) = F(f_i(t), g_i(t), 0) = f_i(t)$$

3.   Set $Q_i(t) = 1$ then

$$K_{Q_i} = \bar{F}(f_i(t), g_i(t), 1) = \bar{g}_i(t).$$

4.   Implement these excitation equations along with the output equations subject to any parts constraint imposed, reducing the logic equations as much as time and resources permit.

5.   Check your work against careless errors by analyzing each flip-flop using the general next state equation

$$Q_i(t + \Delta) = J_{Q_i}(t)\bar{Q}_i(t) + \bar{K}_{Q_i}(t)Q_i(t)$$

and comparing the results against the original equations.

Let us illustrate how we can use this design procedure by means of a few examples.

Example 9.6 Design a clocked sequential circuit using a master-slave JK flip-flop which satisfies the next-state equation

$$X(t + \Delta) = A(t)X(t) + B(t)X(t) + A(t)B(t).$$

Solution. Set $X(t) = 0$ and evaluate $X(t + \Delta)$ in order to determine the J_x excitation equation

$$X(t + \Delta)|_{X(t)=0} = J_x(t) = A(t)B(t).$$

Now set $X(t) = 1$ and evaluate $X(t + \Delta)$ in order to determine the $\bar{K}_x$ excitation equation

$$X(t+\Delta)|_{X(t)=1} = \bar{K}_x = A(t) + B(t) + A(t)B(t)$$
$$= A(t) + B(t).$$

Therefore

$$K_x(t) = \overline{A(t) + B(t)} = \bar{A}(t)\bar{B}(t).$$

The resulting circuit configuration is shown in Fig 9.38.

As a check we have

$$Q(t + \Delta) = J_Q(t)\bar{X}(t) + \bar{K}(t)X(t)$$
$$= A(t)B(t)\bar{X}(t) + \overline{\overline{A(t) + B(t)}}X(t)$$
$$= A(t)B(t)\bar{X}(t) + A(t)X(t) + B(t)X(t)$$
$$= A(t)B(t) + A(t)X(t) + B(t)X(t).$$

Let us now consider an example involving coupled next-state equations.

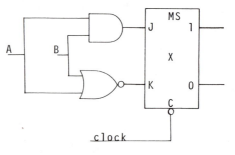

Figure 9.38

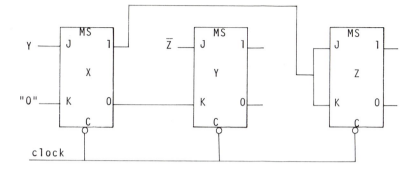

Figure 9.39

Example 9.7 Design a clocked sequential circuit using master-slave JK flip-flops which satisfy the next-state equations:

$$X(t + \Delta) = X(t) + Y(t)$$
$$Y(t + \Delta) = X(t)Y(t) + \bar{Z}(t)\bar{Y}(t)$$
$$Z(t + \Delta) = X(t) \oplus Z(t).$$

Solution. Using the expansion procedure we have:

setting $X(t) = 0$, $X(t + \Delta) = J_x(t) = Y(t)$;

setting $X(t) = 1$, $X(t + \Delta) = \bar{K}_x(t) = 1 \rightarrow K_x(t) = 0$;

setting $Y(t) = 0$, $Y(t + \Delta) = J_y(t) = \bar{Z}(t)$;

setting $Y(t) = 1$, $Y(t + \Delta) = \bar{K}_y(t) = X(t) \rightarrow K_y(t) = \bar{X}(t)$;

setting $Z(t) = 0$, $Z(t + \Delta) = J_z(t) = X(t)$; and

setting $Z(t) = 1$, $Z(t + \Delta) = \bar{K}_z(t) = \bar{X}(t) \rightarrow K_z(t) = X(t)$.

The resulting circuit configuration is shown in Fig. 9.39.

9.10 PRESET AND PRECLEAR INPUTS

There are times when it would be convenient to have a separate set of inputs available on our flip-flop which would allow us to either set the unit to one or clear it to zero without the necessity of a clock pulse.

The circuit shown in Fig. 9.40 shows how these inputs can be added to a simple NAND implemented SR flip-flop.

A zero on the preclear line, $\overline{PC}$, will cause $\bar{Q}$ to transition to a one regardless of the values of the other inputs. A zero on the preset line, $\overline{PS}$, will cause Q to

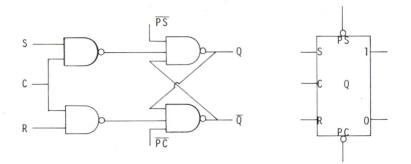

Fig. 9.40 NAND SR Flip-Flop with Preset and Preclear Inputs

transition to a one regardless of the values of the other inputs. The additional inputs act together as a pair of unclocked $\overline{SR}$ inputs. When used, these inputs are subject to the same restrictions as an SR flip-flop.

Since TTL logic circuits reflect a logical one to inputs left open-circuited, the preset and preclear inputs will not affect the operation of the unit if these inputs are not used. Where noise is a problem, these unused inputs would have to be set to a logical one in order to prevent circuit malfunctions. It is also more convenient for the user if these inputs are left complemented. For example, if the unit is to be preset whenever $A(t) \cdot B(t) = 1$, and precleared whenever $C(t) + D(t) = 1$, then a NAND and a NOR gate can be used to provide the added gating. (See Fig. 9.41.) Where more complicated signals are required, the AND-OR-INVERT logic gates can be economically employed. The reader should however be

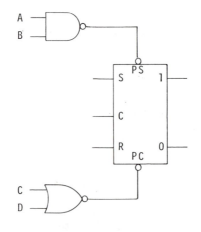

Figure 9.41

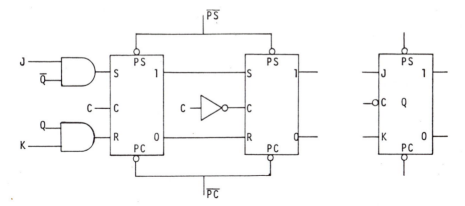

Fig. 9.42. Master Slave Flip-Flop Configuration with Preset and Preclear Inputs.

forewarned that the preset and preclear inputs take precedence over the clocked inputs. Being unclocked, logical hazards may cause erroneous outputs to appear. When operating with master-slave devices, it is necessary to apply the preset and preclear signals to both stages.

If the signals are applied only to the master, they will not affect the slave until the clock signal returns to zero. It is also possible for the feedback lines from the slave, in conjunction with the original J and K inputs, to restore the old value to the master once the preset and preclear signals are removed.

If the signals are only applied to the slave, it is possible for the master to reset the slave to its original value prior to the establishment of a new equilibrium condition in the master. By simultaneously applying the signal to both the master and slave, reliable operation is assured. A typical circuit configuration is shown in Fig. 9.42.

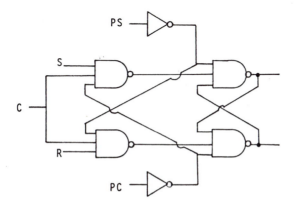

Fig. 9.43 Improved Preset-Clear SR Flip-Flop Configuration

As a final warning, the reader should note that the preset and preclear inputs act as unclocked set-reset inputs and their use must follow all the guidelines given for unclocked SR flip-flops. In general, the preset and preclear input should only be used when the clock signal is equal to zero, thereby eliminating any possible interference from the clocked inputs.

However, if this is not possible, then the circuit shown in Fig. 9.43 should be used in order to isolate the clocked inputs from the preset and preclear lines.

9.11 PROBLEMS

9.1 One method of implementing a toggle flip-flop is to connect the 0 output of a delay flip-flop to its data input. The clock input then becomes the toggle input line. Can the gated latch circuit shown in Fig. 9.3 be successfully converted into a toggle flip-flop? Investigate the timing requirement on the gate input. *Hint*: Construct several timing charts for this unit using varying width pulses.

9.2 Your boss has purchased a large supply of master-slave JK flip-flops. Unfortunately he is not sure whether the units have inverter isolation. Devise a simple test procedure which can quickly determine if the flip-flops are configured like the one in Fig. 9.16 or like the one in Fig. 9.17.

9.3 Construct a timing chart for the modified SR flip-flop shown in Fig. 9.25. The device carries a warning that the S and R inputs are not to be changed while the clock line is a one. Comment on your observed output patterns. Assume all gates have one unit of gate delay. Assume $x(t) = 0$, and $x(t) = 1$ for $t < 0$.

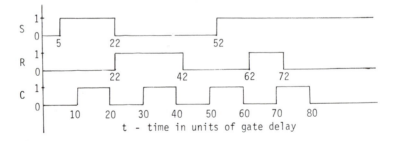

9.4 Construct a timing chart for the edge-triggered JK flip-flop shown in Fig. 9.23. Assume $x(t) = 1$ and $\bar{x}(t) = 0$ for $t < 0$, and that all gates have 1 unit of pure delay.

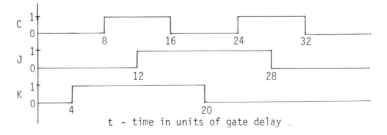

9.5 Construct a timing chart for the edge-triggered delay flip-flop shown in Fig. 9.29d. Assume $x(t) = 1$, and $\bar{x}(t) = 0$ for $t < 0$, and that all gates have one unit of pure delay.

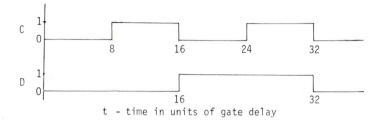

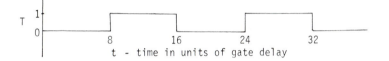

t - time in units of gate delay

9.6 Construct a timing chart for the edge-triggered toggle flip-flop shown in Fig. 9.29e. Assume $x(t) = 1$ and $\bar{x}(t) = 0$ for $t < 0$. Assume all gates have one unit of pure delay.

t - time in units of gate delay

9.7 Repeat problem 9.6 using the edge triggered toggle flip-flop shown in Fig. 9.30.

9.8 Modify the basic SR master-slave flip-flop to obtain
  a) a clocked-toggle flip-flop,
  b) a clocked-delay flip-flop.
  Show an expanded NAND gate configuration for each unit.

9.9 Determine the next-state equations for the following clocked flip-flops.

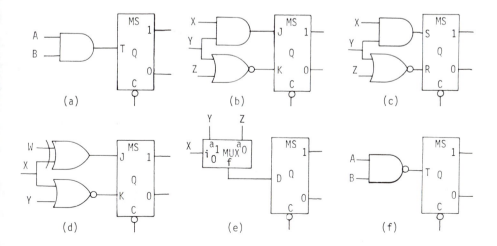

9.10 Synthesize sequential circuits using clocked edge-triggered toggle flip-flops which realize the following next-state equations.

   a) $x(t + \Delta) = A(t) + \bar{x}(t)B(t)$

   b) $y(t + \Delta) = A(t)y(t)$

   c) $z(t + \Delta) = A(t)B(t) + \bar{z}(t)$

9.11 Synthesize sequential circuits using clocked edge-triggered delay flip-flops which realize the following next-state equations.

   a) $x(t + \Delta) = A(t) + x(t)$

   b) $x(t + \Delta) = \bar{A}(t)B(t) + \bar{x}(t)A(t)$

9.12 Determine the next-state equations for the circuit below. Also, construct a present-state/next-state table and a state diagram for this circuit.

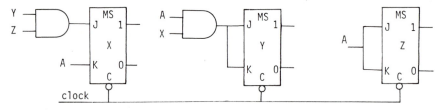

9.13 Implement, using Algorithm 9.2, the following next-state equations.

   a) $x(t + \Delta) = \bar{x}(t)A(t) + B(t)$

   b) $x(t + \Delta) = x(t) \oplus A(t)$

   c) $x(t + \Delta) = x(t)y(t)$

   $y(t + \Delta) = [A(t) + \bar{y}(t)]\bar{x}(t)$

   d) $x(t + \Delta) = y(t) \oplus x(t)$

   $y(t + \Delta) = \bar{A}(t) + \bar{x}(t)$

9.14 Implement the following set of next-state equations using JK flip-flops and Algorithm 9.3.

   a) $x(t + \Delta) = A(t)B(t) + \bar{x}(t)$

   b) $x(t + \Delta) = A(t) + B(t)$

   c) $x(t + \Delta) = A(t) + x(t)$

   d) $x(t + \Delta) = A(t) \oplus x(t)$

   e) $x(t + \Delta) = A(t)$

   $y(t + \Delta) = A(t)x(t)$

   f) $x(t + \Delta) = y(t)z(t)$

   $y(t + \Delta) = A(t) \oplus x(t)z(t)$

   $z(t + \Delta) = \bar{A}(t)\bar{x}(t) + y(t)z(t)$

g) $x(t + \Delta) = \bar{x}(t)$

     $y(t + \Delta) = \bar{x}(t)\bar{y}(t) + z(t)$

     $z(t + \Delta) = x(t)y(t)z(t) + \bar{x}(t)\bar{y}(t)\bar{z}(t)$

9.15   Modify the basic inverter isolated master-slave JK flip-flop shown in Fig. 9.16 by adding reliable preset and preclear inputs that can be operated independent of the clock.

9.16   Add preset and preclear inputs to the basic edge-triggered delay flip-flop (Figure 9.29c) and to the basic edge-triggered toggle flip-flop (Fig. 9.29e).

9.17   A young engineer has been asked to design a sequential circuit to realize the following set of next-state equations

$x(t + \Delta) = \bar{y}(t),$

$y(t + \Delta) = x(t)\bar{y}(t).$

The breadboard circuit is shown below.

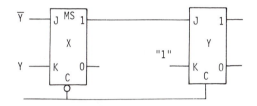

Flip–flop x is a master-slave JK with inverter isolation and flip-flop y is an edge-triggered unit. The clock line is a low-frequency square wave. The circuit when tested acts rather strangely. Can you explain why and show how this circuit can be modified to eliminate the design flaw. *Hint*: Reread the material on the effects of changes to the J and K inputs to a master-slave flip-flop when the clock line is high. Try sketching a timing chart for the unit.

9.18   You have been given the task of designing a sequential circuit to form a two-phase clock for a special control application. The input signal is a square wave. The desired output is shown as $\phi 1$ and $\phi 2$, the two phases of the new clocking signals.

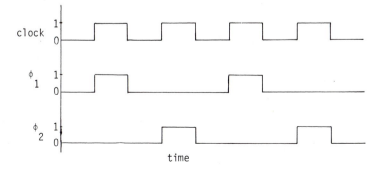

a) Use master-slave JK flip-flops.

b) Use edge-triggered delay flip-flops.

9.19 You have been given the task of designing a wrong direction indicator for a highway exit ramp. Assuming the exiting cars are moving left to right, design a sequential circuit which will output an alarm signal if a vehicle attempts to enter the exit ramp. The alarm should remain on for a period of approximately 5 seconds. State your assumptions and assume that any vehicle detectors you use provide bounceless signals.

Hint. Consider placing two or more sensors across the road, with the space between the sensors being short enough so that a vehicle cannot fit completely between them.

9.20 You have been asked to design a new electronic toy, which operates as follows:

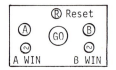

The unit is to consist of two bounceless switches *A* and *B* and a bounceless reset switch *R*.

RULES. When the reset button is depressed the "win" lights and the "go" light are to be turned off. After a delay of approximately 5 seconds, the "go" light is to be turned on.

Once the "go" light is on, the first person to depress (his) (her) switch is the winner.

However, depressing your switch before the "go" light comes on will set your opponent's "win" light on.

The result is to be displayed until the reset button is depressed.

Hint. Use a one shot to generate the "go" signal.

9.12 BIBLIOGRAPHY

Beraru, J., "A one-step process for obtaining flip flop input logic equations." *Computer Design*, Vol. 6, No. 10, pp. 58–63, Oct. 1967.

Booth, T. L., *Digital Networks and Computer Systems.* New York: Wiley, 1971.

Chu, Y., *Digital Computer Design Fundamentals.* New York: McGraw-Hill, 1962.

Larsen, D. G., and P. R. Rony, *The Bugbook I.* Derby, Conn.: E & L Instruments, 1974.

Larsen, D. G., and P. R. Rony, *The Bugbook II.* Derby, Conn.: E & L Instruments, 1974.

Maley, G. A., *Manual of Logic Circuits.* Englewood Cliffs, N.J.: Prentice-Hall, 1970.

Marcus, M. P., *Switching Circuits for Engineers.* Englewood Cliffs, N.J.: Prentice-Hall, 1970.

Millman, J., and H. Taub, *Pulse, Digital and Switching Waveforms.* New York: McGraw-Hill, 1965.

Texas Instruments, *The Integrated Circuits Catalog for Design Engineers.* No. CC401, Dallas: Texas Instruments, Inc., 1971.

Woollons, D. J., *Introduction to Digital Computer Design.* New York: McGraw-Hill, 1972.

Chapter 10

Binary Counters

10.1 INTRODUCTION

In this chapter, digital counting circuits will be introduced. In Section 10.2, we describe the basic asynchronous counting cell. This cell will be interconnected to form ripple counters. In Section 10.3, synchronous binary counters will be discussed. In this section, modular N counters will also be introduced. The design of arbitrary number counters will be discussed in Section 10.4. These units which are particularly important when used as control counters for complex digital systems will be synchronously reset. In Section 10.5, we will discuss direct resetting Modulo-N counters. These units, while synchronously counting, will be asynchronously reset. Section 10.6 will introduce binary down counters. Certain special counters, including switch-tail counters, will be discussed in Section 10.7. Finally, in Section 10.8, we will combine all the material and show how one could design a simple digital stop watch.

10.2 A BASIC BINARY COUNTING CELL

A careful examination of either a table of binary numbers (similar to that in Fig. 10.1) or of the set of idealized wave forms representing a sequence of these numbers (similar to those in Fig. 10.2) will reveal several interesting patterns.

First of all, as one would expect for a binary number system, the frequency of signal changes for variable n_j is exactly half that of variable n_{j-1}. A second noteworthy observation is that variable n_j always changes value whenever variable n_{j-1} changes from a one to a zero.

Lastly, it should be noted that variable n_j never changes value when variable n_{j-1} changes from a zero to a one, or while the variable n_{j-1} remains at a fixed logic level.

$(N)_{10}$	$(n_3\ n_2\ n_1\ n_0)_2$
0	0 0 0 0
1	0 0 0 1
2	0 0 1 0
3	0 0 1 1
4	0 1 0 0
5	0 1 0 1
6	0 1 1 0
7	0 1 1 1
8	1 0 0 0
9	1 0 0 1
10	1 0 1 0
11	1 0 1 1
12	1 1 0 0
13	1 1 0 1
14	1 1 1 0
15	1 1 1 1

$$N_{10} = \sum_{j=0}^{3} n_j \cdot 2^j$$

Fig. 10.1 Table of Binary Numbers

Combining these observations leads one to the conclusion that a simple binary counting cell requires a logic circuit which can discriminate between the leading and trailing edge of an input signal, allowing the output to change value only after the input signal returns from a logical one to a logical zero, that is on the trailing edge of the input signal.

The master-slave flip-flop developed in Chapter 9 inherently exhibits this property. The edge-triggered toggle flip-flop also exhibits this property when modified by adding an inverter to the toggle input line.

A master-slave JK flip-flop can be converted to act as a basic counting cell (as a trailing-edge-sensitive toggle flip-flop) by either feeding back $\bar{x}$ into J and x into K or by placing both J and K inputs at logical one, thereby converting the JK flip-flop directly into a master-slave toggle flip-flop.

A third alternative would be the direct connection of the input A to both J and K as well as to the clock input. The unit is able to correctly operate in this mode because the signal A must return from a one to a zero in order to complete the master-slave clocking cycle. The input A, while at a logical one, causes the master to assume the value of $\bar{x}$. When A is returned to zero the inputs to the master are removed and simultaneously the value of $\bar{x}$ stored in the master is transferred into the slave unit. These toggle flip-flop circuits are shown in Fig. 10.3.

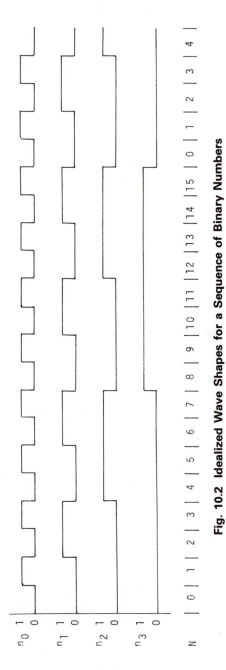

Fig. 10.2 Idealized Wave Shapes for a Sequence of Binary Numbers

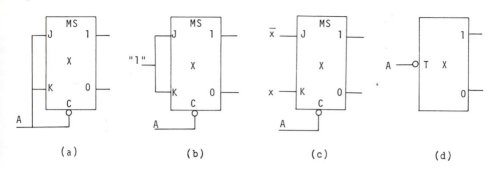

Fig. 10.3 Trailing Edge Toggle Flip-Flop Circuits

In Chapter 9 a six-gate edge-sensitive toggle flip-flop was developed which triggered on the leading edge of the clock signal rather than the trailing edge.

Figure 10.4 shows a three-stage ripple-counter implementation using these units. The propagation delay for this toggle flip-flop was established in Chapter 9 with 2 units on a 0 to 1 transition and 3 units on a 1 to 0 transition. Rather than adding inverters to the T inputs, the zero output of the preceding stage is used.

A second variation noted in this figure is the fact that this unit has been wired to count complete clock cycles as opposed to counting input-level changes. This amounts to adding an additional cell to our unit and ignoring the input X as an output variable. The first cell changes state on the leading edge of the input signal.

10.3 SYNCHRONOUS BINARY COUNTERS

In the following sections, we will focus our attention on a few special design techniques which will allow us to specify and construct reliable binary counting devices. For preliminary discussion purposes we will restrict ourselves to simple binary counters which utilize either master-slave or edge-triggered flip-flops as basic building blocks. We will initially consider only autonomous counters, i.e., we will investigate the restricted class of counters whose next counter state depends only on the present counter state. We will investigate counters without this restriction in the next section. Each clocking signal will cause the counter to change to a new state. The maximum number of states, i.e., the largest number which the counter can retain before repeating a state, will depend on the number of flip-flops and the hardware organization employed.

If all possible states of a sequential circuit with k flip-flops are used as counter states then the counter will have a period of 2^k. However, it is not necessary that all of the internal states of a sequential circuit be used as counter states. For example, if we were to construct a counter with a period of 10, i.e., a decimal counter, using four flip-flops as memory elements, then six of the sixteen internal

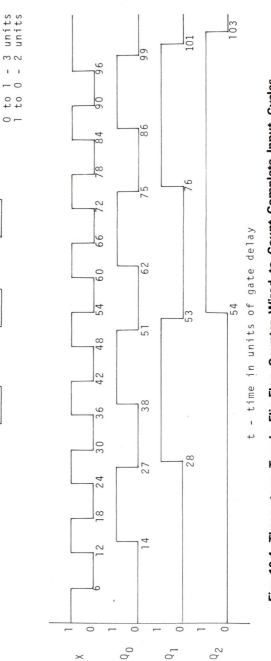

Fig. 10.4 Three-stage Toggle Flip-Flop Counter Wired to Count Complete Input Cycles

states available would not be used as counting states. We will see shortly that these excess internal states must be handled in a special way if one is to construct reliable counters.

Due to the finite counting ability of our hardware-implemented binary counters, a sequential circuit having N counter states is also referred to as being a modulo-N counter or simply a mod-N counter.

The term *molulo N* has been borrowed from modern algebra. For our purposes the modulo-N operator will convey the following meaning. *K modulo N is the positive integer remainder which results when the integer K is divided by the positive integer N.* A modulo-N counter is also commonly referred to as a *divide by N* unit.

A modulo-N binary counter thus has the property that it can unambiguously count a series of events that occurs at most $N-1$ times. The nth-count pulse causes the counter to return to the state representing zero or no occurrences of the event. The event being counted in this case will be the clock signal.

Assuming we are able to place an upper limit, M, on the number of occurrences of an event, then we can also fix a lower limit, K, on the number of flip-flop storage elements required to implement such a counter.

K will be the integer which satisfies the inequalities

$$\log_2 M \le K < \log_2 M + 1,$$

i.e., K is the smallest integer such that $M \le 2^K$. At least ten flip-flops would be required in order to implement a binary counter to count up to $(1000)_{10}$.

As a starting point, let us restrict N to be a power of 2, i.e., $N = 2^K$ for some K. This class of counters will require only K flip-flops.

Let us consider the design of a modulo-16 binary counter with the added restriction that all flip-flops change state simultaneously. The synchronizing signal will be obtained by using a common clock. By restricting the counter to change states only on each complete clock cycle, the counting signal can also be used as the clock input. We will refer to counters which operate with a common clock as being *synchronous* counters.

The starting point for our design can be either a state diagram or a present-state next-state table description of the counter. Using the present-state next-state table, a set of next-state equations can be obtained from which one can apply the sequential-circuit-synthesis procedures developed in Chapter 9.

A present-state/next-state table for a modulo-16 binary counter is shown in Fig. 10.5.

The next-state equations for this counter can be obtained from this table with the aid of K-maps. See Fig. 10.6.

Upon careful examination of these next-state equations a pattern becomes apparent. First, if we number our variables from the least significant to the most significant, then it can be observed that the next-state equations for variable Q_i depend only on Q_i and the variables of lower significance. A general set of

Present State				Next State			
$Q_3(t)$	$Q_2(t)$	$Q_1(t)$	$Q_0(t)$	$Q_3(t+\Delta)$	$Q_2(t+\Delta)$	$Q_1(t+\Delta)$	$Q_0(t+\Delta)$
0	0	0	0	0	0	0	1
0	0	0	1	0	0	1	0
0	0	1	0	0	0	1	1
0	0	1	1	0	1	0	0
0	1	0	0	0	1	0	1
0	1	0	1	0	1	1	0
0	1	1	0	0	1	1	1
0	1	1	1	1	0	0	0
1	0	0	0	1	0	0	1
1	0	0	1	1	0	1	0
1	0	1	0	1	0	1	1
1	0	1	1	1	1	0	0
1	1	0	0	1	1	0	1
1	1	0	1	1	1	1	0
1	1	1	0	1	1	1	1
1	1	1	1	0	0	0	0

Fig. 10.5 Present-State/Next-State Table for a Synchronous Modulo-16 Binary Counter

next-state equations for a modulo-2^N counter following the observed pattern for the modulo-16 counter would be written as

$$Q_0(t + \Delta) = 1 \oplus Q_0(t)$$
$$Q_1(t + \Delta) = Q_0(t) \oplus Q_1(t) \quad (10.1)$$
$$Q_i(t + \Delta) = \left[\bigwedge_{j=0}^{i-1} Q_j(t)\right] \oplus Q_i(t)$$

for

$$1 < i < N \quad (10.2)$$

Each of these equations is of the form $X(t + \Delta) = A(t) \oplus X(t)$, where $A(t)$ does not depend on $X(t)$. Following the synthesis procedure developed for sequential circuits using clocked master-slave JK flip-flops, we have

$$J(t) = X(t + \Delta)|_{X(t)=0} = A(t)$$

and

$$\bar{K}(t) = X(t + \Delta)|_{X(t)=1} = \bar{A}(t) \to K(t) = A(t) = J(t)$$

Thus a general set of excitation equations for a modulo-2^N binary counter can

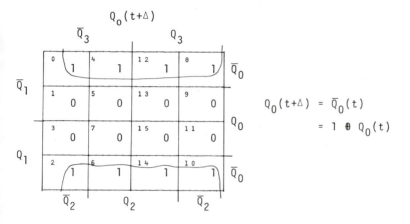

$$Q_0(t+\Delta) = \bar{Q}_0(t)$$
$$= 1 \oplus Q_0(t)$$

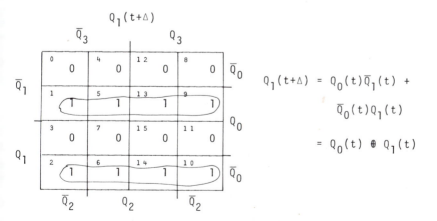

$$Q_1(t+\Delta) = Q_0(t)\bar{Q}_1(t) +$$
$$\bar{Q}_0(t)Q_1(t)$$
$$= Q_0(t) \oplus Q_1(t)$$

Figure 10.6

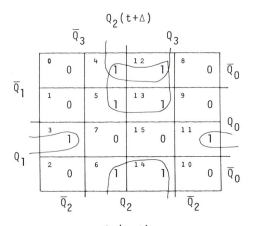

$$Q_2(t+\Delta) = Q_2(t)\overline{Q}_1(t) +$$
$$Q_2(t)\overline{Q}_0(t) +$$
$$\overline{Q}_2(t)Q_1(t)Q_0(t)$$
$$= [Q_1(t)Q_0(t)] \oplus Q_2(t)$$

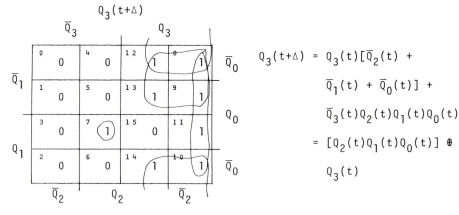

$$Q_3(t+\Delta) = Q_3(t)[\overline{Q}_2(t) +$$
$$\overline{Q}_1(t) + \overline{Q}_0(t)] +$$
$$\overline{Q}_3(t)Q_2(t)Q_1(t)Q_0(t)$$
$$= [Q_2(t)Q_1(t)Q_0(t)] \oplus$$
$$Q_3(t)$$

Figure 10.6 *(Continued)*

be expressed as

$$J_{Q_0}(t) = K_{Q_0}(t) = 1$$
$$J_{Q_1}(t) = K_{Q_1}(t) = Q_0(t)$$
$$J_{Q_i}(t) = K_{Q_i}(t) = \bigwedge_{j=0}^{i-1} Q_j(t)$$

for

$$1 < i < N.$$

In particular, for a modulo-16 counter, we have

$$J_{Q_0}(t) = K_{Q_0}(t) = 1,$$
$$J_{Q_1}(t) = K_{Q_1}(t) = Q_0(t),$$
$$J_{Q_2}(t) = K_{Q_2}(t) = Q_1(t)Q_0(t)$$

and

$$J_{Q_3}(t) = K_{Q_3}(t) = Q_2(t)Q_1(t)Q_0(t).$$

Two sequential circuits realizing these equations are shown in Fig. 10.7.

An additional output M (Max count) has been added to each of the circuits. Its purpose is to indicate when the counter has reached its maximum value.

We can make a relative comparison of the upper frequency limit for these two counters. Let us assume we are using master-slave JK flip-flops similar to those shown in Chapter 9. These units have a setup time Δ_S of 3Δ and a transfer delay Δ_T of 3Δ. The AND gates have a propagation delay Δ_{AG} of 2Δ, Δ being the propagation delay of a single inverting type gate.

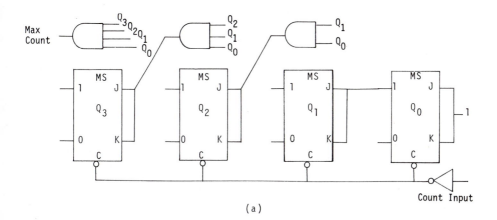

(a)

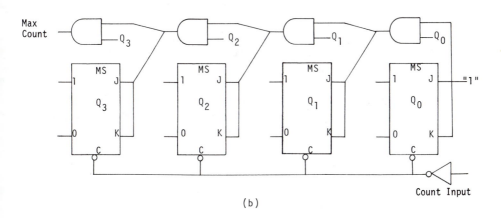

(b)

Fig. 10.7 Modulo-16 Binary Counters. (a) Circuit One; (b) Circuit Two

Using these delay values, counter-circuit one can operate successfully at frequencies up to $1/(\Delta_S + \Delta_T + \Delta_{AG})$ cycles per second, with a periodic-clock wave shape of

Counter-circuit two can operate successfully up to frequencies of $1/(\Delta_S + \Delta_T + (N - 1)\Delta_{AG})$ with a worst-case periodic count wave shape of

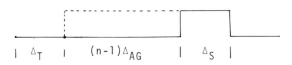

The clock line must be at zero for at least Δ_T seconds and at one for at least Δ_S seconds. The zero-to-one transition can take place any time during the interval marked Δ_{AG}. Using our nominal delay times, circuit 1 has an upper frequency limit of $1/8\Delta$ while circuit 2 has an upper limit of $1/12\Delta$. By using NOR gates with inverted inputs instead of the AND gates, these limits can be increased to $1/7\Delta$ and $1/9\Delta$ respectively. However, it should be remembered that although circuit 1 has a higher input frequency limit, it also requires AND gates (or NOR gates with inverted inputs) with up to N inputs, whereas circuit 2 requires only 2 input gates. It should also be pointed out that flip-flop Q_i in circuit 1 is required to drive $N-i$ gates, while each flip-flop in circuit 2 is required to drive only a single AND gate.

Let us discuss a few of the design problems which occur when implementing an arbitrary modulo-N synchronous binary counter. As an example a modulo-10 counter will be used. The present-state next-state table for such a counter (Fig. 10.8) requires that the next state after state $S_9(1001)$ be state $S_0(0000)$. States S_{10}, S_{11}, S_{12}, S_{13}, S_{14}, and S_{15} are not counting states and thus, in a sense, we don't care what next states follow them. For this reason the next-state entries corresponding to these noncounting states have been designated by d's indicating don't-care conditions.

Using K-maps a reduced set of next-state equations can easily be obtained (Fig. 10.9).

This set of next-state equations can now be implemented using clocked master-slave JK flip-flops. Using the synthesis procedure for JK units

$$Q_3(t + \Delta) = Q_3(t)\bar{Q}_0(t) + Q_2(t)Q_1(t)Q_0(t)$$

Present State				Next State			
$Q_3(t)$	$Q_2(t)$	$Q_1(t)$	$Q_0(t)$	$Q_3(t+\Delta)$	$Q_2(t+\Delta)$	$Q_1(t+\Delta)$	$Q_0(t+\Delta)$
0	0	0	0	0	0	0	1
0	0	0	1	0	0	1	0
0	0	1	0	0	0	1	1
0	0	1	1	0	1	0	0
0	1	0	0	0	1	0	1
0	1	0	1	0	1	1	0
0	1	1	0	0	1	1	1
0	1	1	1	1	0	0	0
1	0	0	0	1	0	0	1
1	0	0	1	0	0	0	0
1	0	1	0	d	d	d	d
1	0	1	1	d	d	d	d
1	1	0	0	d	d	d	d
1	1	0	1	d	d	d	d
1	1	1	0	d	d	d	d
1	1	1	1	d	d	d	d

Figure 10.8

implies

$$J_{Q_3}(t) = Q_2(t)Q_1(t)Q_0(t)$$
$$K_{Q_3}(t) = Q_0(t)[\bar{Q}_2(t) + \bar{Q}_1(t)];$$

$$Q_2(t + \Delta) = Q_2(t) \oplus (Q_1(t)Q_0(t))$$

implies

$$J_{Q_2}(t) = K_{Q_3}(t) = Q_1(t)Q_0(t);$$
$$Q_1(t + \Delta) = Q_1(t)\bar{Q}_0(t) + \bar{Q}_1(t)\bar{Q}_3(t)Q_0(t)$$

implies

$$J_{Q_1}(t) = \bar{Q}_3(t)Q_0(t)$$
$$K_{Q_1}(t) = Q_0(t);$$

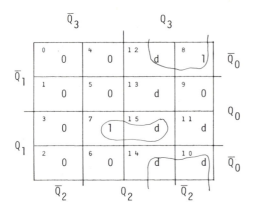

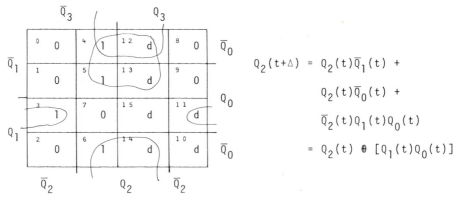

$$Q_3(t+\Delta) = Q_3(t)\bar{Q}_0(t) +$$
$$Q_2(t)Q_1(t)Q_0(t)$$

$$Q_2(t+\Delta) = Q_2(t)\bar{Q}_1(t) +$$
$$Q_2(t)\bar{Q}_0(t) +$$
$$\bar{Q}_2(t)Q_1(t)Q_0(t)$$
$$= Q_2(t) \oplus [Q_1(t)Q_0(t)]$$

Figure 10.9

and

$$Q_0(t + \Delta) = \bar{Q}_0(t)$$

implies

$$J_{Q_0}(t) = K_{Q_0}(t) = 1.$$

The resulting logic circuit wired to count on the leading edge of the clock, and its associated state-transition diagram are shown in Figs. 10.10 and 10.11. An examination of the state-transition diagram for the actual unit reveals that the counter contains a single periodic cycle of ten states and that the remaining states will always transition to one of the states on this cycle within two clock periods. We will refer to counters which exhibit the property of having a single cycle as being *self-starting*.

In general, one will not be lucky enough to obtain a self-starting counter when all the unused states are assigned don't-care next-state values. When this condition does not occur, additional logic will be required to force the unit to one of

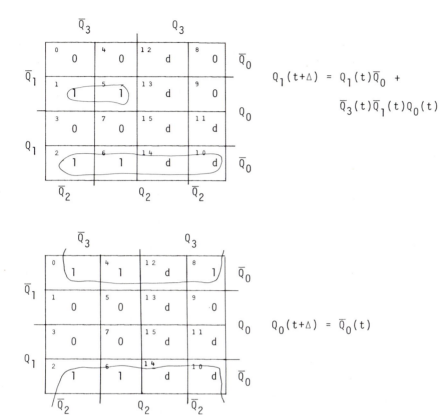

$$Q_1(t+\Delta) = Q_1(t)\overline{Q}_0 +$$
$$\overline{Q}_3(t)\overline{Q}_1(t)Q_0(t)$$

$$Q_0(t+\Delta) = \overline{Q}_0(t)$$

Figure 10.9 (*Continued*)

the assigned counter states. An alternative means of guaranteeing a self-starting counter is to require all noncounting states to have state S_0 (or any arbitrary counting state) as a next-state.

This alternative has an advantage of reducing the transient period to one cycle. However, this reduced transient will be at the expense of additional logic circuitry.

10.4 ARBITRARY MODULO-*N* COUNTER

One disadvantage with the procedure just outlined for the design of a modulo-*N* counter is that, if at a later time it is determined that *N* must be changed, the entire circuit must be scrapped. Let us now consider the design of a general modulo-*N* binary counter with the added requirement that the design be easily modified to handle changes in the value of *N*.

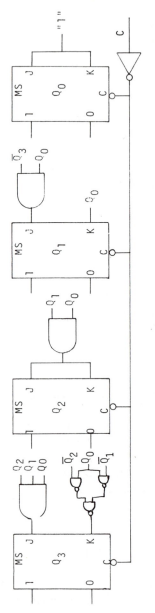

Fig. 10.10 A Self-Starting, Synchronous, Modulo-10 Binary Counter

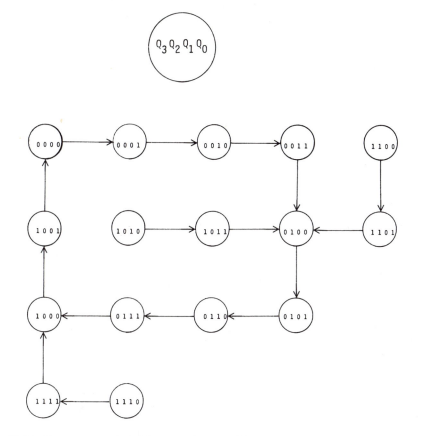

Fig. 10.11 State Transition Diagram for the Modulo-10 Binary Counter Shown in Fig. 10.10

In order to accomplish this objective, a binary counter with an external reset line is required. With the reset line at logical zero, the unit will behave as an ordinary binary modulo-2^k counter where k is the number of flip-flops. However, when the reset line is enabled, the counter will be disabled, and the unit will be reset to zero on the next clocking cycle. This behavioral description is easily translated into a set of next-state equations for such a counting unit, with

$$Q_0(t + \Delta) = \bar{R}(t)(Q_0(t) \oplus 1) \tag{10.3}$$

and

$$Q_i(t + \Delta) = \bar{R}(t)\left[Q_i(t) \oplus \left[\bigwedge_{j=0}^{i-1} Q_j(t)\right]\right]$$

$$1 \le i \le k-1 \tag{10.4}$$

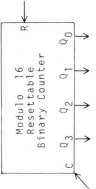

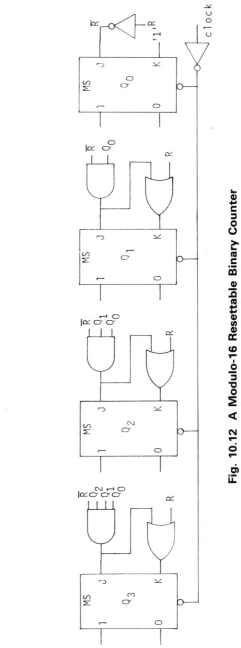

Fig. 10.12 A Modulo-16 Resettable Binary Counter

These equations yield J and K input equations of

$$J_{Q_0}(t) = \bar{R}(t), \; K_{Q_0}(t) = J_{Q_0}(t) + R(t)$$
$$= \bar{R}(t) + R(t) = 1 \qquad (10.5)$$

and

$$\left. \begin{array}{c} J_{Q_i}(t) = \bar{R}(t) \cdot \bigwedge\limits_{j=0}^{i-1} Q_j(t) \\[2mm] K_{Q_i}(t) = R(t) + J_{Q_i}(t) \\[2mm] = R(t) + \bigwedge\limits_{j=0}^{i-1} Q_j(t) \end{array} \right\} \qquad 1 \le i \le k-1 \quad (10.6)$$

A 4-stage resettable binary counter circuit is shown in Fig. 10.12. Using this basic unit, a modulo-10 counter would be obtained by externally connecting an AND gate such that $R = Q_3 \bar{Q}_2 \bar{Q}_1 Q_0$, as is shown in Fig. 10.13.

This counter is self starting and synchronous with a worst-case transient of $(2^k - N)$ cycles. Variations of this basic design could easily provide for an inverted input $\bar{R}$ and the outputting of the complements of the flip-flops. This would allow a single 4-input NAND gate to control the maximum count, with the NAND gate being either internally available or externally provided. A single-cell unit suitable for use in constructing large counters is shown in Fig. 10.14, along with a modulo-11 binary counter illustrating the method of cell interconnection.

For prototype designs and test equipment, it is convenient to have available a binary counter whose cycle length (period) can be quickly changed. Figures 10.15, 10.16, and 10.17 show three schemes for accomplishing this goal.

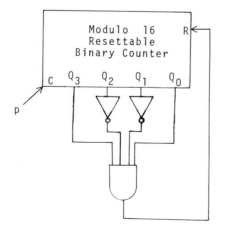

Fig. 10.13　Modulo-10 Binary Counter Unit

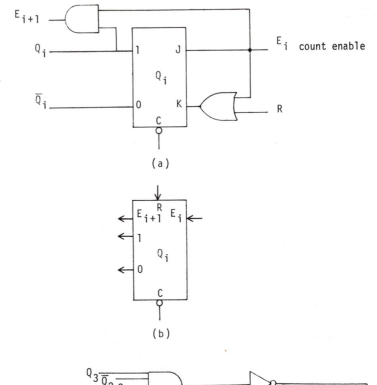

(a)

(b)

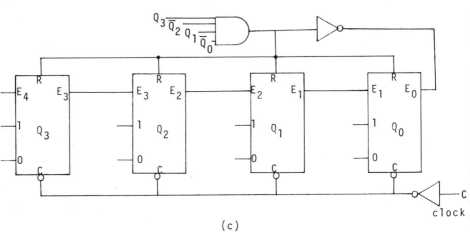

(c)

Fig. 10.14 Modulo-11 Binary Counter Unit Wired for Transition on Leading Edge of the Clock. (a) Basic Cell; (b) Cell Symbol; (c) Modulo-11 Binary Counter Using Resetting Counter Cell

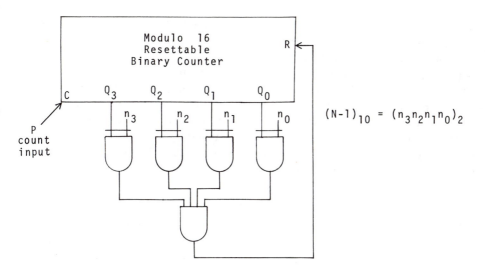

Fig. 10.15 Modulo-N Binary Counter Using Coincidence Gates

$$(N-1)_{10} = (n_3 n_2 n_1 n_0)_2$$

The maximum count for the circuit shown in Fig. 10.15 may be changed either by placing 0's or 1's on the input lines n_3, n_2, n_1, and n_0 directly or by using some external logic circuit to vary the 4-input lines. The second circuit, Fig. 10.16, requires a decoding network consisting of a set of 16 NAND gates along with the appropriate input gating. The maximum count for this circuit can be varied by means of a rotary switch or by the direct connection of the inverter to the proper

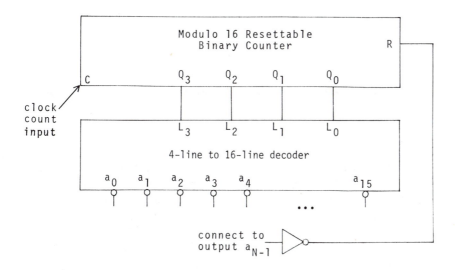

Fig. 10.16 Modulo-N Binary Counter Using a Decoding Network

output. However, for large counters, this circuit will require a considerable amount of excess hardware in the decoder.

The third circuit, Fig. 10.17, uses only a single NAND gate but requires four switches or input line changes in order to vary the maximum counting period. The final circuit choice will depend on the application and the convenience required.

Figure 10.18 shows a modulo-5 binary counter along with its associated timing chart obtained by resetting the modulo-8 binary counter when in state S_4 (100). The logic gates in this figure were assumed to have the following units of gate delay: NOT, NAND, 1 unit; AND, OR, 2 units; MS-JK's, 3 units; 0-to-1 transition, 4 units; 1-to-0 transition.

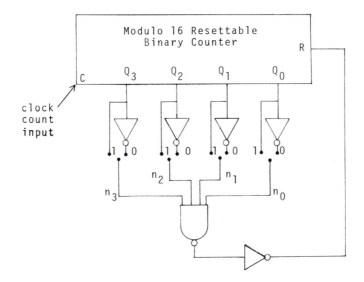

Fig. 10.17 Modulo-N Binary Counter Using NAND-Gate Input Switching

The reader should be cautioned that in many cases, particularly for small values of N, the amount of hardware required by the general resettable counter organization is more than would be required by an optimum logic circuit.

By carefully selecting the next states for the invalid counter states, reduced logic equations can be obtained, while still preserving the self-starting property. In the case of the modulo-5 counter, a simple circuit is obtained by requiring all invalid states to return to a counting state on the next clock pulse. Figure 10.19 shows the resulting reduced-logic circuit.

For each N there exists a reduced-logic circuit configuration; finding it, however, may be very time consuming.

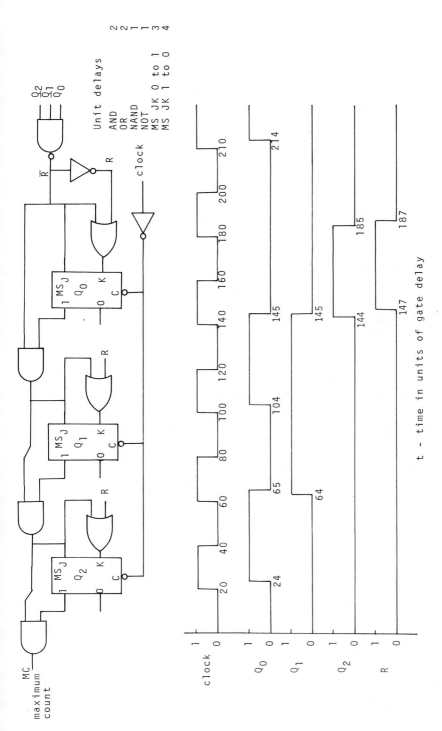

Fig. 10.18 Timing Chart for a Modulo-5 Binary Counter

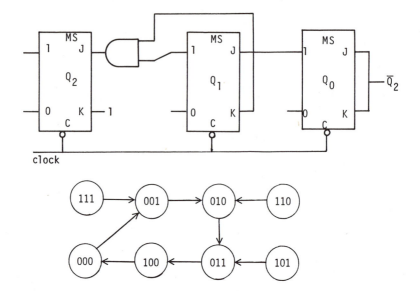

Fig. 10.19 Alternate-Reduced-Logic Self-Starting Modulo-5 Binary Counter Wired to Count on the Trailing Edge of the Clock

10.5 DIRECT RESETTING COUNTER CIRCUITS

Due to the availability of JK flip-flops with preclear and preset inputs, these units are often utilized in the construction of modulo-N counting units. We can classify these counters as being hybrid units in that the binary counting operation is synchronously controlled, whereas the resetting operation is not clocked but rather asynchronously controlled.

An examination of a typical flip-flop cell similar to that shown in Fig. 10.20, leads one to conclude that the preclear and preset input lines will be sensitive to short-duration logic hazards. This inherent problem is caused by the absence of any internal clocking control on these auxiliary inputs.

The difficulty with using the preclear inputs occurs when a state-decoding hazard, which is inherent in a binary-decoding unit, prematurely resets either the entire unit or only some of the cells to zero causing erratic circuit behavior. Fortunately, these unwanted decoding hazards can be successfully eliminated if the clock signal is utilized to control the preclear inputs. Figure 10.21 shows the organization of a typical modulo-N direct-resetting binary counter. The direct reset signal is obtained by using a NAND gate wired to detect state S_N. Note, this is one more state than was required by the synchronous resetting unit discussed in the preceding section.

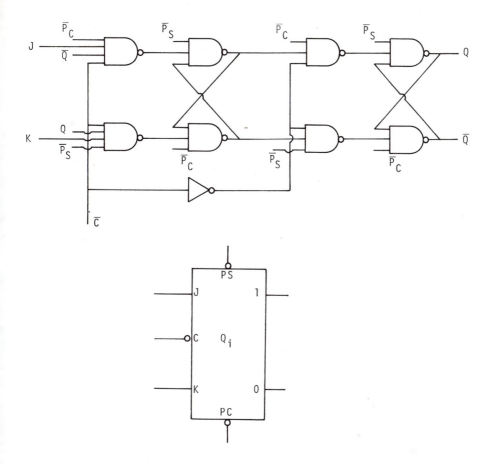

Fig. 10.20 Basic Master-Slave JK Flip-Flop with Preset and Preclear Inputs

Figure 10.22 shows a modulo-3 direct-resetting binary counter along with its associated timing chart. This particular circuit uses a single NAND gate to provide a reset pulse when Q_1 and Q_0 are both at one. However, there is an inherent decoding hazard on this reset line during the transition from state S_1 (01) to state S_2 (10). This hazard will probably cause this unit to be prematurely reset to zero, as indicated by the dotted lines on the Q_0 waveform.

In order for us to obtain a reliable modulo-N counter, these inherent decoding spikes must be prevented from reaching the reset input. Figures 10.23 and 10.24 illustrate two circuit modifications which accomplish this task. Both circuits AND the clock signal with the basic resetting signal. In Fig. 10.23 the clock signal is inverted and delayed by five units, thereby filtering the decoding hazards from the resetting signal line. The penalty which must be paid for this is a longer stay in state S_N and a shorter stay in the reset state S_0. For low-frequency

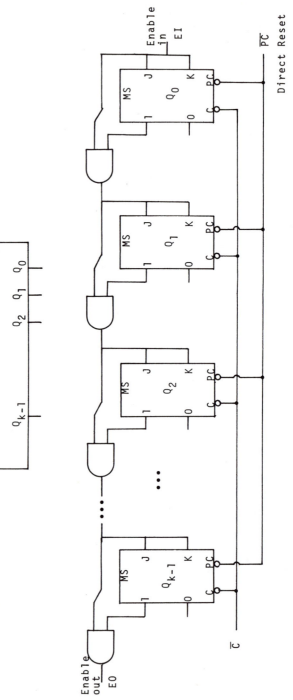

Fig. 10.21 Basic Circuit Organization for a Direct-Resetting-Modulo 2^K Binary Counter

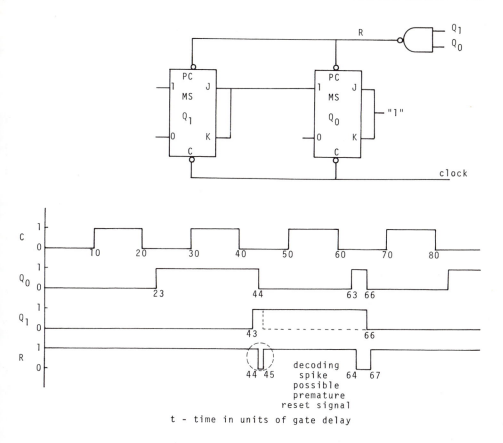

Fig. 10.22 Modulo-3 Direct-Resetting Binary Counter With Decoding Spikes

counters, this modification is ideal. However, for high-frequency counters, the short duration stay in state S_0 may not be tolerable if this state is being used to trigger other digital systems which require a fixed minimum setup time.

The circuit shown in Fig. 10.24 AND's the clock signal directly with the resetting signal, thereby delaying the resetting signal until the start of the latter half of the clocking cycle. This results in a sort of modulo $(N + \frac{1}{2})$ counter. This modification will not always work, due to the internal configuration of the JK flip-flops used to construct the counter. As was pointed out in Chapter 9, many flip-flops restrict the use of the preclear and preset inputs to those times when the clock input is a zero, i.e., when the master is off and the slave is on. This restriction is necessary to prevent the combination of external J and K inputs along with the internal feedback paths from the slave from improperly setting the master once the reset input is removed.

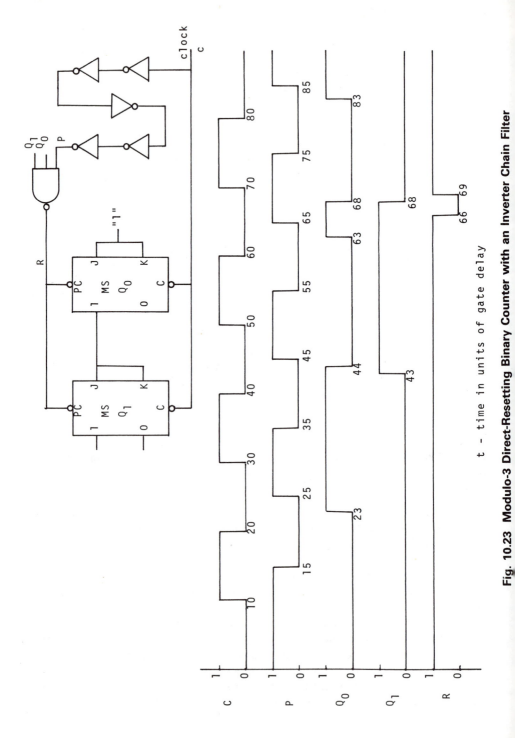

Fig. 10.23 Modulo-3 Direct-Resetting Binary Counter with an Inverter Chain Filter

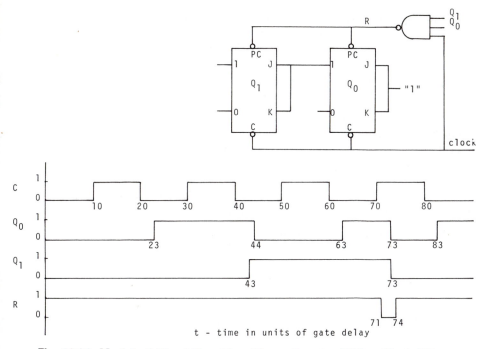

Fig. 10.24 Modulo-3 Direct-Resetting Binary Counter With a Clock Filter

The master-slave JK flip-flop shown in Fig. 10.25 does not need this restriction since the $\overline{\text{PRECLEAR}}$ (PC) input not only resets the master and slave unit but also effectively blocks any input or feedback signal on the J inputs from affecting the master unit by holding points A and B at 1 for the duration of the reset pulse.

The only restriction on the flip-flop shown in Fig. 10.25 is that the JK input and $\overline{\text{PC}}$ not be changed within three units of gate delay of a change in the clock input

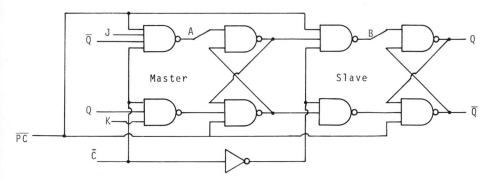

Fig. 10.25 Master-Slave JK Flip-Flop With a Preclear Input

from a one to a zero and that all signals on these input lines be maintained for at least three units of gate delay.

10.6 BINARY DOWN COUNTERS

In the preceding sections of this chapter our attention has been focused on counters which count in an ascending order. There are times when it is necessary to design counters which operate in a descending order, i.e., counters which count backwards or down.

The basic operation of such a counter is illustrated in Fig. 10.26.

Scanning this table from top to bottom a similar pattern of operation to that of a binary up counter is evident. Thus it is apparent from this table that:

1. Q_0 changes state on every count operation;

$(\text{Count})_{10}$	Q_3	Q_2	Q_1	Q_0
15	1	1	1	1
14	1	1	1	0
13	1	1	0	1
12	1	1	0	0
11	1	0	1	1
10	1	0	1	0
9	1	0	0	1
8	1	0	0	0
7	0	1	1	1
6	0	1	1	0
5	0	1	0	1
4	0	1	0	0
3	0	0	1	1
2	0	0	1	0
1	0	0	0	1
0	0	0	0	0

Fig. 10.26 Table of Successive States on a Modulo-16 Down Counter

2. Q_1 changes state only when Q_0 equals zero;
3. Q_2 changes state only when both Q_1 and Q_0 are equal to zero;
4. Q_3 changes states only when Q_2, Q_1, and Q_0 are all equal to zero.

These observations lead us to conclude that a general set of next-state equations for a binary down counter would be

$$Q_0(t+\Delta) = Q_0(t)\oplus 1 \tag{10.7}$$

$$Q_i(t+\Delta) = Q_i(t)\oplus\left[\bigwedge_{j=0}^{i-1} \bar{Q}_j(t)\right]$$

for

$$1 \leq i \leq n-1. \tag{10.8}$$

These equations are easily implemented using JK flip-flops. Our synthesis procedure yields

$$J_{Q_0}(t) = K_{Q_0}(t) = 1$$

and

$$J_{Q_i}(t) = K_{Q_i}(t) = \bigwedge_{j=0}^{i-1} \bar{Q}_j(t)$$

for

$$1 \leq i \leq n-1$$

A modulo-8 down counter circuit is illustrated in Fig. 10.27 along with its associated timing chart.

Where dual-mode operation is required, a synchronous modulo-2^N up-down counter can be easily constructed.

If a mode-control line M is provided with $M=0$, implying count down and $M=1$ implying count up, the next-state equations of an up counter (Eqs. 10.1 and 10.2) can be combined with those of a down counter (Eqs. 10.7 and 10.8). Doing so, we have

$$Q_0(t+\Delta) = M(t)\cdot(Q_0(t)\oplus 1) + \bar{M}(t)\cdot(Q_0(t)\oplus 1) = Q_0(t)\oplus 1 \tag{10.9}$$

$$Q_i(t+\Delta) = M(t)\cdot(Q_i(t)\oplus \bigwedge_{j=0}^{i-1} Q_j(t)) + \bar{M}(t)\cdot(Q_i(t)\oplus \bigwedge_{j=0}^{i-1} \bar{Q}_j(t))$$

$$= Q_i(t)\oplus(M(t)\cdot \bigwedge_{j=0}^{i-1} Q_j(t) + \bar{M}(t) \bigwedge_{j=0}^{i-1} \bar{Q}_j(t))$$

for

$$1 \leq i \leq n-1. \tag{10.10}$$

Figure 10.28 shows a modulo-8 binary synchronous up-down counter.

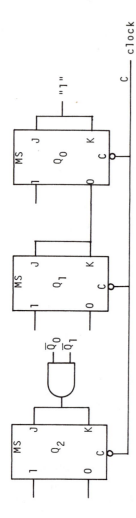

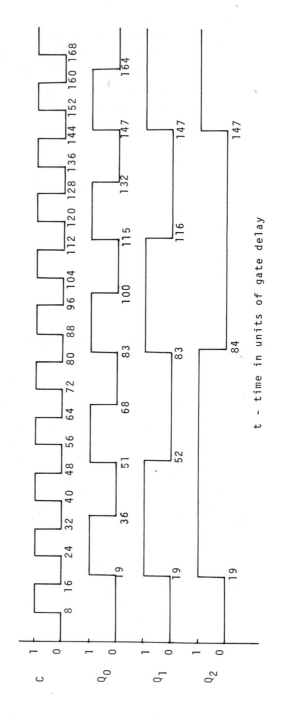

Fig. 10.27 A Modulo-8 Down Counter

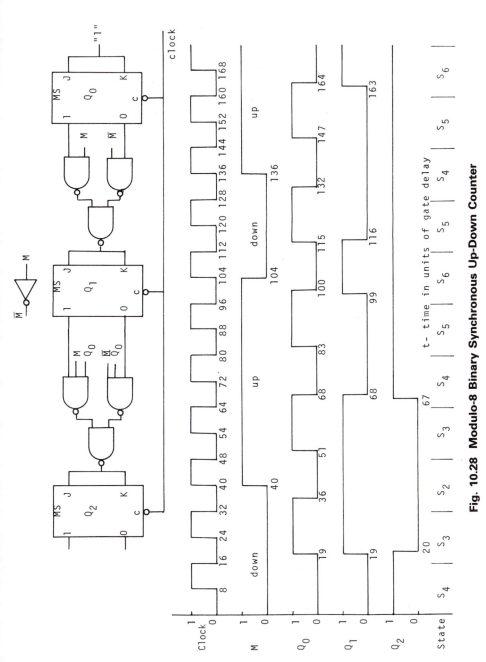

Fig. 10.28 Modulo-8 Binary Synchronous Up-Down Counter

One place where up-down counters are widely used is in the design of analog-to-digital converters. In this application, the output of the binary counter is fed to a digital-to-analog converter. The output of this unit is compared with the input signal. If the difference is positive, the counter is counted down; if the difference is negative, the counter is counted up. In this way the counter tends to track the input signal. A block diagram of such a system is shown in Fig. 10.29.

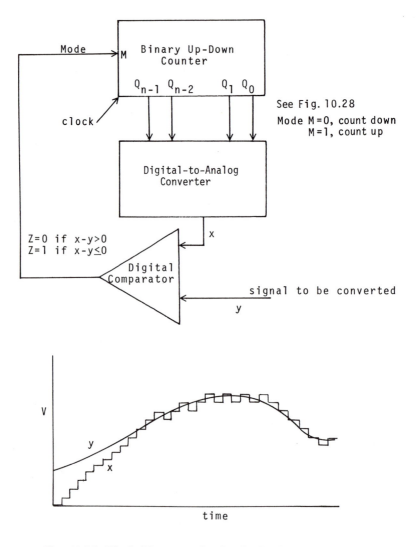

Fig. 10.29 Block Diagram of a Crude Analog-to-Digital Unit

10.7 SPECIAL BINARY COUNTING UNITS

In this section, four special counting units will be studied. Each unit has been chosen to illustrate alternative approaches to the design of binary counting units.

10.7.1 Gray Code Counters

Decoding spikes are one of the problems which are inherent in the standard binary synchronous modulo-N up counters. These spikes are due to the simultaneous change of several of the state variables. One method of eliminating decoding hazards is to use a Gray code instead of a binary code. Since, with a Gray code, only one variable changes at a time, there will be no unwanted decoding spikes.

    A present-state next-state table for a modulo-8 Gray-code synchronous counter is shown in Fig. 10.30. Using our standard synthesis procedure, we can construct a sequential circuit to satisfy the table. As a first step we will determine the next-state equations using the K-maps in Fig. 10.31.

    These equations lead to the JK flip-flop realization shown in Fig. 10.32.

10.7.2 A Modulo-10 Excess-3 Binary Counter

A modulo-10 binary counter using an excess-3 coding can be constructed by making a slight modification to the synchronous resettable modulo-N counter organization. Instead of resetting the entire counter to 0 when state S_{12} is reached, the counter must be reset to state S_3. This can be accomplished by placing OR gates on the J input of the first two stages and the K inputs of the last two stages of a standard binary up counter.

    The state-transition diagram for this counter is shown in Fig. 10.33. The logic circuit for the counter is shown in Fig. 10.34. It is easily seen from Fig. 10.34 that this counter is self-starting.

10.7.3 Ring Counters

An alternate method for eliminating decoding spikes is to assign a coding scheme such as a 1-out-of-N code. A modulo-4 binary counter using this code would count as follows:

	Q_3	Q_2	Q_1	Q_0
A_0	0	0	0	1
A_1	0	0	1	0
A_2	0	1	0	0
A_3	1	0	0	0

Present State			Next State			Output Section							
$Q_2(t)$	$Q_1(t)$	$Q_0(t)$	$Q_2(t+\Delta)$	$Q_1(t+\Delta)$	$Q_0(t+\Delta)$	$A_0(t)$	$A_1(t)$	$A_2(t)$	$A_3(t)$	$A_4(t)$	$A_5(t)$	$A_6(t)$	$A_7(t)$
0	0	0	0	0	1	1	0	0	0	0	0	0	0
0	0	1	0	1	1	0	1	0	0	0	0	0	0
0	1	1	0	1	0	0	0	1	0	0	0	0	0
0	1	0	1	1	0	0	0	0	1	0	0	0	0
1	1	0	1	1	1	0	0	0	0	1	0	0	0
1	1	1	1	0	1	0	0	0	0	0	1	0	0
1	0	1	1	0	0	0	0	0	0	0	0	1	0
1	0	0	0	0	0	0	0	0	0	0	0	0	1

Fig. 10.30 Present-State/Next-State Table for a Modulo-8 Gray Code Counter

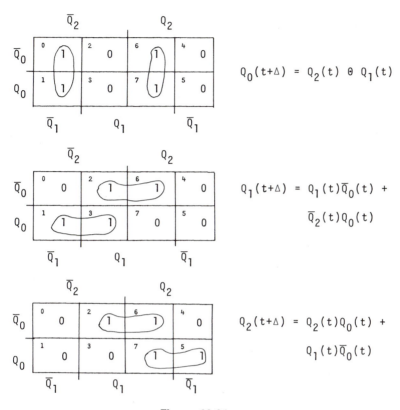

$$Q_0(t+\Delta) = Q_2(t) \oplus Q_1(t)$$

$$Q_1(t+\Delta) = Q_1(t)\overline{Q}_0(t) +$$
$$\overline{Q}_2(t)Q_0(t)$$

$$Q_2(t+\Delta) = Q_2(t)Q_0(t) +$$
$$Q_1(t)\overline{Q}_0(t)$$

Figure 10.31

In general, with a modulo-N counter, counter state A_i would have stage Q_i equal to one and all other stages equal to zero.

One simple method of implementation for this type of counter is to use an n-stage binary shift register, with the output of stage Q_{n-1} used as the shift input into stage Q_0. To insure that the unit will start in a valid counter state, the counter can be provided with an initial state input which would preset the unit into state A_0.

Because of the movement of the single 1 down the register, this type of counter organization is commonly referred to as a ring counter.

A modulo-4 ring counter circuit is shown in Fig. 10.35.

This unit, while having a trivial decoding scheme, requires an exorbitant number of flip-flops. However, for small values of N this type of counter is very popular.

An alternative method of insuring that this counter will cycle through the correct set of N states is to feed back a 1 into the first flip-flop, Q_0, whenever the first $(n-1)$ flip flops are all equal to zero. Thus as long as there are any ones in

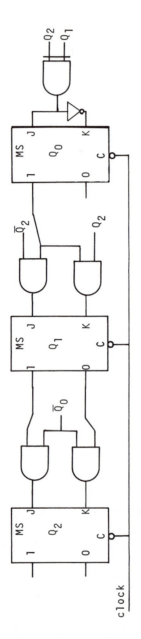

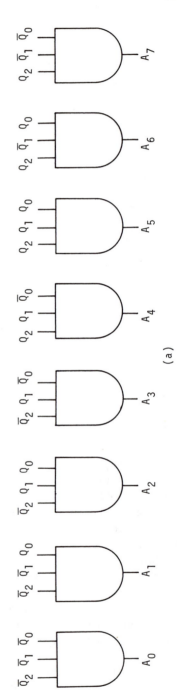

(a)

clock

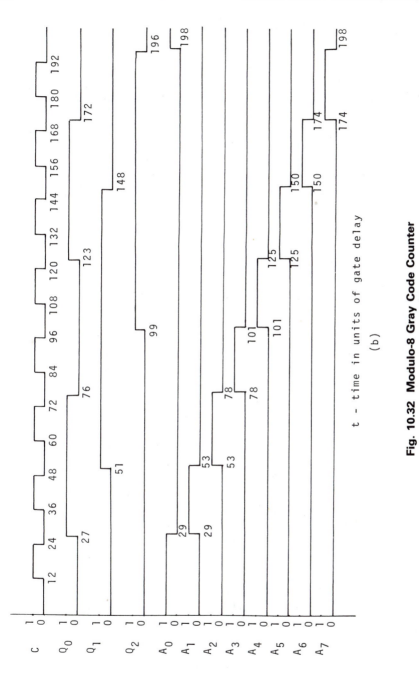

t – time in units of gate delay

(b)

Fig. 10.32 Modulo-8 Gray Code Counter

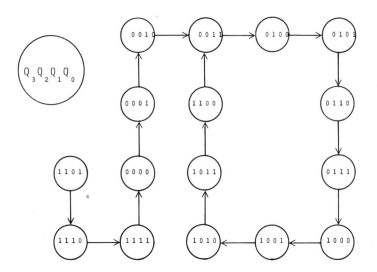

Fig. 10.33 State Transition Diagram for an Excess-3 Modulo-10 Counter

this position, zero will be fed back, thereby forcing the unit into the correct counting cycle. Figure 10.36 illustrates this modification with a modulo-4 circuit.

The state transition for this unit is shown in Fig. 10.37.

10.7.4 Switch-Tail Counters

If instead of feeding back Q_{n-1} into the shift input, as was done to form a ringcounter, we feed $\bar{Q}_{n-1}$ back, a switch-tail counter results. This counter, when correctly initialized to state S_0, $(00 \cdots 00)$, alternately shifts in 1's until the register reaches the all-one's state S_{2^n-1}, $(111 \cdots 11)$, at which point zeros are shifted in until the unit returns to the all-zero state.

Figure 10.38 shows the logic circuit for a modulo-8 switch-tail counter. The state coding for this unit is shown in Fig. 10.39. Again, because at most one variable changes at a time, this unit will not have any decoding spikes.

As is apparent from the coding, a general n-stage switch-tail counter has $2n$ coded states. These states can be uniquely decoded using 2 input gates following the method indicated on the coding chart for the modulo-8 counter above. This counter is very popular as a control counter for small values of N. The feedback function, $F(Q_{n-1}, \ldots, Q_0) = \bar{Q}_{n-1} + \overline{(Q_{n-2} \cdot Q_{n-2} \cdots Q_2 \cdot Q_1)} \cdot Q_0$, will insure that the counter will quickly reach the desired counting cycle if the unit is initially started in an invalid counter state.

A self-starting modulo-8 switch-tail counter is shown in Fig. 10.40 with the feedback function $F(Q_3, Q_2, Q_1, Q_0) = \bar{Q}_3 + \overline{Q_2 \cdot Q_1} \cdot Q_0$. The timing chart for this

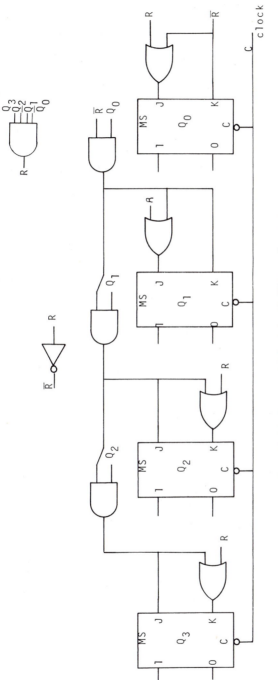

Fig. 10.34 Excess-3 Modulo-10 Binary Counter

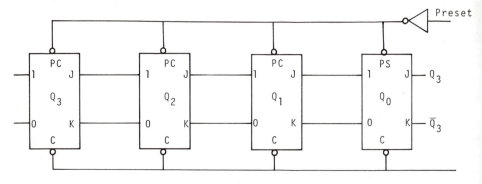

Fig. 10.35 Modulo-4 Ring Counter

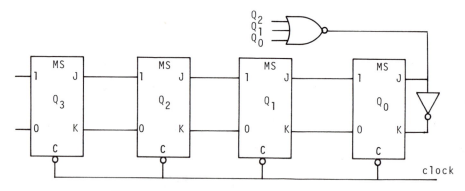

Fig. 10.36 A Self-Starting Modulo-4 Ring Counter

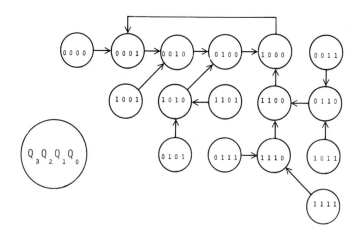

Fig. 10.37 State-Transition Diagram for a Modulo-4 Self-Starting Ring Counter

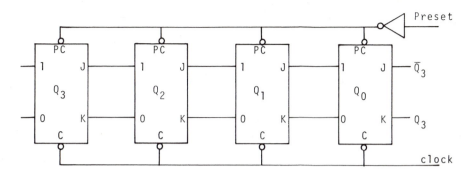

Fig. 10.38 Modulo-8 Switch-Tail Counter

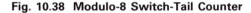

Figure 10.39

unit is shown in Fig. 10.41 and the state-transition diagram for this unit is shown in Fig. 10.42.

10.8 DESIGN OF A DIGITAL STOP WATCH

Let us put the material on counters to use by considering in this section the design of a simple digital stop watch similar to that shown in Fig. 10.43. The unit will operate up to 59 minutes, 59.99 seconds using seven segment indicators.

This hand-held unit will use a common start-stop push button and a separate reset push button to control the unit manually. Separate logic inputs for start-stop and reset are also provided to allow for electronic control of the stop watch. An accuracy requirement, when electronically controlled, on the unit will be $\pm 0.001\%$ or ± 0.04 seconds at the maximum timing interval.

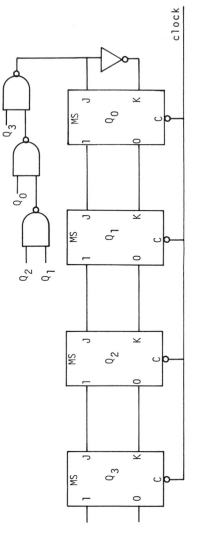

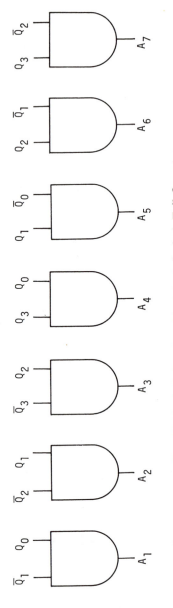

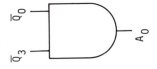

Fig. 10.40 Self Starting Modulo-8 Switch-Tail Counter

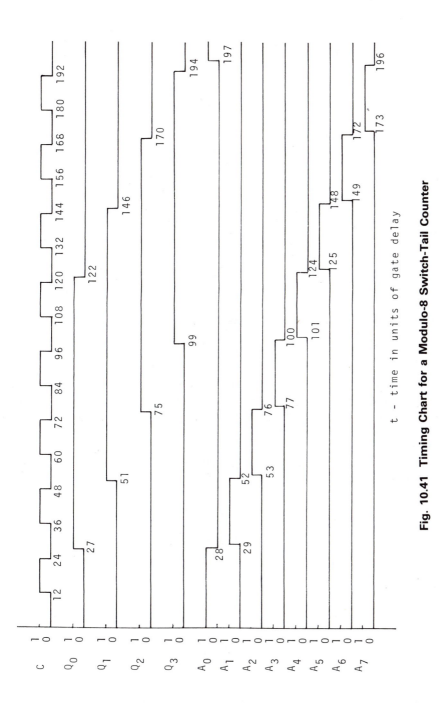

Fig. 10.41 Timing Chart for a Modulo-8 Switch-Tail Counter

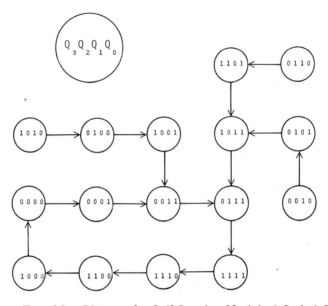

Fig. 10.42 State Transition Diagram for Self-Starting Modulo-8 Switch-Tail Counter

This accuracy requirement implies that the internal clock used by the control system must also be accurate to within ± 0.001%. To maintain this clock stability, a crystal oscillator will be used. Once a clock frequency is selected, the stop watch consists of a series of binary counters used to hold the number of clock pulses between the start and stop signals. For simplicity, let us select a clock frequency of 10,000 cycles per second.

For this clock frequency, six decade counters and two modulo-6 binary counters will be required by the stop watch. A block diagram of the counter system is shown in Fig. 10.44.

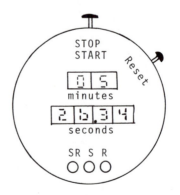

Fig. 10.43 Digital Stop Watch

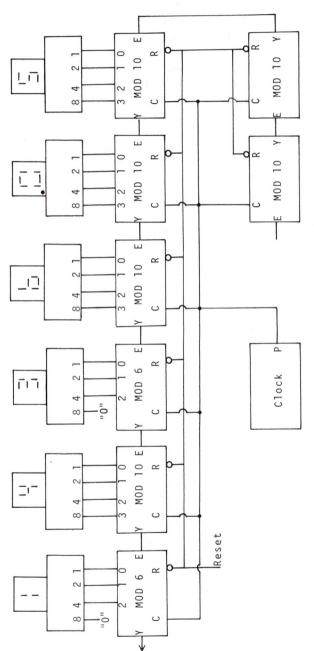

Fig. 10.44 Block Diagram of Counter Units and 7-Segment Display Units Needed for the Stop Watch

Detailed logic circuits for the modulo-6 and modulo-10 self-starting counters are shown in Figs. 10.45 and 10.46. The output of the minute and second counters (top row of counters) is fed directly to a seven segment indicator driver to give a visual display of the time.

All that remains is the design of the control system interface with the manual and electronic input commands. The register section requires only two signals: an enable E for the first counter unit and a reset signal $\bar{R}$ for resetting the count to zero. Let us consider initially the manual input signals from the two momentary contact switches. The reset signal is used to clear all of the counters back to zero and also turn off the count enable. The start-stop push button is to alternately enable and disable the clock signal to the counters. This alternate operation implies the use of a toggle flip-flop clocked by the action of the push button. Using bounce-free switches as input signals, the inverted start-stop signal can be connected directly to the JK and clock inputs of a master slave JK flip-flop, thereby forming a toggle flip-flop which triggers each time the button is depressed. The inverted reset signal can also be connected to the $\overline{PC}$ input on the flip-flop. The electronic commands can be ORed together with the manual input to complete the circuit. In this case $(START)_e$ and $(STOP)_e$ are ORed with the

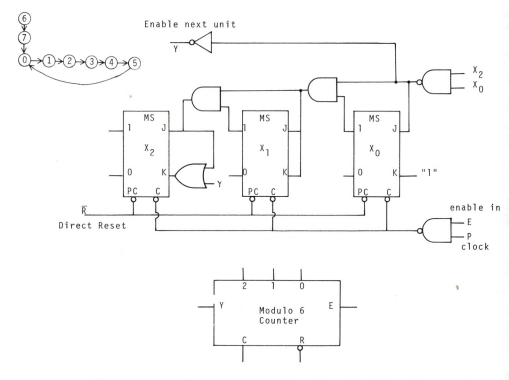

Fig. 10.45 Self-Starting Modulo-6 Counter with a Direct Reset

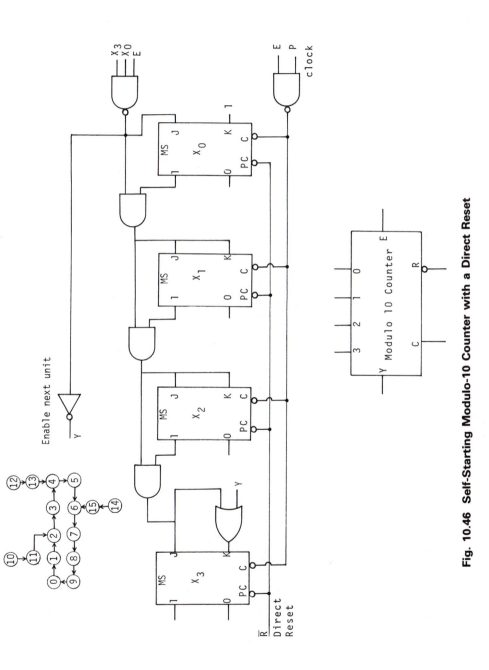

Fig. 10.46 Self-Starting Modulo-10 Counter with a Direct Reset

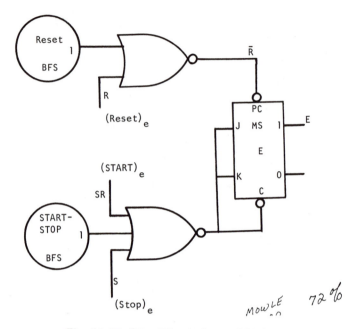

MowLE 72%

Fig. 10.47 Stop Watch Control Unit

start-stop push button, and the $(RESET)_e$ ORed with the manual reset signal. The final control circuit is shown in Fig. 10.47.

10.9 PROBLEMS

10.1   You have been given the task of designing a set of special-purpose, synchronous, self-starting, modulo-N binary counters using the approach outlined in Section 10.3. The value of N is:

a) 3   b) 5   c) 7
d) 9   e) 11   f) 12
g) 13   h) 15

These units are to count the leading-edge clock line only when an enable signal E is a one. Your unit should provide an output signal when in state $N - 1$.

10.2   You have been charged with the task of designing a group of synchronous self-starting modulo-N binary counters using the procedure outlined in Section 10.4. Each of your counters is to operate with four values of N. The selection of the value of N is determined by two control lines x and y as indicated below. Assume these units are to count the trailing edge of a clock line only when an enable signal E is a one. Ignore transients which might occur during a change in the values of x and y.

Control lines		Values of N selected			
x	y	(a)	(b)	(c)	(d)
0	0	2	1	3	5
0	1	4	3	6	10
1	0	8	5	9	15
1	1	16	7	12	20

10.3   Design a special-purpose, synchronous, self-starting decimal counter (modulo 10) using the following codes from Table 7.2. Show a state diagram for your unit.
      a)   excess-6   b)   5–4–2–1
      c)   2–4–2–1   d)   4–3–2–1

10.4   Modify the basic design of the Gray code counter discussed in Section 10.7.1 so that it can operate as either an up counter ($M = 1$) or as a down counter ($M = 0$). Show a state diagram for your final design.

10.5   Design a self-starting modulo-10 switch-tail counter, including the state decoding circuitry. Demonstrate that your design is self-starting by constructing a state diagram for your final design.

10.6   Design a modulo-8 synchronous binary counter which has the following state diagram:

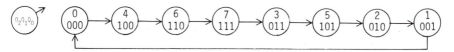

The unit is to be designed to count the leading edges of the clock line.

10.7   You have been asked to design a digital calendar for a table clock. Assume you have available a 60-Hz square-wave clock.

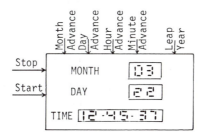

Your unit must be capable of operating automatically over a year. The leap-year switch must be set some time prior to Feb. 29 for the unit to work correctly. The manual advance push buttons will operate only if the stop signal has been previously pressed. Depressing the minute advance will also automatically clear the seconds on the clock to zero.

Block diagram your unit showing only the details of the control unit (use bounceless switches) and any counters used. You may assume the availability of seven-segment decoders and indicator lights using standard 8–4–2–1 code and a special blanking signal to inhibit the lights.

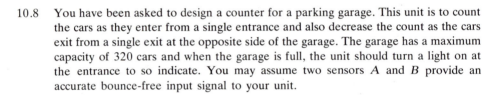

10.8   You have been asked to design a counter for a parking garage. This unit is to count the cars as they enter from a single entrance and also decrease the count as the cars exit from a single exit at the opposite side of the garage. The garage has a maximum capacity of 320 cars and when the garage is full, the unit should turn a light on at the entrance to so indicate. You may assume two sensors A and B provide an accurate bounce-free input signal to your unit.

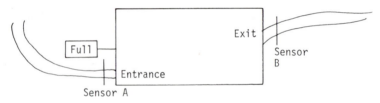

Your unit should have a manual increment and decrement button and a reset signal in order to allow an operator to adjust the count when unexplained errors occur (a boy on a bicycle rides back and forth on the entrance or exit ramp, etc.).

10.9   Design a pair of electronic dice, using either 8–4–2–1 segment indicators or an

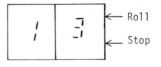

array of leds (light-emitting diodes) for outputs.

The roll push button should cause the internal logic to count a high-frequency clock signal. The stop push button should disable the internal clock and also prevent the roll push button from again activating the unit for a period of approximately 3 seconds.

a) Design your dice to be unbiased.

b) Comment on how you could design loaded dice (hopefully not easy to detect).

10.10 Design a simple digital alarm clock. Your unit is to operate by counting the 60-cycle AC line frequency which has been converted into a clean square wave clock (You do not have to design the conversion unit).

Explain how your unit works; show a block diagram for your unit.

You may also assume that you have available, as inputs for the alarm time, a group of miniature rotary switches which can be set with a screw driver. These units yield a binary-coded output and are available with 1, 2, 3, and 4 output lines.

10.11 You have been asked to design a special frequency-division unit. An input square wave of frequency F is to be inputted to your unit. The output from your unit is to be a square wave of frequency $F/(2(N+1))$ with $0 \le N \le 15$. The value of N is entered by means of four toggle switches with $(N)_{10} = (n_3 n_2 n_1 n_0)_2$.

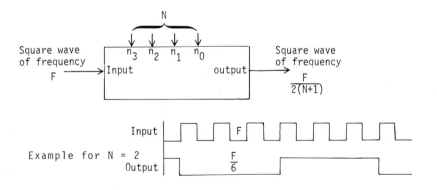

Hint: Reread the section on arbitrary number counters.

10.12 You have been asked to design a special frequency-division unit. An input square wave of frequency F is to be inputted to your unit. The output from your unit is to be a square wave of frequency $F/(N+1)$ with $0 \le N \le 15$. The value of N is to be entered by means of four toggle switches with $(N)_{10} = (n_3 n_2 n_1 n_0)_2$.

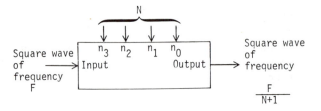

Hint: Draw a timing chart for the unit shown below and then use it to modify the circuit you used in Problem 10.11.

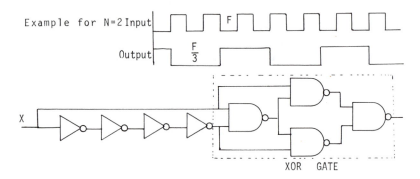

Assume all gates have one unit of delay and that x is a square wave with a period of at least 20 units of gate delay.

10.10 BIBLIOGRAPHY

Boyce, J. C., *Digital Logic and Switching Circuits*. Englewood Cliffs, N.J.: Prentice-Hall, 1975.

Maley, G. A., *Manual of Logic Circuits*. Englewood Cliffs, N.J.: Prentice-Hall, 1970.

Morris, R. L. and J. R. Miller (Eds.), *Designing with TTL Integrated Circuits*. New York: McGraw-Hill, 1971.

Nagle, H. T., B. D. Carroll, and J. D. Irwin, *An Introduction to Computer Logic*. Englewood Cliffs, N.J.: Prentice-Hall, 1975.

Peatman, J. B., *The Design of Digital Systems*. New York: McGraw-Hill, 1972.

Rhyne, V. T., *Fundamentals of Digital Systems Design*. Englewood Cliffs, N.J.: Prentice-Hall, 1973.

Wickes, W. E., *Logic Design with Integrated Circuits*. New York: Wiley, 1968.

Chapter 11

Register Design
Techniques

11.1 INTRODUCTION

In this chapter we will discuss a variety of design procedures for data-storage registers. In particular, in Section 11.2, the design of one-step and two-step data-storage registers will be considered. The design of a shift-register unit will also be included in this section. Multifunction binary counting and complementing registers will be the subject material for Section 11.3.

In particular, the concept of logical superposition as a design aid will be discussed. Finally, the design of arbitrary periodic sequence generators will be introduced in Section 11.4. Two alternate techniques will be presented, the first using feedback shift registers and the second using resettable modulo-N binary counters.

11.2 DATA-STORAGE REGISTERS

There will be many occasions when we will find it either necessary or convenient to store the results of our digital calculations. In this section we will explore several different circuit configurations suitable for the temporary storage of binary data.

Before proceeding into the logical organization of data-storage devices, let us define what we mean by the term *register*.

Definition 11.1   A *register* is a collection of flip-flop storage elements sharing a common timing network and a common purpose.

Using this definition, a storage register would consist of a collection of flip-flops whose purpose is to hold data temporarily. A shift register would consist of a collection of clocked flip-flops interconnected in such a manner that, at the end of each clocking sequence, the data contained in the register is shifted one bit to the left or right.

11.2.1 Design of Binary Storage Registers

Let us examine a few of the design considerations which must be taken into account by a design engineer given the task of specifying a simple n-bit binary storage register, i.e., a register such that each flip-flop satisfies a next-state equation of the form

$$Q_i(t + \Delta) = D_i(t).$$

Let us assume that our design engineer has at his or her disposal both clocked and unclocked SR flip-flops, master-slave JK flip-flops, and edge-triggered delay flip-flops.

For the moment let us suppose that our design engineer selects a basic unclocked SR flip-flop as the main storage device.

One approach to the design of the data register using unclocked SR flip-flops would be to recognize the next state equation for each flip-flop as being that of a standard delay flip-flop. Hence by simply placing an inverter between the S and R inputs of each flip-flop a data register will result. See Fig. 11.1.

At first glance this approach looks reasonable and it certainly is inexpensive. However, a second look reveals that there is no way to retain the old value of the data while new data is being made ready. As soon as new data appears on an input line, it is immediately transferred into the register. There is also no way that the register contents can be cleared except by setting each data input to zero.

If we are willing to use two signals to help control the flow of data into the register, then we could first clear all the flip-flops to zero by setting all of the S inputs to zeros and all of the R inputs to ones. The data could then be stored by placing ones on the S inputs of those flip-flops which have ones on their respective data lines. The resulting circuit is usually referred to as a data register with a two-step read operation (clear before setting) and is shown in Fig. 11.2.

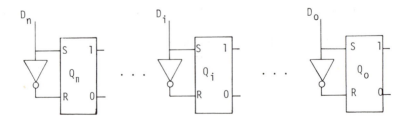

Fig. 11.1 Unclocked SR Storage Register

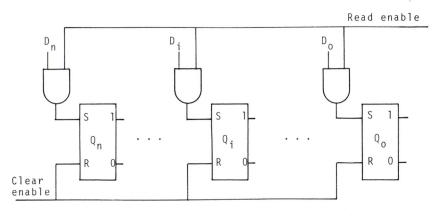

Fig. 11.2 Data Register Wired with a Two-Step Read

To assure proper operation of this register configuration, the clear-enable line and the read-enable line should not be simultaneously equal to one. This condition could cause the S and R inputs to be simultaneously equal to one, leading to an inherent failure of the data register. It is also apparent that the data lines should be allowed to change only while the read-enable line is a logical zero since once a register bit has become a one it can only be reset by means of the clear-enable. It should also be clear from an examination of the circuit that once data has been successfully loaded into the register, it will remain there until the clear-enable is again used. Figure 11.3 shows a simplified timing chart for the ith register bit.

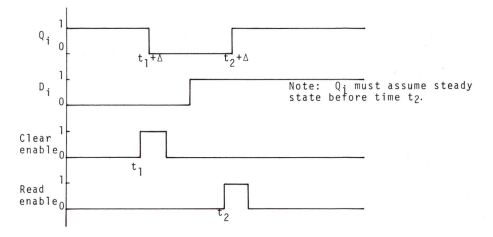

Fig. 11.3 Timing Signals for Storage Register in Fig. 11.2

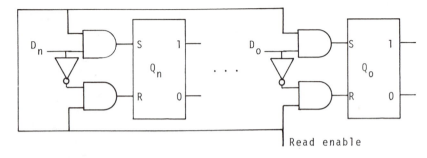

Fig. 11.4 One-Step-Read Data Register

Instead of first clearing the register and then setting the required bits to ones, we could just as well have first set the entire register to all ones and then cleared the bits corresponding to zero information.

Both of the above approaches require two separate timing signals to store the new data. A one-step approach (i.e., a single timing signal) can be obtained by gating the true and complement data inputs into the flip-flop as is shown in Fig. 11.4.

When the read enable line is at zero, the old data is retained in the register. By applying a read enable signal, the data on the input line is transferred into the register. A closer look at this circuit configuration also reveals that we have effectively converted our unclocked SR flip-flops into simple clocked SR devices with the read enable line serving as the clock line. The resulting clocked circuit is shown in Fig. 11.5.

As long as the read-enable line is a one, the SR flip-flop will tend to track the data inputs. Hence, the read-enable line should be pulsed only during the time interval when the input data is known to be stabilized.

Instead of using simple SR flip-flops, we could just as well have constructed the above clocked circuit using master-slave JK flip-flops or simple edge-triggered delay flip-flops.

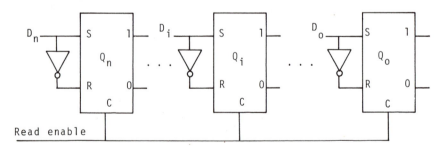

Fig. 11.5 Clocked SR Flip-Flop Storage Register

11.2.2 Shift Registers

We can look at a shift register as a data-storage register which has as its inputs the output of a previous stage to its right or left. If we number the flip-flops from right to left, the equations describing a right-shifting register become:

$$Q_n(t + \Delta) = \text{SRI}(t), \qquad (11.1)$$

$$Q_i(t + \Delta) = Q_{i+1}(t); \qquad (11.2)$$

where the Shift Right Input, $\text{SRI}(t)$, is the input data to be shifted into the left-most flip-flop. Figure 11.6 shows a 4-bit right-shifting register using clocked JK flip-flops which transitions on the leading edge of the shift signal. A timing chart for this register is also shown in Fig. 11.7.

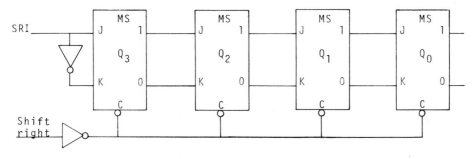

Fig. 11.6 A JK Implemented Shift Register

The reader may be wondering why clocked SR flip-flops haven't been used as building blocks for shift registers as they were used for data registers. The reason lies in the fact that as long as the clock signal is a one, the data on the input will be read continually into the unit and then passed from stage to stage as quickly as each flip-flop can change state. Thus, on one clock signal, it is difficult to prevent the data from shifting more than one stage down the line.

As was the case with data registers, shift registers can be obtained prewired in varying bit lengths on integrated-circuit chips.

11.2.3 Multifunction Register Design

Let us combine some of the ideas expressed in the preceding sections and consider the logical design of a small 4-bit multifunction register. The register we will consider has two control inputs, X and Y, which are used to select one of four

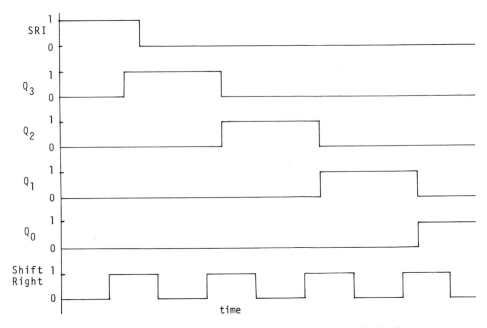

Fig. 11.7 A Timing Chart for a JK Implemented Shift Register

register functions on each clock cycle. A third control, input E, is used to enable the four functions specified by X and Y. The basic operations which the register is to perform are listed in Table 11.1.

From Table 11.1 one can easily write out a set of next-state equations for this unit:

$$Q_3(t + \Delta) = [\bar{X}(t)\bar{Y}(t)B_3(t) + \bar{X}(t)Y(t)\text{SRI}(t) + X(t)\bar{Y}(t)Q_2(t)$$
$$+ X(t)Y(t)A_3(t)]E(t) + \bar{E}(t)Q_3(t) \quad (11.3)$$

$$Q_2(t + \Delta) = [\bar{X}(t)\bar{Y}(t)B_2(t) + \bar{X}(t)Y(t)Q_3(t) + X(t)\bar{Y}(t)Q_1(t)$$
$$+ X(t)Y(t)A_2(t)]E(t) + \bar{E}(t)Q_2(t) \quad (11.4)$$

$$Q_1(t + \Delta) = [\bar{X}(t)\bar{Y}(t)B_1(t) + \bar{X}(t)Y(t)Q_2(t) + X(t)\bar{Y}(t)Q_0(t)$$
$$+ X(t)Y(t)A_1(t)]E(t) + \bar{E}(t)Q_1(t) \quad (11.5)$$

$$Q_0(t + \Delta) = [\bar{X}(t)\bar{Y}(t)B_0(t) + \bar{X}(t)Y(t)Q_1(t) + X(t)\bar{Y}(t)\text{SLI}(t)$$
$$+ X(t)Y(t)A_0(t)]E(t) + \bar{E}(t)Q_0(t). \quad (11.6)$$

Using delay flip-flops, this register could be implemented without further work since the excitation input equation for an edge-triggered delay flip-flop is $D_i(t) = Q_i(t+\Delta)$. One such realization is shown in Fig. 11.8.

TABLE 11.1
Multifunction Register Function Specifications

Control Inputs			Register Function to be Done on Next Clock Pulse	Next State of the Register			
$E(t)$	$X(t)$	$Y(t)$		$Q_3(t+\Delta)$	$Q_2(t+\Delta)$	$Q_1(t+\Delta)$	$Q_0(t+\Delta)$
1	0	0	Load register with the data on the parallel B input lines	$B_3(t)$	$B_2(t)$	$B_1(t)$	$B_0(t)$
1	0	1	Shift register data right one place with SRI (shift right input) shifted into Q_3	$SRI(t)$	$Q_3(t)$	$Q_2(t)$	$Q_1(t)$
1	1	0	Shift register data left one place with SLI (shift left input) shifted into Q_0	$Q_2(t)$	$Q_1(t)$	$Q_0(t)$	$SLI(t)$
1	1	1	Load register with the data on the parallel A input lines	$A_3(t)$	$A_2(t)$	$A_1(t)$	$A_0(t)$
0	d	d	Chip disabled, retain old data	$Q_3(t)$	$Q_2(t)$	$Q_1(t)$	$Q_0(t)$

d - don't care

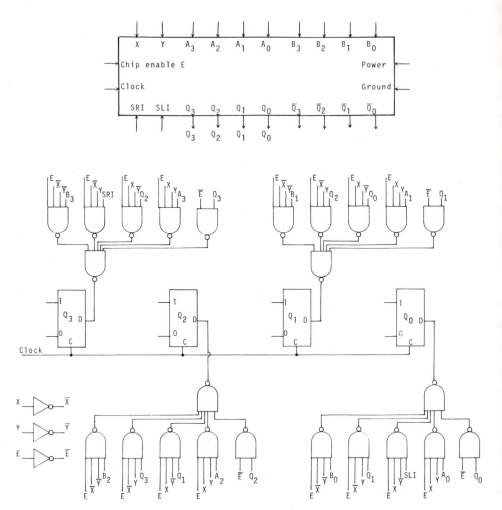

Fig. 11.8 Parallel Load Bidirectional Shift Register Using *D* Flip-Flops

An alternative approach using JK flip-flops requires the determination of a pair of excitation equations for each flip-flop. (Note: The addition of an inverter between JK thereby converting the JK into a delay flip-flop is not really an alternative approach.)

A little thought about when the register changes state reveals that the present state is retained as long as the register is not enabled ($E = 0$). The JK or delay flip-flops can also be wired to retain their data by gating in the clock signal. Thus, if we AND the register-enable signal with the clock input, we need only consider those state changes which take place when the unit is enabled, i.e., when $E = 1$.

Using this information, the next-state equations can be reduced to the following set.

When $E(t) = 1$ then:

$$Q_3(t + \Delta) = \bar{X}(t)\bar{Y}(t)B_3(t) + \bar{X}(t)Y(t)SRI(t) + X(t)\bar{Y}(t)Q_2(t)$$
$$+ X(t)Y(t)A_3(t) \qquad (11.7)$$

$$Q_2(t + \Delta) = \bar{X}(t)\bar{Y}(t)B_2(t) + \bar{X}(t)Y(t)Q_3(t) + X(t)\bar{Y}(t)Q_1(t)$$
$$+ X(t)Y(t)A_2(t) \qquad (11.8)$$

$$Q_1(t + \Delta) = \bar{X}(t)\bar{Y}(t)B_1(t) + \bar{X}(t)Y(t)Q_2(t) + X(t)\bar{Y}(t)Q_0(t)$$
$$+ X(t)Y(t)A_1(t) \qquad (11.9)$$

$$Q_0(t + \Delta) = \bar{X}(t)\bar{Y}(t)B_0(t) + \bar{X}(t)Y(t)Q_1(t) + X(t)\bar{Y}(t)SLI(t)$$
$$+ X(t)Y(t)A_0(t). \qquad (11.10)$$

Applying the synthesis procedure developed in Chapter 9 to these equations, we can obtain the $J_{Q_i}(t)$ exicitation equation by calculating $Q_i(t + \Delta)$ with $Q_i(t) = 0$. The $\bar{K}_{Q_i}(t)$ equations can be obtained by calculating $Q_i(t+\Delta)$ with $Q_i(t) = 1$. For this design problem, the synthesis procedure yields for the ith stage:

$$J_{Q_i}(t) = \bar{X}(t)\bar{Y}(t)B_i(t) + \bar{X}(t)Y(t)Q_{i+1}(t) + X(t)\bar{Y}(t)Q_{i-1}(t)$$
$$+ X(t)Y(t)A_i(t) \qquad (11.11)$$

$$\bar{K}_{Q_i}(t) = \bar{X}(t)\bar{Y}(t)B_i(t) + \bar{X}(t)Y(t)Q_{i+1}(t) + X(t)\bar{Y}(t)Q_{i-1}(t)$$
$$+ X(t)Y(t)A_i(t) \qquad (11.12)$$

$$\bar{K}_{Q_i}(t) = J_{Q_i}(t) \qquad (11.13)$$

$$K_{Q_i}(t) = \bar{J}_{Q_i}(t) \qquad (11.14)$$

For $i = 0$, $Q_{i-1}(t)$ is replaced by SLI(t), and, for $i = 3$, $Q_{i+1}(t)$ is replaced by SRI(t). In this case, an inverter between J and K is the best solution.

A register configuration using these excitation equations is shown in Fig. 11.9. In this circuit, AND-OR-INVERT (NAND-WIRED-AND) gates were selected for the implementation of the K_{Q_i} equations in order to illustrate an alternative gating approach. As stated earlier, the enable line E is used to gate the clock signal to each stage of the register. This same clocking technique can also be applied to the delay flip-flop unit.

In this example it is clear from an examination of the two register configurations shown in Fig. 11.8 and 11.9, that the second delay unit organization is the least complex, only because of special handling of the enable signal.

Let us consider the design of a data register with slightly different requirements as a second example of the type of design procedure one could follow when dealing with multifunction registers. The chip input and output lines are shown in

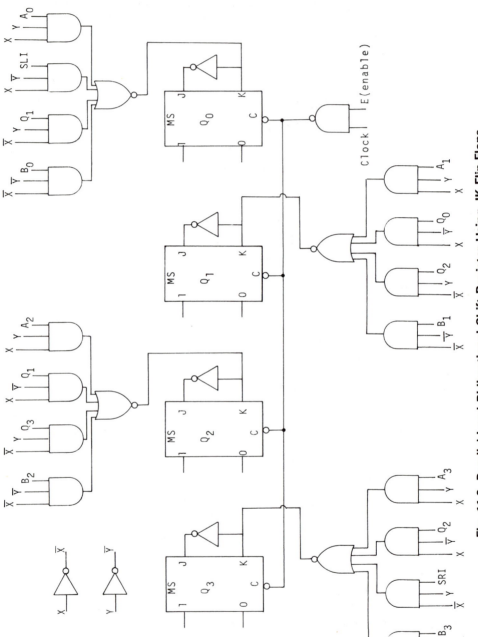

Fig. 11.9 Parallel-Load Bidirectional Shift Register Using JK Flip-Flops

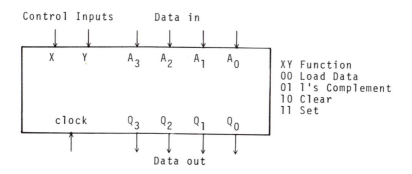

Fig. 11.10 Chip Input and Output Lines for a Multifunction Data Register

Fig. 11.10. The function specifications for this second example are shown in Table 11.2.

From Table 11.2 we can write the next-state equation for the ith register stage as:

$$Q_i(t + \Delta) = \bar{X}(t)\bar{Y}(t)A_i(t) + \bar{X}(t)Y(t)\bar{Q}_i(t) + X(t)\bar{Y}(t) \cdot 0$$
$$+ X(t)Y(t) \cdot 1.$$

Reducing this yields:

$$Q_i(t + \Delta) = \bar{X}(t)\bar{Y}(t)A_i(t) + Y(t)\bar{Q}_i(t) + X(t)Y(t) \qquad (11.15)$$

for $0 \le i \le 3$.

Using a delay flip-flop, the $D_i(t)$ excitation equation would be $D_i(t) = Q_i(t + \Delta)$.

A logic-circuit realization for a typical stage using delay flip-flops is shown in Fig. 11.11.

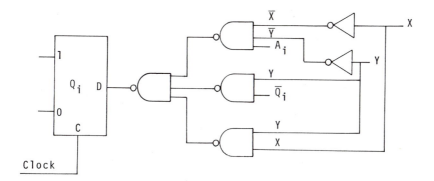

Fig. 11.11 Delay Flip-Flop Realization for a Multifunction Register

TABLE 11.2
Multifunction Register Function Specifications

Control Inputs		Register Function to be Done on the Next Clock Pulse	Next State of the Register			
$X(t)$	$Y(t)$		$Q_3(t+\Delta)$	$Q_2(t+\Delta)$	$Q_1(t+\Delta)$	$Q_0(t+\Delta)$
0	0	Load register with the data on the parallel input lines	$A_3(t)$	$A_2(t)$	$A_1(t)$	$A_0(t)$
0	1	1's complement the data held in the register	$\overline{Q}_3(t)$	$\overline{Q}_2(t)$	$\overline{Q}_1(t)$	$\overline{Q}_0(t)$
1	0	Clear the register	0	0	0	0
1	1	Set the register	1	1	1	1

11.3 SPECIAL REGISTER UNITS

In Chapter 10, a few binary counting circuits were introduced. In this section we will discuss the design of multifunction counting registers as well as data-complementing register units. Finally, we will discuss a logical superposition technique for *JK* flip-flop circuits.

11.3.1 Binary Counting Registers

A very useful counting register is a unit which can either accept external data, count up or down, or be cleared to zero. A logic symbol for such a register unit is shown in Fig. 11.12.

In order to allow this 4-bit unit to be easily interconnected to form larger counting units, a count-enable input line E and a count-enable output line Z have been provided.

In general, the best coding of the function-select lines will depend upon the particular application for the unit, with the most used function assigned the code $XY = 00$, and the least used, the code $XY = 11$. For this design example we will assign a sort of natural code to the function select variables.

X	Y	Function
0	0	Clear to zero
0	1	Count up in binary
1	0	Count down in binary
1	1	Load external data

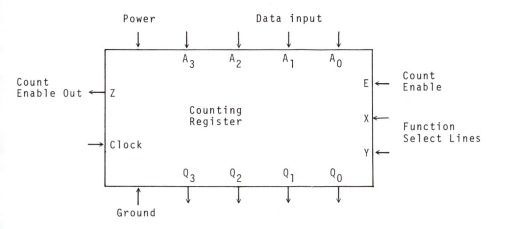

Fig. 11.12 Multifunction Counting Register Symbol

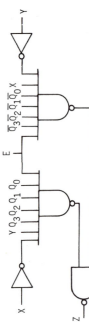

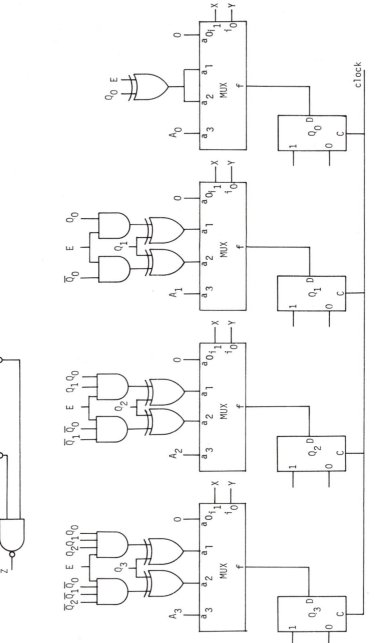

Fig. 11.13 Circuit for a Multifunction Counting Unit (Clear—Count up—count down—load) Using Delay Flip-Flops

Once a function assignment is made, the next step in the design procedure is to write out a set of next-state equations.

$$Q_0(t + \Delta) = \overset{\text{clear}}{\overbrace{\bar{X}(t)\bar{Y}(t)0}} + \overset{\text{count up}}{\overbrace{\bar{X}(t)Y(t)\bar{Q}_0(t)E(t)}}$$

$$+ \overset{\text{count down}}{\overbrace{X(t)\bar{Y}(t)E(t)\bar{Q}_0(t)}} + \overset{\text{load}}{\overbrace{X(t)Y(t)A_0(t)}}$$

$$= \bar{X}(t)Y(t)E(t)\bar{Q}_0(t) + X(t)\bar{Y}(t)E(t)\bar{Q}_0(t) + X(t)Y(t)A_0(t)$$

$$(11.16)$$

and

$$Q_i(t + \Delta) = \overset{\text{clear}}{\overbrace{\bar{X}(t)\bar{Y}(t) \cdot 0}} + \overset{\text{count up}}{\overbrace{\bar{X}(t)Y(t)E(t)}} \cdot \left[Q_i(t) \oplus \bigwedge_{j=0}^{i-1} Q_j(t) \right]$$

$$+ \overset{\text{count down}}{\overbrace{X(t)\bar{Y}(t)E(t)}} \cdot \left[Q_i(t) \oplus \bigwedge_{j=0}^{i-1} \bar{Q}_j(t) \right] + \overset{\text{load}}{\overbrace{X(t)Y(t)A_i(t)}}$$

$$(11.17)$$

for $1 \leq i \leq 3$

$$Z(t) = \overset{\text{count up enable out}}{\overbrace{\bar{X}(t)Y(t)E(t)Q_3(t)Q_2(t)Q_1(t)Q_0(t)}}$$

$$+ \overset{\text{count down enable out}}{\overbrace{X(t)\bar{Y}(t)E(t)\bar{Q}_3(t)\bar{Q}_2(t)\bar{Q}_1(t)\bar{Q}_0(t)}}$$

$$(11.18)$$

When these counting cells are used by themselves, the count enable input E is set to a one. When these units are interconnected to form larger counters, the Z output of each unit serves as the count enable input E for the next most significant counting unit.

Using delay flip-flops and multiplexers, this set of next-state equations (Eqs. (11.16) to (11.18)) can be easily implemented. The control inputs X and Y are used as the selection inputs for each multiplexer. One such circuit implementation is shown in Fig. 11.13.

The implementation of these equations using master-slave JK flip-flops is left as an exercise for the reader.

11.3.2 A Complementing Register

When operating in either two's complement or one's complement arithmetic, there are many occasions when it is necessary to negate the contents of a register. Let us consider the design of a 4-bit complementing register.

DESIGN SPECIFICATION

A 4-bit data register is required which will, on command, either accept data, clear itself to zero, take the one's complement of the data in the register, or take the two's complement of the data in the register. The unit is to be synchronous with all operations being completed in one clock cycle. An additional requirement is that the unit must be able to be interconnected to form longer registers.

Solution. First we must determine both the number and the type of inputs and outputs required by this register specification. The inputs include: four data lines A_3, A_2, A_1, A_0, for external data; two control inputs, X and Y; a clock signal line, C; and a two's complement enable line, T_{in}; and power and ground lines. The outputs include the four register variables, Q_3, Q_2, Q_1, Q_0; and a two's complement output enable line, T_{out}. See Fig. 11.14. Let us assume the control input selection code to be

X	Y	Function	Mnemonic
0	0	Clear register to zero	E_{CL}
0	1	One's complement the register	E_{CM}
1	0	Two's complement the register	E_{TC}
1	1	Load in new data	E_{LD}

Using a decoder unit (Fig. 11.15) we can convert the selection inputs into unique function-enable signals. We will also require that only a single enable line be equal to a one during a clocking interval. With these enable signals, we can write out the next state equations for the unit as

$$Q_0(t + \Delta) = E_{CL} \cdot 0 + E_{CM} \cdot \bar{Q}_0(t) + E_{TC} \cdot (T_{in}(t) \oplus Q_0(t)) + E_{LD} \cdot A_0(t) \quad (11.19)$$

$$Q_i(t + \Delta) = E_{CL} \cdot 0 + E_{CM} \cdot \bar{Q}_i(t) + E_{TC} \cdot \left[\left(T_{in}(t) + \bigvee_{j=0}^{i-1} Q_j(t) \right) \oplus Q_i(t) \right] + E_{LD} \cdot A_i(t) \quad (11.20)$$

For $1 \le i \le 3$,

$$T_{out}(t) = T_{in}(t) + \bigvee_{j=0}^{3} Q_i(t). \quad (11.21)$$

In this case, the two's complement operation is being accomplished by scanning the register from right to left until a nonzero entry is found. All digits to the left of this entry are complemented. Thus, when these units are interconnected to form longer registers, the rightmost unit must have T_{in} set to zero. The output T_{out} of each unit then is fed into the T_{in} input of the next most-significant unit.

Applying the standard synthesis procedures in Chapter 9 to Eqs. (11.19) and (11.20), we obtain the following excitation equations for the ith stage of the register:

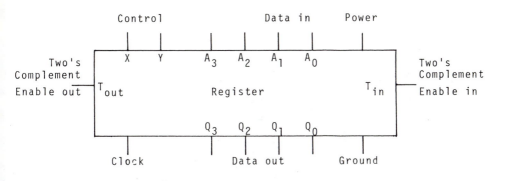

Figure 11.14

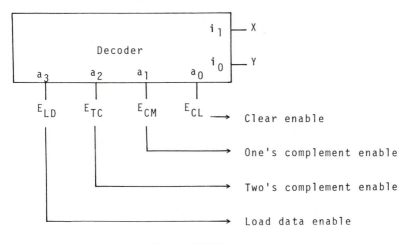

Figure 11.15

$$J_{Q_i}(t) = Q_i(t + \Delta)\big|_{Q_i(t)=0} = E_{CM} + E_{TC}\left[T_{in}(t) + \bigvee_{j=0}^{i-1} Q_j(t) \right] + E_{LD} \cdot A_i(t) \qquad (11.22)$$

and

$$\overline{K_{Q_i}(t)} = Q_i(t + \Delta)\big|_{Q_i(t)=1} = E_{TC}\overline{\left[T_{in}(t) + \bigvee_{j=0}^{i-1} Q_j(t) \right]} + E_{LD} \cdot A_i(t). \qquad (11.23)$$

These equations suggest the circuit implementation shown in Fig. 11.16.

A direct implementation of the next-state equation using a delay flip-flop with a multiplexer input is shown in Fig. 11.17. If we include the complexity of the multiplexer into our logic cost, then the JK flip-flop organization is less complex for this register configuration.

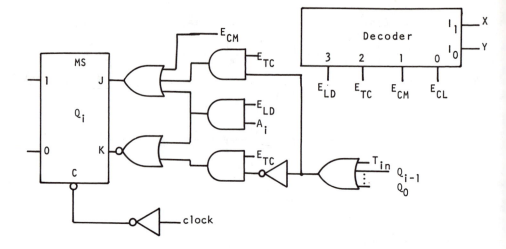

Fig. 11.16 Typical Logic Circuit for a Multifunction Complementing Register

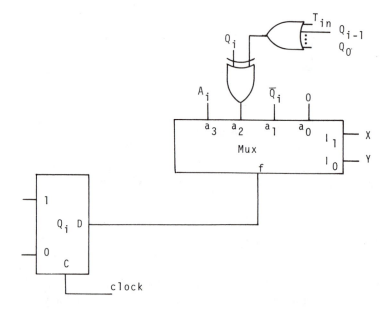

Fig. 11.17 Alternate Multifunctioning Complementing Register Stage Implementation Using a Delay Flip-Flop

11.3.3 Logical Superposition

In the preceding example, an alternate set of excitation equations can be obtained by using a logical superposition approach. Since it is known by the circuit designer that at most only one of the function-enable lines E_{f_j} can be 1 during a given clocking interval, the overall register design can be broken into a series of smaller problems. By treating each function separately, we obtain simpler J and K excitation equations. The final set of JK equations will be the logical OR of the separate function-excitation equations. Since each of the function-excitation equations will be ANDed with the function enable signal during a given clocking interval, only the required function will affect the flip-flop. Thus the excitation equations for stage Q_i become

$$J_{Q_i}(t) = \bigvee_{j=0}^{n} E_{f_j}(t) \cdot J_{f_{j_{Q_i}}}(t) \tag{11.24}$$

and

$$K_{Q_i}(t) = \bigvee_{j=0}^{n} E_{f_j}(t) \cdot K_{f_{j_{Q_i}}}(t) \tag{11.25}$$

where $E_{f_j}(t)$ is the enable for function j and $J_{f_{j_{Q_i}}}(t)$ and $K_{f_{j_{Q_i}}}(t)$ are the excitation equations for this single function.

Using the preceding example, the next-state equation

$$Q_i(t + \Delta) = E_{CL} \cdot 0 + E_{CM} \cdot \bar{Q}_i(t)$$

$$+ E_{TC} \cdot \left[\left(T_{in}(t) + \bigvee_{j=0}^{i-1} Q_j(t) \right) \oplus Q_i(t) \right] + E_{LD} A_i(t)$$

can be broken into four separate next-state equations.

1. If $E_{CL}(t) = 1$, then $Q_i(t + \Delta) = 0$. This function yields the excitation equations

$$J_{CL_{Q_i}}(t) = 0 \quad \text{and} \quad K_{CL_{Q_i}}(t) = 1.$$

2. If $E_{CM}(t) = 1$, then $Q_i(t + \Delta) = \bar{Q}_i(t)$. This function yields the excitation equations

$$J_{CM_{Q_i}}(t) = 1 \quad \text{and} \quad K_{CM_{Q_i}}(t) = 1.$$

3. If $E_{TC}(t) = 1$, then $Q_i(t + \Delta) = \left[T_{in}(t) + \bigvee_{j=0}^{i-1} Q_j(t) \right] \oplus Q_i(t)$.

This function yields the excitation equations

$$J_{TC_{Q_i}}(t) = K_{TC_{Q_i}}(t) = \left[T_{in}(t) + \bigvee_{j=0}^{i-1} Q_j(t) \right].$$

4. If $E_{LD}(t) = 1$, then $Q_i(t + \Delta) = A_i(t)$. This function yields the excitation equations

$$J_{LD_{Q_i}}(t) = A_i(t) \quad \text{and} \quad K_{LD_{Q_i}}(t) = \bar{A}_i(t).$$

Applying logical superposition to this problem yields the desired overall excitation equations for stage Q_i

$$J_{Q_i}(t) = E_{CL}(t) \cdot J_{CL_{Q_i}}(t) + E_{CM}(t) \cdot J_{CM_{Q_i}}(t) + E_{TC}(t) \cdot J_{TC_{Q_i}}(t) + E_{LD}(t) \cdot J_{LD_{Q_i}}(t)$$

and

$$K_{Q_i}(t) = E_{CL}(t) \cdot K_{CL_Q}(t) + E_{CM}(t) \cdot K_{CM_Q}(t) + E_{TC}(t) \cdot K_{TC_Q}(t) + E_{LD}(t) \cdot K_{LD_Q}(t).$$

Hence, for this example we have

$$J_{Q_i}(t) = E_{CM}(t) + E_{TC}(t) \cdot \left[T_{in}(t) + \bigvee_{j=0}^{i-1} Q_j(t) \right] + E_{LD} \cdot A_i(t) \tag{11.26}$$

and

$$K_{Q_i}(t) = E_{CL}(t) + E_{CM}(t) + E_{TC}(t) \cdot \left[T_{in}(t) + \bigvee_{j=0}^{i-1} Q_j(t) \right] + E_{LD} \cdot \bar{A}_i(t). \tag{11.27}$$

An implementation of these equations is shown in Figure 11.16.

The overall complexity of both circuit configurations (Fig. 11.16 and Fig. 11.18) is about the same. However, the latter circuit configuration lends itself to easier analysis of the register operations.

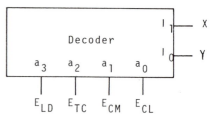

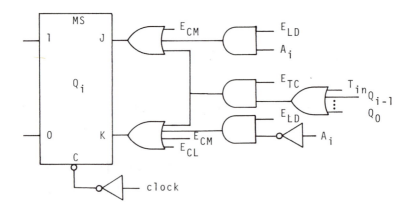

Fig. 11.18 Complementing-Register Stage Resulting from Superposition Approach

TABLE 11.3
Summary of JK Flip-Flop Operations

Operation	Enable	Next State Equation $Q(t + \Delta)$	Excitation Inputs $J_Q(t)$	$K_Q(t)$
no operation	E_{NOP}	$Q(t)$	0	0
Clear	E_{CL}	0	0	1
Set	E_{ST}	1	1	0
Complement	E_{CM}	$\bar{Q}(t)$	1	1
Load	E_{LD}	$W(t)$	$W(t)$	$\bar{W}(t)$
OR	E_{OR}	$Q(t) + X(t)$	$X(t)$	0
AND	E_{AND}	$Q(t) \cdot Y(t)$	0	$\bar{Y}(t)$
XOR	E_{XOR}	$Q(t) \oplus Z(t)$	$Z(t)$	$Z(t)$

A little thought reveals that there are only eight basic register functions which must be considered. These are: No operation, clear to zero, set to all ones, one's complement the register contents, load new data, and OR, AND, or XOR the register contents with data. Table 11.3 summarizes the requirements for each of these basic operations. Shifting operations would use the load function and counting operations of the XOR function.

The general next-state equation for such a register would be:

$$
\begin{aligned}
Q(t + \Delta) = {} & E_{NOP}(t)Q(t) + &&\text{no operation} \\
& E_{CL}(t) \cdot 0 + &&\text{clear} \\
& E_{ST}(t) \cdot 1 &&\text{set} \\
& E_{CM}(t) \cdot \bar{Q}(t) &&\text{complement} \\
& E_{LD}(t) \cdot W(t) &&\text{load} \\
& E_{OR}(t) \cdot (Q(t) + X(t)) &&\text{OR} \\
& E_{AND}(t) \cdot (Q(t) \cdot Y(t)) &&\text{AND} \\
& E_{XOR}(t) \cdot (Q(t) \oplus Z(t)). &&\text{XOR} && (11.28)
\end{aligned}
$$

The set of excitation equations for this general register using logical superposition are

$$J_Q(t) = E_{ST}(t) + E_{CM}(t) + E_{LD}(t) \cdot W(t) + E_{OR}(t) \cdot X(t) + E_{XOR}(t) \cdot Z(t) \quad (11.29)$$

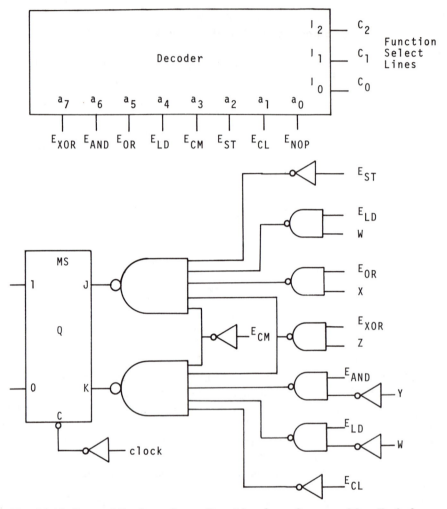

Fig. 11.19 General-Register Stage Resulting from Superposition Technique

and

$$K_Q(t) = E_{CL}(t) + E_{CM}(t) + E_{LD}(t) \cdot \bar{W}(t) + E_{AND}(t) \cdot \bar{Y}(t) + E_{XOR}(t) \cdot Z(t).$$
$$(11.30)$$

A NAND gate implementation of these equations is shown in Fig 11.19.

The conventional design procedure (i.e., without using any prior knowledge about the enable signals) yields the following set of excitation equations.

$$J_Q(t) = Q(t + \Delta)\Big|_{Q(t)=0}$$

$$= E_{ST}(t) + E_{CM}(t) + E_{LD}(t) \cdot W(t) + E_{OR}(t)X(t) + E_{XOR}(t)Z(t) \qquad (11.31)$$

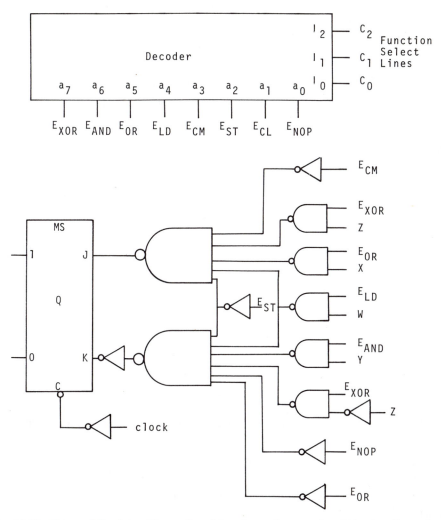

Fig. 11.20 General-Register Stage Resulting from Conventional Design Procedure

and

$$K_Q(t) = \overline{Q(t + \Delta)}\Big|_{Q(t)=1}$$

$$= \overline{E_{NOP}(t) + E_{ST}(t) + E_{LD}(t)W(t) + E_{OR}(t) + E_{AND}(t) \cdot Y(t) + E_{XOR}(t)\bar{\bar{Z}}(t)}$$

$$(11.32)$$

A NAND-gate implementation of these equations is shown in Fig. 11.20.

A comparison of the two sets of excitation equations and their corresponding circuit implementation reveals that these two design techniques yield configurations with about equal circuit complexity.

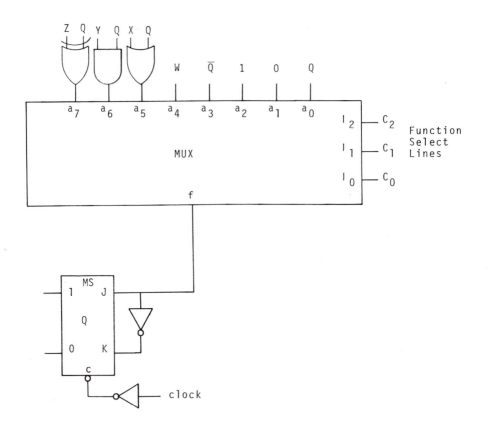

Fig. 11.21 General-Register Stage Resulting from a Delay Flip-Flop Design Procedure

Finally, let us directly implement the next-state equation (11.28) by converting a JK flip-flop into a delay flip-flop. A multiplexer input unit will be used to provide the function selection. The resulting D flip-flop circuit configuration is shown in Fig. 11.21.

The final selection of design techniques for register units is left to the reader. However, the delay-unit approach generally yields the least amount of hardware when more than four functions are involved.

11.4 TIMING PATTERN GENERATORS

There are occasions when special periodic sequences of ones and zeroes are required. One application for these units is the generation of testing patterns used for diagnosing logic circuit faults.

In general these special patterns can be obtained in two ways. The first approach is to use a feedback shift register wired in such a manner that it automatically generates the desired pattern. The second approach is to use a modulo-N binary counter and assign zeroes and ones to the output function in

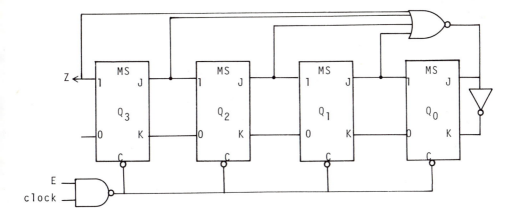

Fig. 11.22 Feedback Shift Register Realization of the Periodic Pattern 00001

such a way as to satisfy the pattern specification. The circuit shown in Fig. 11.22 illustrates the first approach and the circuit shown in Fig. 11.23 illustrates the second. In both cases the desired sequence is the pattern 00001.

Both of these sequence generators are self-starting and synchronously clocked, with the counter unit requiring the least amount of hardware. In this section we will explore briefly the synthesis procedures used to obtain these two circuit configurations. However, the interested reader is referred to the books

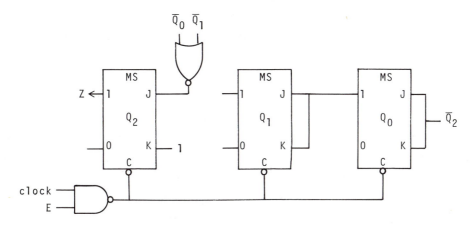

Fig. 11.23 Counter Generation of the Periodic Sequence 00001

listed at the end of this chapter for a more in-depth treatment of this interesting subject.

11.4.1 Feedback Shift Register Sequence Generators

Given a finite periodic sequence of zeroes and ones, we would like to construct the shortest self-starting feedback shift register which will generate the sequence. The general form of the feedback shift register is shown in Figure 11.24.

A k-stage feedback shift register for an n-bit periodic sequence of the form $I_0 I_1 I_2 I_3 \cdots I_{n-2} I_{n-1}$ will require n unique states, each k bits wide, with the restriction that

$$\lceil \log_2 n \rceil \le k \le n.$$

It is understood, of course, that the sequence given is not a multiple of a shorter sequence.

The state coding is obtained by collecting together each consecutive group of k inputs from the sequence:

$$S_0 = I_0 I_1 I_2 \cdots I_{k-1}$$
$$S_1 = I_1 I_2 I_3 \cdots I_k$$
$$S_2 = I_2 I_3 I_4 \cdots I_{k+1}$$

$$\vdots$$

$$S_{n-k} = I_{n-k} I_{n-k+1} I_{n-k+2} \cdots I_{n-2}$$
$$S_{n-k+1} = I_{n-k+1} I_{n-k+2} I_{n-k+3} \cdots I_{n-1}$$

$$\vdots$$

$$S_{n-2} = I_{n-2} I_{n-1} I_0 \cdots I_{k-3}$$
$$S_{n-1} = I_{n-1} I_0 I_1 \cdots I_{k-2}.$$

The time-dependent next-state equations for this feedback shift register unit are:

$$Q_i(t + \Delta) = Q_{i+1}(t); \qquad 0 \le i \le k - 2$$

and

$$Q_{k-1}(t + \Delta) = F(Q_0(t), Q_1(t) \cdots, Q_{k-2}(t), Q_{k-1}(t)) = F(S_t)$$

with

$$(S_t) = (I_{t \bmod N}, I_{(t+1) \bmod N}, \ldots, I_{(t+k-2) \bmod N}, I_{(t+k-1) \bmod N})$$

with the subscripts on the I's all evaluated modulo N. The time t is expressed in clock cycles.

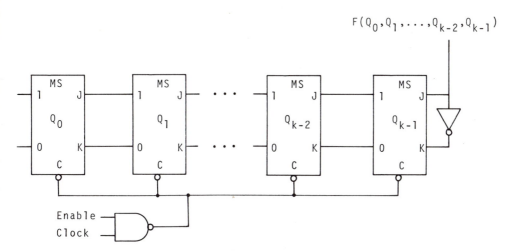

Fig. 11.24 General Feedback Shift Register Organization

The value of the feedback function for the $2^k - n$ states not included in the sequence can be selected in any way, provided no additional cycles are created.

For example, the periodic sequence

0	1	1	0	1	0
I_0	I_1	I_2	I_3	I_4	I_5

can be realized by a three-stage feedback shift register, with the feedback function specification shown in Fig. 11.25.

An inspection of the state diagram for this unit, shown in Fig. 11.26, reveals that the states (000) and (111) which are not included in the desired sequence must feedback a 1 and 0 respectively in order to preserve the self-starting (single cycle) characteristic of the unit.

Using a K-map as an aid, the feedback function for this example is

$$f(Q_0(t), Q_1(t), Q_2(t)) = \bar{Q}_0(t)\bar{Q}_1(t) + Q_0(t) \cdot \bar{Q}_2(t).$$

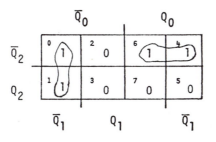

j	$S_j(Q_0,Q_1,Q_2)$			$F(S_j)$	
	I_{j+0}	I_{j+1}	I_{j+2}	I_{j+3}	
0	0	1	1	0	
1	1	1	0	1	
2	1	0	1	0	states on sequence
3	0	1	0	0	
4	1	0	0	1	
5	0	0	1	1	

Q_0	Q_1	Q_2	$F(Q_0,Q_1,Q_2)$	
0	0	0	1	states not on sequence
1	1	1	0	

Figure 11.25

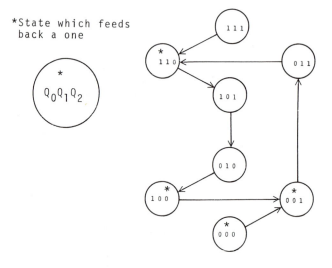

Fig. 11.26 State Diagram for the Sequence 011010 Using a Three-Stage Feedback Shift Register

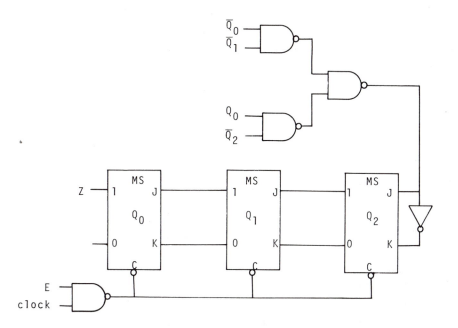

Fig. 11.27 Feedback Shift-Register Generator for the Sequence 011010

The resulting shift-register configuration is shown in Fig. 11.27.

The key to understanding the synthesis procedure for these feedback shift-register generators is the fact that when we partition the input sequence into n states there can be no duplicate states. In other words, there cannot be two k-bit subsequences with the same bit pattern. Violating this condition will lead to a conflicting feedback function specification with $F(S_i) = \bar{F}(S_j)$ and $S_i = S_j$. For example, given the sequence 0001, a 2-bit state assignment will yield

$$S_0 = 00 \qquad \text{with } F(S_0) = 0$$
$$S_1 = 00 \qquad \text{with } F(S_1) = 1$$
$$S_2 = 01 \qquad \text{with } F(S_2) = 0$$
$$S_3 = 10 \qquad \text{with } F(S_3) = 0$$

In this case, we have $S_0 = S_1 = 00$ and

$$F(S_0) = \bar{F}(S_1).$$

This conflict implies that a larger value of k is required to implement this sequence using a feedback shift register. With this conflict possibility in mind, we can state a simple synthesis procedure.

ALGORITHM 11.1

Synthesis procedure for the generation of an N-bit periodic sequence using feedback shift registers

Given the periodic sequence $I_0 I_1 \cdots I_{N-1}$,

STEP	OPERATION
1.	Pick $k = \lceil \log_2 N \rceil$
2.	Partition the sequence into N states with $S_j = I_j I_{j+1} \cdots I_{j+k-2} I_{j+k-1}$ where the subscripts are evaluated modulo N.
3.	If two states S_i and S_j have the same assignment, increase k by 1 and go back to Step 2. Otherwise, proceed to Step 4.
4.	For the state assignments which form a subsequence of the original sequence, assign the feedback function $F(S_j) = I_{(j+k) \bmod N}$. For those state assignments which do not form a subsequence of the original sequence, assign the feedback function so as to preserve the self-starting (single cycle) characteristic of the unit.
5.	Using the general feedback shift-register circuit shown in Fig. 11.22, implement the feedback function obtained in Step 5 and STOP.

Let us illustrate this algorithm using the periodic sequence

$$I_0 \quad I_1 \quad I_2 \quad I_3 \quad I_4 \quad I_5$$
$$0 \quad \ 0 \quad \ 0 \quad \ 0 \quad \ 1 \quad \ 1$$

STEP	OPERATION
1.	$k = \lceil \log_2 6 \rceil = 3$
2.	$k = 3; \quad S_0 = I_0 I_1 I_2 = 000$
	$S_1 = I_1 I_2 I_3 = 000$
	$S_2 = I_2 I_3 I_4 = 001$
	$S_3 = I_3 I_4 I_5 = 011$
	$S_4 = I_4 I_5 I_0 = 110$
	$S_5 = I_5 I_0 I_1 = 100$
3.	Since $S_0 = S_1 = 000$, we must increase k to 4 and return to Step 2.
2.	$k = 4; \quad S_0 = I_0 I_1 I_2 I_3 = 0000$
	$S_1 = I_1 I_2 I_3 I_4 = 0001$
	$S_2 = I_2 I_3 I_4 I_5 = 0011$
	$S_3 = I_3 I_4 I_5 I_0 = 0110$
	$S_4 = I_4 I_5 I_0 I_1 = 1100$
	$S_5 = I_5 I_0 I_1 I_2 = 1000$

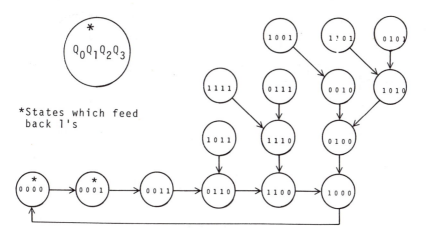

*States which feed
back 1's

Figure 11.28

3.   Since all the state assignments are unique, we can proceed to Step 4.
4.   The feedback function values for the six states in the original sequence are

$$f(S_0) = I_4 = 1$$
$$f(S_1) = I_5 = 1$$
$$f(S_2) = I_6 = I_0 = 0$$
$$f(S_3) = I_7 = I_1 = 0$$
$$f(S_4) = I_8 = I_2 = 0$$
$$f(S_5) = I_9 = I_3 = 0$$

Since state (0000) is on the original sequence, we can allow the feedback function for all the remaining states not in the original sequence to feed back a zero without causing any additional cycles. This can be verified by checking the state diagram in Fig. 11.28.

5.   The feedback function specified in Step 4 is

$$F(Q_0(t), Q_1(t), Q_2(t), Q_3(t)) = \bar{Q}_0(t)\bar{Q}_1(t)\bar{Q}_2(t).$$

The feedback shift register in Fig. 11.29 will generate the periodic sequence 000011.

11.4.2 Resettable Counter Sequence Generators

The above feedback shift-register sequence-generator procedure may require a large number of iterations before a valid generator can be found. An alternate approach to the generation of an arbitrary periodic sequence does exist.

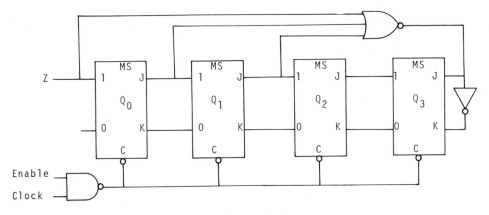

Figure 11.29

Given an N-bit periodic sequence

$$I_0 I_1 \cdots I_{n-1},$$

we can construct a modulo-N binary counter using the techniques outlined in Chapter 10. The outputs of this counter can then be decoded using a standard decoding circuit such that

$$D_I = 1; \quad \text{if and only if the counter is in}$$
$$\text{state } S_I = (i_{k-1}, i_{k-2}, \ldots, i_1, i_0)$$
$$\text{where } (I)_{10} = (i_{k-1} i_{k-2} \cdots i_1 i_0)_2.$$

The required periodic sequence is obtained by setting

$$Z = \bigvee_{j=0}^{N-1} (D_j \cdot I_j).$$

Any output hazards can be eliminated by using Z as the input to a delay flip-flop. However, this change will result in the output sequence being delayed by one unit of time. The general circuit configuration for this unit is shown in Fig. 11.30. The periodic sequence 011010 can be realized as shown in Fig. 11.31.

The preceding discussion leads to the following synthesis procedure.

ALGORITHM 11.2

Synthesis procedure for the generation of an N-bit periodic sequence using resettable counters.

Given the periodic sequence

$$I_0 I_1 \cdots I_{n-1},$$

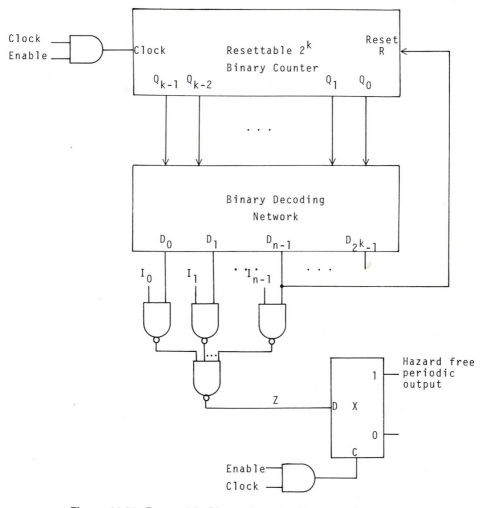

Figure 11.30 Resettable Binary-Counter Sequence Generator

STEP   OPERATION

1.   Calculate $k = \lceil \log_2 N \rceil$.

2.   Design a modulo-N self-starting synchronous binary counter using k flip-flop stages.

3.   Decode the binary counter with $D_t = 1$ if and only if the counter is in the counter state S_t. Form the output function

$$Z = \bigvee_{j=0}^{N-1} (D_j \cdot I_j).$$

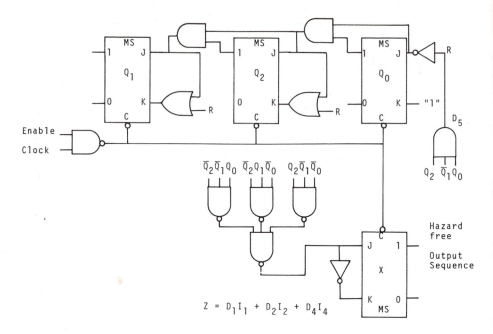

Fig. 11.31 Sequence Generator for 011010

4.   If the logical hazards in the output Z are no problem, stop here. Otherwise use Z as the input to a delay flip-flop synchronized with the counter and then STOP.

Let us illustrate this algorithm using the same periodic sequence as we used with the feedback shift register.

Given the 6-bit sequence

$$I_0 \quad I_1 \quad I_2 \quad I_3 \quad I_4 \quad I_5$$
$$0 \quad \ 0 \quad \ 0 \quad \ 0 \quad \ 1 \quad \ 1,$$

we apply Algorithm 11.2.

STEP   OPERATION

1. $k = \lceil \log_2 6 \rceil = 3$.

2.   Design a modulo-6 binary counter using three flip-flop stages.

3.   Form the function

$$Z = \bigvee_{j=0}^{5} D_j I_j = D_4 + D_5$$
$$= Q_2 \cdot \bar{Q}_1.$$

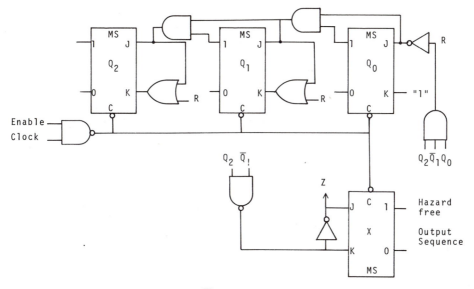

Figure 11.32

4.   Feed this function into a delay flip-flop X in order to eliminate possible output hazards, and STOP.

Figure 11.32 shows the resulting circuits.

A comparison of these two methods reveals that the latter method is easier to implement and, in general, will use the least amount of hardware. This reasoning is based on the fact that the feedback function for the shift-register generator will be more complex than the decoding network used by the counter. However, for long sequences a hardware approach will always be expensive.

11.5 PROBLEMS

11.1   Using the procedure outlined in Section 11.2.1, design a 4-bit data register using unclocked SR flip-flops and a two-step read operation (set before clearing).

11.2   Design a 5-bit bidirectional shift register described by the following set of equations:

$$Q_5(t + \Delta) = Q_4(t) \cdot M(t) + \text{SRI}(t)\bar{M}(t)$$

$$Q_i(t + \Delta) = Q_{i-1}(t) \cdot M(t) + Q_{i+1}(t) \cdot \bar{M}(t) \qquad i = 2, 3, 4$$

$$Q_1(t + \Delta) = \text{SLI}(t) \cdot M(t) + Q_2(t) \cdot \bar{M}(t)$$

where $M(t)$ is the direction enable, with $M = 0$ implying a right shift and $M = 1$ a left shift. SRI and SLI are the shift right and shift left data inputs respectively.

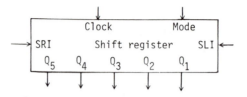

The register is to shift left or right on the leading edge of the clock line. You may assume that M, SRI, and SLI will only change when the clock line is zero.

11.3   Design the multifunction register specified in Fig. 11.33.

Control Inputs			Register Function to be Done on the Next Clock Pulse	Next State of the Register			
$E(t)$	$x(t)$	$y(t)$		$Q_3(t+\Delta)$	$Q_2(t+\Delta)$	$Q_1(t+\Delta)$	$Q_0(t+\Delta)$
1	0	0	Load register with the data on the parallel data input lines	$A_3(t)$	$A_2(t)$	$A_1(t)$	$A_0(t)$
1	0	1	Shift the register data right one place with SRI (shift right input) shifted into Q_3	$SRI(t)$	$Q_3(t)$	$Q_2(t)$	$Q_1(t)$
1	1	0	AND the data in the register with the data on the input lines	$A_3(t)Q_3(t)$	$A_2(t)Q_2(t)$	$A_1(t)Q_1(t)$	$A_0(t)Q_0(t)$
1	1	1	Clear the register to zero	0	0	0	0
0	d	d	Chip disabled, retain old data	$Q_3(t)$	$Q_2(t)$	$Q_1(t)$	$Q_0(t)$

d - don't care

Figure 11.33

a) Use JK flip-flops (do not convert into delay units).
b) Use delay flip-flops with multiplexers on the data inputs.
c) Use delay flip-flops and AND the chip enable with the clock line.

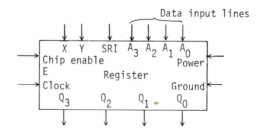

11.4   Design a 4 bit multifunction data register which satisfies the specifications in Fig. 11.34.

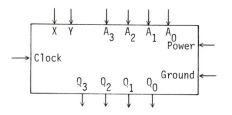

a) Write out a set of next-stage equations for this unit.
b) Implement the set of next-state equations using JK flip-flops (do not convert to delay units).

| Control Inputs | | Register Function to be Done |
$x(t)$	$y(t)$	on the next clock pulse
0	0	Retain old data in the register
0	1	Load in the data on the data input lines
1	0	Count up in binary by one
1	1	Clear the register to zero

Register (1)

| Control Inputs | | Register Function to be Done |
$x(t)$	$y(t)$	on the Next Clock Pulse
0	0	Retain old data
0	1	Clear the unit to zero
1	0	XOR the data in the register with the data on the data input lines
1	1	AND the data in the register with the data on the data input lines

Register (2)

Figure 11.34

Control Inputs x(t) y(t)	Register Function to be Done on the Next Clock Pulse
0 0	Retain old data
0 1	Load in data on the data in-put lines
1 0	Count up in binary by one Register(t+Δ) = Register(t) plus 1
1 1	Count up in binary by two Register(t+Δ) = Register(t) plus 2

Register (3)

Figure 11.34 (*Continued*)

    c) Implement the set of next-state equations using D flip-flops with multiplexers on the inputs.

    d) Compare your designs for (b) and (c).

11.5   Redesign the multifunction registers specified in Problem 11.4 using the concept of logical superposition discussed in Section 11.3.3.

11.6   Using Algorithm 11.1 as outlined in Section 11.4.1 design a set of feedback shift registers which output the following periodic sequences:

$I_0 I_1 \cdots I_{n-1}$

a) 0 1 0 1 0 $n = 5$
b) 1 1 1 0 1 0 0 0 $n = 8$
c) 0 0 0 1 1 1 $n = 6$
d) 0 0 1 1 1 $n = 5$
e) 1 0 1 1 1 0 0 0 $n = 8$

11.7   Design a set of periodic sequence generators using resettable binary counters, as outlined in Algorithm 11.2, which output the following periodic sequences:

$I_0 I_1 \cdots I_{n-1}$

a) 0 1 1 1 0 $n = 5$
b) 1 1 1 0 1 0 0 0 $n = 8$
c) 1 0 1 0 1 1 $n = 6$
d) 1 0 1 $n = 3$

11.8   You have been asked to design a clocked synchronous sequential circuit which can be used to examine a serial input line outputting a one each time that the sequence 110 occurs.

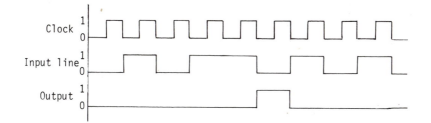

Your unit should include a reset line to initially clear the unit to zero. Assume the input line can only be changed when the clock line is at zero.

11.9   You have been asked to design a clocked synchronous sequential circuit which will accept eight bits of serial data being sent, using the following 11-bit format.

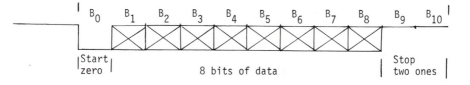

a) The first bit must be a zero.
b) The next eight bits contain the data sent.
c) Bits B_9 and B_{10} must both be ones.

The data is sent over at irregular intervals with the line maintained at one during intervals between data words. Your task is to design a unit which will accept the data and indicate when a valid data word has been received.

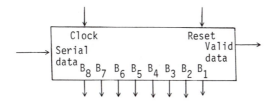

A reset line should be included to allow the valid data time to be externally cleared. The valid data line should automatically return to zero at the start of each new data sequence.

11.10 You have been given the task of designing an alarm system of a control system. The control unit outputs a series of pulses on two separate lines, x and y. Over a sample period of 100 clock times, the same number of pulses must be used by both control lines, otherwise an alarm should be set. A reset signal is used to synchronize the counting period with the control systems. The reset signal is used to start each sample period.

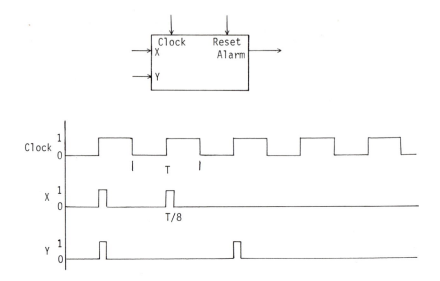

Note: The pulses are sent during the interval when the clock line is high in synch with the 0 to 1 transition. These pulses are wide enough to be used as a clock line for either a master-slave or a JK flip-flop. The pulse width is only $\frac{1}{8}$ the duration of the clocking period, T.

11.6 BIBLIOGRAPHY

Bartee, T. C., *Digital Computer Fundamentals*. 3rd ed. New York: McGraw-Hill, 1972.

Gill, A., *Introduction to the Theory of Finite State Machines*. New York: McGraw-Hill, 1962.

Golomb, S. W., *Shift Register Sequences*. San Francisco: Holden Day, 1967.

Huffman, D. A., "The synthesis of sequential switching circuits," *J. Franklin Institute*, Vol. 257, pp. 275–303, Apr. 1954, and pp. 161–190, Mar. 1954.

Kautz, W. W., *Linear Sequential Switching Circuits*. San Francisco: Holden Day, 1965.

Morris, R. L., and J. R. Miller (Eds.), *Designing with TTL Integrated Circuits*. New York: McGraw-Hill, 1971.

Peatman, J. B., *The Design of Digital Systems*. New York: McGraw-Hill, 1972.

Rhyne, V. T., *Fundamentals of Digital Systems Design*. Englewood Cliffs, N.J.: Prentice Hall, 1973.

Chapter 12

Advanced
Arithmetic Units

12.1 INTRODUCTION

Up until now we have been working with digital sub-systems. In the two preceding chapters, we have developed techniques for designing binary counting and register circuits. We have also discussed elementary control structure using monostable multivibrators or one shots.

In this chapter, we will be concerned with the development of control strategies for some basic digital systems.

Section 12.2 will be devoted to the design of a serial adder-subtracter. In this section, we will use a binary counting circuit as our primary control device.

Section 12.3 will be devoted to the development of binary multiplication circuits using a serial-parallel technique. The concept and use of a control matrix will be presented as a design aid.

Finally, in Section 12.4, algorithms for binary division will be introduced. The exercises at the end of this chapter are intended to extend the material contained in this chapter. In particular, algorithms for signed binary multiplication and signed binary division are included in the problem set.

12.2 A BINARY SERIAL ADDER

Let us combine some of the components we have been discussing in the last few chapters into a useful computation circuit. In particular, let us design a serial adder capable of adding or subtracting two binary numbers. We will restrict our example to four binary digits, i.e., we are going to add two 4-bit binary numbers together yielding a 4-bit sum, a carry out, and an overflow signal. Figure 12.1 shows a simple block diagram for a serial adder. Two's complement arithmetic will be used in this adder-subtracter unit.

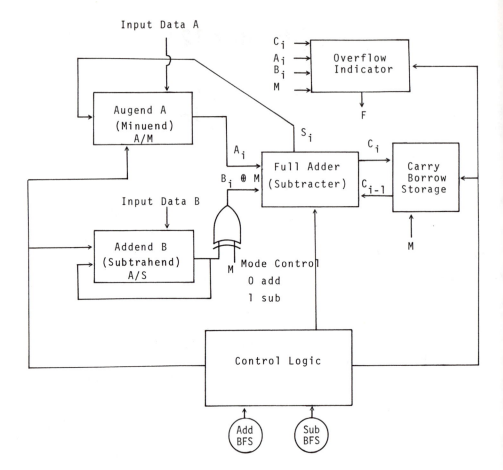

Fig. 12.1 Block Diagram of a Serial Adder

Because this is the first control-system example, we will make a few simplifying assumptions in the interest of keeping the control system as uncomplicated as possible.

The two's complement required for subtraction will be obtained by taking the one's complement of each bit entering the serial adder and setting the initial carry/borrow storage register to a one. We will also assume that the required data will be transmitted to the registers in parallel.

The starting point for the design of an arithmetic unit is a basic algorithm describing the function to be implemented by the hardware unit. Algorithm 12.1 for binary addition ($M = 0$) and subtraction ($M = 1$) will be used. Careful examination of this algorithm will reveal the necessary control hardware requirements.

ALGORITHM 12.1

An algorithm for a 4-bit serial addition and subtraction unit

STEP	OPERATION	Mode, $M = 0$ Add $M = 1$ Sub	Augend/ minuend A/M	Addend/ subtrahend A/S	Carry/ borrow
1.	Start. Turn on event clock.		$xxxx$	$xxxx$	x
2.	Clear overflow indicator F. Read A into the augend/ minuend register. Jam the value of M into carry/borrow register setting c_{-1} to M. $M = 0$ Addition. $M = 1$ Subtraction.		$a_3 a_2 a_1 a_0$	$xxxx$	M
3.	Read B into the addend/ subtrahend register.		$a_3 a_2 a_1 a_0$	$b_3 b_2 b_1 b_0$	M
4.	Set $i = 0$.				
5.	Add a_i, $(b_i \oplus M)$ and c_{i-1}. Shift the A/M and A/S registers right one place, jamming s_i into the A/M register from the left. Shift b_i back into the A/S register and replace c_{i-1} with c_i,		$s_0 a_3 a_2 a_1$	$b_0 b_3 b_2 b_1$	c_0
6.	Set $i = 1$ and repeat 5,		$s_1 s_0 a_3 a_2$	$b_1 b_0 b_3 b_2$	c_1
7.	Set $i = 2$ and repeat 5,		$s_2 s_1 s_0 a_3$	$b_2 b_1 b_0 b_3$	c_2
8.	If $a_3(b_3 \oplus M)\bar{c}_2$ $+ \bar{a}_3(\overline{b_3 \oplus M})c_2 = 1$, then turn on overflow indicator F.				
9.	Set $i = 3$ and repeat 5.		$s_3 s_2 s_1 s_0$	$b_3 b_2 b_1 b_0$	c_3
10.	STOP. The answer is in the augend/minuend register, $s_3 s_2 s_1 s_0$. It will be correct if the overflow indicator is not on.				

Example: Add 0101 to 0010

Step	Action taken	Augend	Addend	Carry
1	Start	xxxx	xxxx	x
2	Read Augend	0010	xxxx	0
3	Read Addend	0010	0101	0
4	Set $i = 0$			
5	Add and shift	1001	1010	0
6	Add and shift	1100	0101	0
7	Add and shift	1110	1010	0
8	No overflow			
9	Add and shift	0111	0101	0
10	STOP			

Answer is 0111

12.2.1 The Augend/Minuend Register

An examination of the serial adder-subtracter algorithm reveals that the augend/minuend register must be capable of shifting right one, reading information in parallel, and shifting the sum bit into the leftmost bit position on each shift signal. Either clocked master-slave JK or edge-triggered delay flip-flops can be used for this application. Let us use JK flip-flops in this example. We will assume that only the true variables are available to be read into the register. Figure 12.2 shows one possible circuit diagram for this augend/minuend register.

The addend/subtrahend register is the same as the augend/minuend register with the exception of the S_i input. Since we don't care what is left in the addend/subtrahend register when our calculation is complete, let us simply restore the addend/subtrahend to its original value, i.e., replace S_i by B_i in Fig. 12.2.

12.2.2 The Full Adder/Subtracter Logic and Overflow Circuitry

Since we have both the true and complement variables available from our registers, let us use a simple two-level NAND-NAND circuit to generate the sum S_i and use a JK flip-flop, wired as a delay flip-flop, to generate and store the new carry C_i. One such circuit is shown in Fig. 12.3.

The overflow calculation can be accomplished using the rule that the carry into and out of the sign position must be the same in order for the result to be valid. However, this will require additional time, since this calculation cannot begin until the carry out is generated. An alternate overflow calculation will be used. It is based on the rule which states that for positive numbers the carry into the sign position must be a zero and for negative numbers the carry into the sign position must be a one.

The overflow calculation becomes:

$$\text{overflow} = \bar{a}_3\overline{(b_3 \oplus M)}c_2 + a_3(b_3 \oplus M)\bar{c}_2.$$

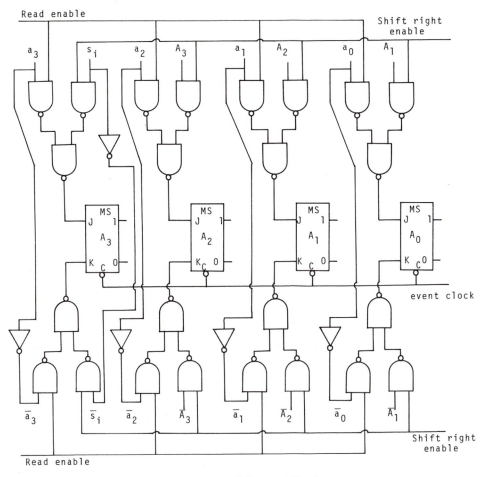

Fig. 12.2 Augend/Minuend Register

This calculation can be carried out at the same time as the calculations of the sum s_3 and carry c_3.

Using two's complement arithmetic for subtraction requires that we take the one's complement of the subtrahend and force the initial carry into the adder, c_{-1}, to be a one. The equations for the arithmetic unit can be modified to

$$S_i = a_i \oplus (b_i \oplus M) \oplus c_{i-1}$$
$$C_i = a_i(b_i \oplus M) + c_{i-1}(b_i \oplus M) + a_i c_{i-1}.$$

However, we would like to store the new value of the carry into a carry/borrow flip-flop and therefore we must provide for initialization of the flip-flop to the value of M during the read/augend step. With this in mind, we can modify the carry equation to

$$C^* = \mathrm{LIC} \cdot M + \mathrm{SHR} \cdot [a_i(b_i \oplus M) + c_{i-1}(b_i \oplus M) + a_i c_{i-1}]$$

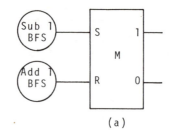

(a)

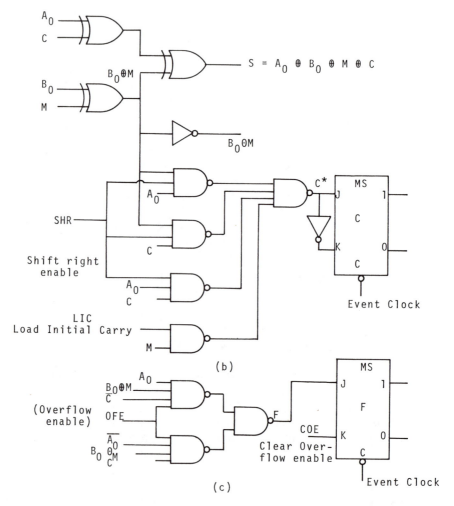

$$S = A_0 \oplus B_0 \oplus M \oplus C$$

(b)

(c)

Fig. 12.3 Adder-Subtracter Logic and Overflow Circuitry. (a) Operation Flip-Flop $M = 1$ for Subtraction, $M = 0$ for Addition; (b) Sum-and-Carry Logic; (c) Overflow Logic

where LIC stands for load initial carry and SHR stands for shift-right enable.

The overflow equation for the adder subtracter is

$$F = \text{OFE} \cdot [a_3(b_3 \oplus M)\bar{c}_2 + \bar{a}_3(\overline{b_3 \oplus M})c_2]$$

where OFE stands for overflow enable. The overflow enable OFE must not be set to a one until carry c_2 has been generated and stored in the carry/borrow flip-flop.

Lastly we note that since we are shifting the information right one place after each operation, we can replace a_i by A_0, b_i by B_0, and c_{i-1} by the output of the carry/borrow flip-flop C. Therefore, the equations governing the operation of the arithmetic unit become:

$$S = A_0 \oplus (B_0 \oplus M) \oplus C$$

$$C^* = \text{LIC} \cdot M + \text{SHR} \cdot [A_0 \cdot (B_0 \oplus M) + C \cdot (B_0 \oplus M) + A_0 C]$$

$$F = \text{OFE} \cdot A_0(B_0 \oplus M)\bar{C} + \bar{A}_0(\overline{B_0 \oplus M})C].$$

An implementation of these equations is shown in Fig. 12.3. Figure 12.3(a) shows that the value of the operation M is stored in a simple unclocked SR flip-flop with the add command connected to the reset side and the subtract command connected to the set side. The overflow signal F is simply connected to the J input of a JK flip-flop and the clear-overflow enable, COE, is connected to the K input to provide for clearing the overflow flip-flop at the start of each new add or subtract operation. The overflow enable must not be turned on until the values of a_3 and b_3 have been shifted into A_0 and B_0 and the carry/borrow flip-flop has been loaded with c_2.

12.2.3 The Control Unit Timing

The control unit for the adder-subtracter unit must provide the overall timing for the unit. It must supply the event clock, EC, and the enable signals: read augend/minuend, RAM; read addend/subtrahend, RAS: shift right enable, SHR; overflow enable, OFE; a stop command; and those commands needed to correctly initialize the carry unit.

Before one can effectively design a control system, an estimate of the time required for the various operations should be obtained. The calculation of course must be based on the actual propagation delays for the gates and flip-flops which will be used in the controller. In order to estimate the time required for this paper design, a few assumptions will be necessary.

Let Δ be the propagation delay time of an inverting type gate such as NAND, or NOR, 2Δ be the propagation delay time of AND and OR gates, and 3Δ be the propagation delay time of an exclusive OR gate. Let us also assume a propagation delay of 5Δ for a JK or SR flip-flop, and let us allow for a worst possible case of 5Δ for the enable time for setting the master section of each master-slave JK flip-flop. Using these estimates, we see that the clock period for reliable operation of the adder-subtracter circuit would be 32Δ. This worst case includes 5Δ

propagation delay for the values of A_0, B_0, and C to change, 6Δ for carry signal C^* to be valid, and 5Δ enable time or the carry/borrow flip-flop. Since we are using a square-wave clock, doubling this delay will give us a reasonable clock period.

An inspection of Fig. 12.2 shows that a square-wave clock with a period of 32Δ will not be satisfactory, since the propagation delay and settling time for the sum input to the left-most stage is 19Δ. Thus a clock period of at least 38Δ is required to reliably operate the adder/subtracter and register units together. If we add a safety margin to allow for wiring delay and circuit aging, a reasonable choice would be at least 40Δ for the clock period. Thus if Δ is approximately 5 ns, the clock period would be 200 ns requiring a 5 megahertz square-wave clock. This would be an upper frequency limit. However, any lower frequency clock would work satisfactorily.

12.2.4 Synchronous Counter Control Unit

One simple method for providing control signals to the various registers and control flip-flops is to use the output of the stages of a binary counter for the enable signals. This type of circuit does not require a precise control on the clock frequency variation or component tolerances and thus, in most cases, will be very reliable. When using a binary counter for the enable signals in our circuits, a synchronous unit is generally preferred to a ripple type counter as we saw in Chapter 10.

The choice of a ring, switch tail, feedback, or ordinary binary counter will depend on the number of counts required. For a small number of counts, five or less, or counters which must provide different enable signals on each count, a self-starting ring counter is a good choice due to its ease in counter state decoding. For counters with greater than five stages or counters which must provide fixed enable signals over several counter stages, feedback counters or ordinary synchronous binary counters are good choices. In all cases, the counter should be self-starting or at least have provision of presetting or preclearing the counter before it is used as a control unit. The control counter for the four-bit serial adder requires only five counts; the RAM, RAS, LIC, and COE enable signal is provided by count 1, while the 4-shift-right signals, SHR, are provided by counts 2, 3, 4, and 5. Count 5 also provides for the overflow enable OFE and the stop command.

The 5-stage self-starting ring counter shown in Fig. 12.4 is controlled by the inverted main clock. As long as the control flip-flop X is zero, the ring counter will shift in only zeroes and thereby initialize the counter. As soon as X is set to a one, the counter is converted into a five-unit ring counter. In this control unit the event clock is not inverted but rather is in phase with the main clock. An idealized timing chart for this unit is shown in Fig. 12.5. The clock period needed to operate the serial adder was previously determined to be 200 ns.

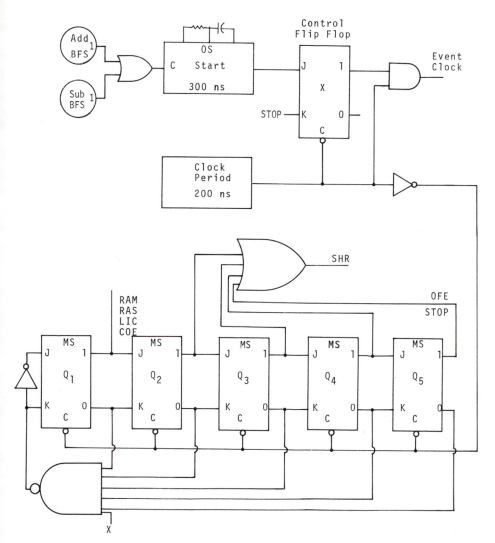

Fig. 12.4 A Synchronous Counter Control Unit for a 4-Bit Serial Adder-Subtracter

12.3 SERIAL-PARALLEL MULTIPLICATION TECHNIQUES

In Chapter 6 we examined a hardware configuration suitable for high-speed parallel multiplication. However, these combinational logic arrays require a large number of gates. For those applications where high speed is not a critical design factor, a considerable saving in hardware can be obtained by using sequential methods. We will classify these algorithms which sequentially scan the bits of the

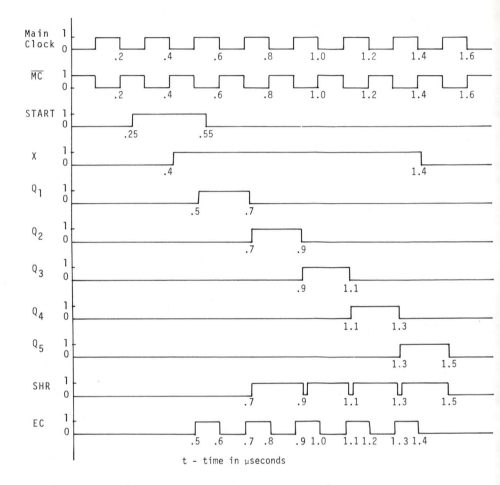

Fig. 12.5 Idealized Time Chart for a 4-Bit-Serial Adder-Subtracter Count-Control Unit

multiplier and add the properly aligned multiplicand to the partial product as being *serial-serial* multipliers, if the addition operation is implemented serially, or *serial-parallel* multipliers, if a parallel addition operation is implemented.

In this section we will examine two of the many algorithms applicable to serial-parallel multiplication of two n-bit binary numbers. In each case studied, a general algorithm will be presented in flow-chart format. A few examples will follow each algorithm. Finally, suggestions for hardware implementation of the algorithm will be given. We will assume unsigned numbers (positive integers) leaving the modifications necessary for the multiplication of signed numbers as an exercise.

12.3.1 Serial-Parallel Multiplication of Binary Integers

Before proceeding to write out a series of multiplication algorithms, let us establish a few simple conventions. The multiplier A is an n-bit binary integer

$$A = a_{n-1}a_{n-2} \cdots a_0.$$

The multiplicand B is also an n-bit binary integer

$$B = b_{n-1}b_{n-2} \cdots b_0.$$

The product $P = A \times B$ will be a $2n$-bit binary integer

$$P = p_{2n-1}p_{2n-2} \cdots p_n p_{n-1} \cdots p_1 p_0.$$

Initially the multiplier A will be loaded into an n-bit register called the MultiplieR register (MR). The multiplicand B will also be loaded into an AuXiliary storage register (AX) where it will remain throughout the multiplication operation. An additional register called the ACcumulator register (AC) will be required to hold temporary results. A single-bit flip-flop called a Link (L) will be used in some cases to temporarily hold the carry out resulting from arithmetic operations. The symbol "←" will be used as a replacement operator in our algorithms. Thus $X \leftarrow Y$ will mean replace the contents of register X with the value or contents of Y.

Link (L) 1 bit	Accumulator Register (AC) n bits	Multiplier Register (MR) n bits	Auxiliary Register (AX) n bits

Let us start with a detailed multiplication algorithm which can be easily converted into a set of instructions for a digital computer.

ALGORITHM 12.2

Add and then shift binary multiplication algorithm

STEP   OPERATION

1.   Clear accumulator register (AC) to zero.

2.   Load multiplier A into the multiplier register (MR).

3.   Load multiplicand B into the auxiliary register (AX).

4.   Repeat steps 5 to 9 N times and then go to step 10.

5.   If the least significant bit of the MR register (MR_0) is 1, go to step 6. Otherwise go to step 7.

6.   Add the contents of the AX to the contents of the AC, storing the carry out in the link and the sum in the AC; then go to step 8.

8.   Rotate the contents of the link and AC one place right, with Link $\leftarrow AC_0$ and $AC_{n-1} \leftarrow$ link.

9.   Rotate the contents of the link and MR one place right, with link$\leftarrow MR_0$ and $MR_{n-1} \leftarrow$ link (old value of AC_0).

10.   STOP. The product is now stored in the AC and MR registers with the most significant half $(P_{2n-1} \cdots P_n)$ in the AC and the least significant half $(P_{n-1} \cdots P_0)$ in the MR.

A flow chart for this algorithm is shown in Fig. 12.6.

Let us illustrate this algorithm by working out an example.

Example 12.1   Multiply $A = 1011$ by $B = 0101$ using Algorithm 12.2.

Step	Link	AC	MR	AX	Count	Comment
Start	d	dddd	dddd	dddd	d	d − don't care
1		0000				$AC \leftarrow 0$
2			1011			$MR \leftarrow A$
3				0101		$AX \leftarrow B$
4					0	Count $\leftarrow 0$
5	d	0000	1011	0101	0	$MR_0 = 1$, go to 6.
6	0	0101				$AC \leftarrow AC$ plus AX, Link $\leftarrow C_{out}$.
8	1	0010				Rotate Link and AC right 1.
9	1		1101			Rotate Link and MR right 1.
4						Increment count by one.
4					1	Since count <3, go to 5.
5	1	0010	1101	0101	1	$MR_0 = 1$, go to 6.
6	0	0111				$AC \leftarrow AC$ plus AX, Link $\leftarrow C_{out}$.
8	1	0011				Rotate Link and AC right 1.
9	1		1110			Rotate Link and MR right 1.
4						Increment count by one.
4					2	Since count <3, go to 5.
5	1	0011	1110	0101	2	$MR_0 = 0$, go to 7.
7	0	0011				$AC \leftarrow AC$ plus zero, Link $\leftarrow C_{out}$.
8	1	0001				Rotate Link and AC right 1.
9	0		1111			Rotate Link and MR right 1.
4						Increment count by one.
4					3	Since count <3, go to 5.
5	0	0001	1111	0101	3	$MR_0 = 1$ go to 6.
6	0	0110				$AC \leftarrow AC$ plus AX, Link $\leftarrow C_{out}$.
8	0	0011				Rotate Link and AC right 1.
9	1		0111			Rotate Link and MR right 1.
4						Since count = 3, go to 10.
10		0011	0111			STOP

Thus $P = A \times B = (1011) \times (0101) = 00110111$

$$11 \times 5 = 55$$

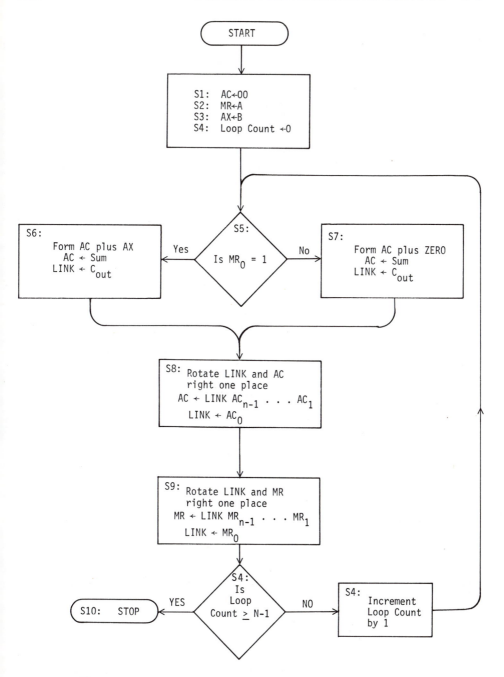

Fig. 12.6 Flow Chart for Binary-Multiplication Algorithm 12.2

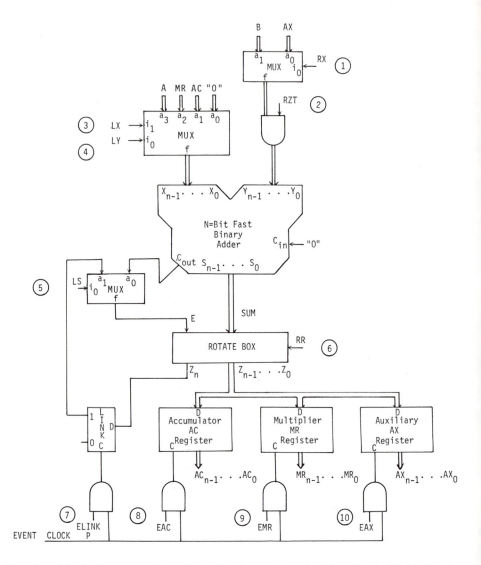

Fig. 12.7 Block Diagram of the Basic Hardware Required for Binary-Multiplication Algorithm 12.2

In Example 12.1 only those values which are involved in the operation are shown at each step except on those lines corresponding to step 5 where all the data is shown. Let us implement this algorithm with a special purpose hardware unit. From an examination of the algorithm it is clear that the link (L), accumulation (AC), multiplier register (MR), and auxiliary registers (AX) can be constructed using D flip-flops. New input data can be transferred to these registers by using a multiplexer on the adder and then passing the data through the adder with one input held at zero. The output from the adder can then be fed into a rotate unit which will either pass the input directly to the output or rotate it right one place. A typical hardware configuration suggested by these comments is shown in Fig. 12.7. Ten numbered control lines are indicated on this figure. These lines cause the following action to take place:

1. RX   Select line for the right-side multiplexer on the input of the adder. $RX = 0$ selects AX, $RX = 1$ selects the multiplicand B.

2. RZT   Right-side zero or true input select line. $RZT = 0$ selects a zero input. $RZT = 1$ selects the output of the right-side input multiplexer.

3. LX $\rceil$
4. LY $\rfloor$ Select lines for the left-side input multiplexer.

LX	LY	Input selected
0	0	Zero
0	1	AC
1	0	MR
1	1	A - Multiplier

5. LS   Link select line. Setting $LS = 0$ causes the carry out of the adder to be passed to the rotate box input E. Setting $LS = 1$ causes the old values of the link to be transferred to the rotate box input E.

6. RR   Rotate right control line.

RR	Input	Output
0	$E\ S_{n-1}\ \cdots\ S_1 S_0$	$E\ S_{n-1}\ \cdots\ S_1 S_0$
1	$E\ S_{n-1}\ \cdots\ S_1 S_0$	$S_0\ E\ S_{n-1}\ \cdots\ S_1$

7. ELINK   Enable the link flip-flop to read data.
8. EAC   Enable the AC register to read data.
9. EMR   Enable the MR register to read data.
10. EAX   Enable the AX register to read data.

    In order to control the registers and arithmetic unit we will need a sequential control unit. Examination of the algorithm reveals that we will need at least two counters and a multiply event flip-flop. The multiply event flip-flop, M, will be set to a one by the start signal and remain at one throughout the multiply operation. As long as M is a one it will enable an event clock P. The multiply event flip-flop, M, will be turned off at the completion of the algorithm by setting the STOP command to a one. A loop counter (LCNT) will be needed to keep track of the number of times the multiply loop is executed. This counter must have at least N

counting states. A control counter (CCNT) will also be required in order to sequence through the steps in the algorithm. This control counter must be able to be preset to zero, to count up in binary, and to be reset to a previous state.

However, before we can design this control counter, we need to know how many counting states will be required. This can be determined by examining the algorithm and the basic register hardware. By using direct reset inputs on each of the counters, controlled by the zero output of the event flip-flop M, we can be sure that the loop counter will be correctly set at the start. The data initialization operations, steps S1, S2, and S3 will each require one clock period (C_0, C_1, and C_2). By using the multiplier bit MR_0 to control the right-side zero-true control line, RZT, steps S5, S6, S7, and S8 can be combined into one clock period (C_3). This can be done by forming the sum [AC plus (MR_0 times AX)] at the output of the adder, and passing this sum along with any carry out to a rotate box where the sum can be rotated right one place prior to being stored in the link and AC registers.

The MR register and link can then be passed to the rotate box through the adder during the next clock cycle. While the MR data is being rotated in Step S9, the loop counter can also be incremented and tested. When the loop count equals $N - 1$, indicating that the loop has been executed N times, the STOP command can be issued. Thus steps S9 and S4 can also be combined and executed during one clock period (C_4) with the control counter (CCNT) looping back to state (C_3) if the loop count is less than $N - 1$. The control unit circuitry suggested by this discussion is shown in Fig. 12.8.

The event clock P will cause the registers to be triggered on a positive or leading edge of the clock signal, while the multiply flip-flop is clocked on the trailing edge of the clock signal. However, because of the inverting action of the NAND gate, these clocking actions will take place within a few gate delays of each other.

When M is set to a one, the NAND gate will not change until the start of the next clocking period. When the STOP signal is given, M will be turned off on the trailing edge of the clocking signal while P is being held at one. Thus this clocking arrangement will not output any split pulses from P. The six additional control lines indicated in this figure cause the following action to take place.

11. CCD2 ⎫
12. CCD1 ⎬ Data inputs to the control counter (used for looping).
13. CCD0 ⎭
14. CI   Control counter inhibit signal. As long as CI = 0, the counter, CCNT, counts up in binary. When CI = 1, the next counter state is read into the counter by means of the data input lines CCD2, CCD1, and CCD0.
15. LCE   Loop counter enable signal.
16. STOP   Turn off the multiply event flip-flop, M, and terminate the algorithm.

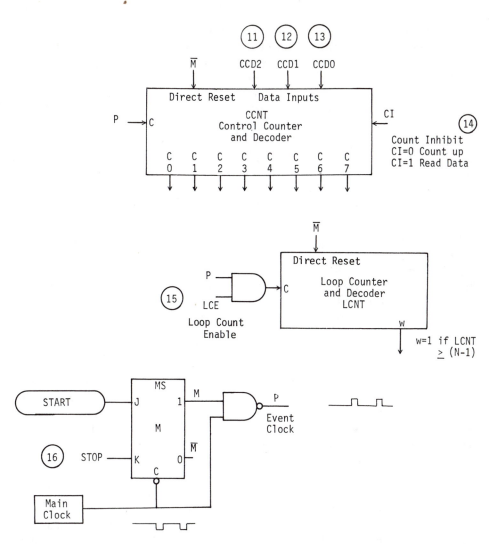

Fig. 12.8 Basic Control-Unit Configuration for Algorithm 12.2

All that remains is for us to determine the sequence of command signals which will be fed to these sixteen control points.

When the control counter is in state C_0, the AC register must be set to zero and the control counter incremented. An examination of the register and control section reveals that this can be accomplished by enabling the AC register, setting EAC to one, and reading in the zero result from the rotate box.

When the control counter is in state C_1, the MR register must be loaded with the multiplier A and the control counter incremented. This can be accomplished by setting both left-side multiplexer inputs LX and LY to ones, and enabling the MR register to read in the output of the rotate box, by setting EMR to a one.

When the counter is in state C_2, the multiplicand B must be read into the AX register and the control counter incremented. This can be accomplished by setting the right-side multiplexer-select input, RX, to a one and simultaneously setting the right-side zero-true control line, RZT, to a one. The output of the rotate box can then be read into the AX register by setting EAX to a one.

When the control counter is in state C_3, the sum [AC plus (MR$_0$ times AX)] is obtained from the adder by setting LY to a one, selecting the AC as the left input to the adder and using MR$_0$ to control the right-side zero-true unit. The result can then be rotated right by setting RR to a one. The rotated results are returned to the link and AC register by setting the control lines ELINK and EAC to ones. The control counter is also incremented during this cycle.

When the control counter is in state C_4, the link and MR register must be rotated right one place, the loop counter incremented, and the next control counter state set to C_3 if the loop count is less than $N - 1$ or to C_5 (by default) if the loop counter is equal to $N - 1$. Also, if the loop counter is greater than or equal to $N - 1$, a STOP command must be issued. These tasks can be accomplished by setting control lines LS and LX to ones, thereby passing the link and MR registers through the adder to the rotate box. The rotate-box input, RR, is set to a one and the rotated result returned to the link and MR register by setting ELINK and EMR to ones.

By placing the complemented output of the loop-counter output signal w on the control-counter data inputs CCD1 and CCD0, and the control-counter inhibit input, CI, the looping operation back to state C_3 can be accomplished. Finally, when the loop counter output w is equal to 1, indicating the loop has been executed N times, the STOP signal can be given. Counter states C_5, C_6, and C_7 are not used in this algorithm, hence they should cause a STOP signal to be issued, indicating an error condition.

The above control operations can be easily summarized and visualized by means of a control matrix like that shown in Fig. 12.9.

The counter states are placed along the lefthand side of the matrix; the control lines are along the top. The intersection points contain the control information for each counter state.

The overall logic equations for the control lines can be easily determined from the control matrix. The entries in each row are ANDed with the control state at the left side of the row and the resulting logical products are ORed down each column.

The 16 control input equations for the control matrix shown in Fig. 12.9 are:

1. $RX = C_2$	2. $RZT = C_2 + C_3 \cdot MR_0$
3. $LX = C_1 + C_4$	4. $LY = C_1 + C_3$
5. $LS = C_4$	6. $RR = C_3 + C_4$

Counter State \ Control Line	1 RXT	2 RZT	3 LX	4 LY	5 LS	6 RR	7 FLINK	8 FEAC	9 FEMR	10 FEAX	11 CCD2	12 CCD1	13 CCD0	14 CI	15 LCE	16 STOP	Comment
C_0								1									$AC \leftarrow 0$; $CCNT \leftarrow 1$
C_1			1	1					1								$MR \leftarrow A$; $CCNT \leftarrow 2$
C_2	1	1															$AX \leftarrow B$; $CCNT \leftarrow 3$
C_3		MR_0		1		1	1	1		1							Form AC plus (MR_Q times AX) then rotate result right; $Link \leftarrow S_0$; $AC \leftarrow Cout_{n-1}\cdots S_1$; $CCNT \leftarrow 4$
C_4			1		1	1	1		1			$\bar{w}$	$\bar{w}$	$\bar{w}$	1	w	Rotate right Link and MR; $Link \leftarrow MR_0$; $MR \leftarrow Link\ MR_{n-1}\cdots MR_1$; If loop count $<N-1$ $CCNT \leftarrow 3$, otherwise $CCNT \leftarrow 5$; If loop count $=N-1$ issue STOP signal
C_5																1	Issue STOP signal
C_6																1	Issue STOP signal
C_7																1	Issue STOP signal

Fig. 12.9　Control Matrix for Algorithm 12.2

7. ELINK $= C_3 + C_4$ 8. EAC $= C_0 + C_3$

9. EMR $= C_1 + C_4$ 10. EAX $= C_2$

11. CCD2 $= 0$ 12. CCD1 $= C_4 \cdot \bar{w}$

13. CCD0 $= C_4 \cdot \bar{w}$ 14. CI $= C_4 \cdot \bar{w}$

15. LCE $= C_4$ 16. STOP $= C_4 \cdot w + C_5 + C_6 + C_7$

In general, this hardware multiplier will require $2N + 3$ clock pulses to complete the multiplication once the start signal is issued. The clock pulse used

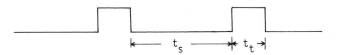

should be asymmetric with a long setup time, t_s, and a relatively short transfer time, t_t. The setup time, t_s, should include the following propagation delay times:

a) the time required for the control counter decoder outputs to decode the new control state (4 units of gate delay).

b) the time required to select a new input for the left side multiplexer (2 units of gate delay).

c) the time required to pass the right side input through the right-side zero-true unit (2 units of gate delay).

d) the n-bit fast-adder time (12 units of gate delay assuming a first-order carry-lookahead unit).

e) the time required to select the extra bit for the rotate box (2 units of gate delay).

f) the time required to pass the data through the rotate box (2 units of gate delay).

g) the input setup time for the registers link, AC, MR, AX (3 units of gate delay) assuming D flip-flops.

h) wiring delay and safety factor as required to account for environmental changes (3 units for wiring and 5 units safety margin).

Typical delay estimates in terms of units of gate delay are shown in parentheses.

The transfer time, t_t, must be long enough to allow the new data to be properly transferred into all of the registers and counters (5 units of gate delay should be adequate). t_t must be short enough so that false clock signals are not issued due to changes in clock enables prior to the removal of the clock signal.

This estimate of 40 units of gate delay for the clock period is not unreasonable for this hardware configuration. This estimate yields a multiply time of *approximately* $40(2N + 3)$ units of gate delay as compared with *approximately*

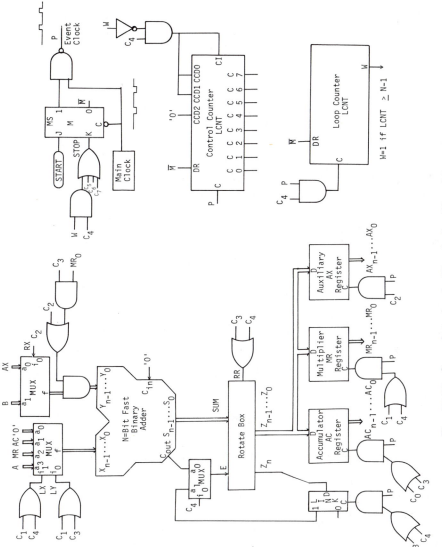

Fig. 12.10 Complete Logic Circuit for Algorithm 12.2

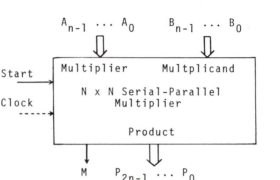

Figure 12.11

$(4N - 2)$ units of gate delay for the parallel multiplier designed in Section 6.8. Therefore for reasonable values of N, this serial-parallel multiplier will be approximately 20 to 30 times slower than a parallel multiplier using equivalent logic gates. However where high speed is not a critical consideration, this sequential circuit may prove more than satisfactory as it will also require less hardware. The complete block diagram for this serial-parallel multiplier is shown in Fig. 12.10.

This unit can be constructed as the large-scale integrated circuit shown in Fig. 12.11.

Here, the multiply-event flip-flop M can be used to determine when the product is completed. As long as M is equal to one, the unit is considered busy. When M is returned to zero, the multiply operation is complete. Either an internal or external clock signal can be used to control the unit. The start signal should be removed prior to the time M returns to zero to prevent repeated multiplication.

Example 12.2 Multiply $A = 1011$ by $B = 0101$. Refer to Fig. 12.12.

12.3.2 A Stored Shifted-Integer Multiplication Algorithm

The hardware configuration used to implement Algorithm 12.2 placed the hardware-rotate box at the output of the adder. This placement caused the multiplication loop to take two full clock periods, one to add and shift the AC, and one to shift the MR. By removing the rotate box and increasing the complexity of the AC and MR registers slightly, we can carry out the multiplication loop in a single clock cycle. This can be accomplished by simultaneously storing the shifted sum [AC plus (MR_0 times AX)] in the AC and simultaneously shifting the MR register. Let us incorporate these suggestions into an alternate integer multiplication algorithm.

Multiply $A = 1011$ by $B = 0101$

Present Control State	M	CCNT	LCNT	E	SUM	LINK	AC	MR	AX	Comment
					Value of register at end of clocking period					
START	1	0	0	–	–	–	–	–	–	Turn on Multiply flip flop
C_0	1	1	0	0	0000	–	0000	–	–	AC ← 0
C_1	1	2	0	0	1011	–	0000	1011	–	MR ← A
C_2	1	3	0	0	0101	–	0000	1011	0101	AX ← B
C_3	1	4	0	0	0101	1	0010	1011	0101	(AC plus AX) rotate right
C_4	1	3	1	1	1011	1	0010	1101	0101	Rotate right Link and MR
C_3	1	4	1	0	0111	1	0011	1101	0101	(AC plus AX) rotate right
C_4	1	3	2	1	1101	1	0011	1110	0101	Rotate right Link and MR
C_3	1	4	2	0	0011	1	0001	1110	0101	(AC plus zero) rotate right
C_4	1	3	3	1	1110	0	0001	1111	0101	Rotate right Link and MR
C_3	1	4	3	0	0110	0	0011	1111	0101	(AC plus AX) rotate right
C_4	0	5	4	0	1111	0	0011	0111	0101	Rotate right Link and MR STOP

AC   MR

$P = 0011 \quad 0111$

Fig. 12.12 Register Contents for the Hardware Circuit Realization of Algorithm 12.2

ALGORITHM 12.3

A stored shifted-binary-integer multiplication algorithm

STEP   OPERATION

1.   a) Clear the accumulator register to zero.
      b) Load the multiplier A into the multiplier register.
      c) Load the multiplicand B into the auxiliary register.

2.   Repeat step 3 N times and then STOP.

3.   a) Form $C_{out} S_{n-1} \cdots S_0 = $ AC plus (MR$_0$ times AX).
      b) Replace the AC with $C_{out} S_{n-1} \cdots S_1$, and
      c) Simultaneously replace the MR with $S_0 MR_{n-1} \cdots MR_1$.

A flow chart for this algorithm is shown in Fig. 12.13. Let us illustrate this algorithm by means of an example.

Example 12.3 Multiply $A = 0110$ by $B = 0111$ using Algorithm 12.3.

Step	C_{out}	Sum	AC	MR	AX	Loop Count	Comment
START	d	dddd	dddd	dddd	dddd	d	d – don't care
1	d	dddd	0000	0110	0111	0	AC ← 0. MR ← A. AX ← B. Loop count ← 0.
2,3	0	0000	0000	0011	0111	1	AC plus zero. AC ← $C_{out} S_3 S_2 S_1$. MR ← $S_0 MR_3 MR_2 MR_1$. Loop count ← 1.
2,3	0	0111	0011	1001	0111	2	AC plus AX. AC ← $C_{out} S_3 S_2 S_1$. MR ← $S_0 MR_3 MR_2 MR_1$. Loop count ← 2.
2,3	0	1010	0101	0100	0111	3	AC plus AX. AC ← $C_{out} S_3 S_2 S_1$. MR ← $S_o MR_3 MR_2 MR_1$. Loop count ← 3.
2,3	0	0101	0010	1010	0111	3	AC plus zero. AC ← $C_{out} S_3 S_2 S_1$. MR ← $S_0 MR_3 MR_2 MR_1$. Since loop count = 3. STOP.

The product $P = A \times B = (0110) \times (0111) = 00101010$

$6 \times 7 = 42$

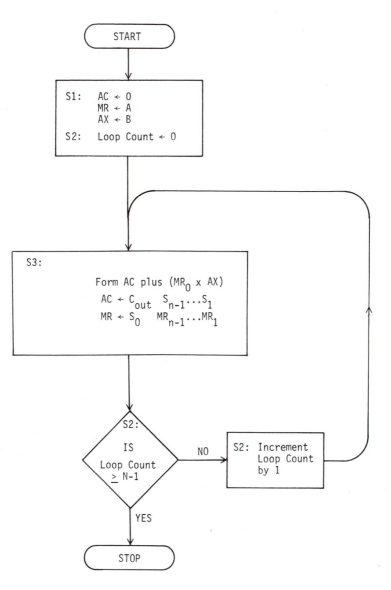

Fig. 12.13 Flow Chart for Binary Multiplication as Shown in Algorithm 12.3

Let us implement this algorithm using a special-purpose hardware unit. Examination of the algorithm reveals that the link flip-flop is no longer needed. There is also no need for a special loop counter in the control section, since the multiply loop requires only a single clock pulse. The complete hardware unit is shown in Fig. 12.14. Let us discuss how we arrived at this configuration. The accumulator register requires two data inputs, the output of the adder and a zero

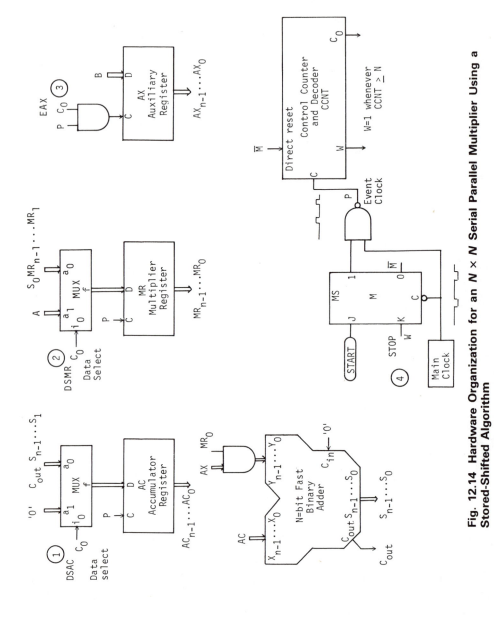

Fig. 12.14 Hardware Organization for an $N \times N$ Serial Parallel Multiplier Using a Stored-Shifted Algorithm

input for initially clearing the register. These inputs can be supplied by using a data multiplexer on the data input to the accumulator.

The multiplier register must initially be loaded with the multiplier A. During each clock pulse of the multiply loop, the MR-register contents must be shifted right. These two operations can be accomplished by also using a data multiplexer on the data input to the multiplier register. Because both of these registers are involved in each clock cycle, there will be no need to provide a separate enable signal. Instead, the event clock P can be fed directly to the clock inputs of these two registers. Since the event clock P is only active as long as the multiply flip-flop is a one, there is no danger of extra clock pulses. Given a choice of data input positions on the multiplexers, the zero input and multiplier input A should be placed on the a_1 inputs with the output of the adder and the shifted MR data placed on the a_0 inputs. This configuration will require a nonzero input on the data select lines DSAC and DSMR only during the first clocking period when the control counter is in state C_0.

The auxiliary register requires only a single data input in order to initially load the multiplicand B. This read operation can be done by using counter state C_0 to gate the event clock P to the clock input of the AX register.

An n-bit fast adder is needed to form the sum [AC plus (MR$_0$ times AX)].

The control counter (CCNT) requires at least $(N + 1)$ counting states with two outputs, the first output C_0 indicating that the counter is in state zero and the second output W indicating that the multiplication loop has been completed. When W is equal to one, the STOP command is enabled.

This stored shifted-multiply algorithm requires only $(N + 1)$ clock pulses as compared with the $(2N + 3)$ clock pulses required by the preceding "add-and-then-shift" algorithm. Thus by transferring the shifting operations from a separate rotate box to inputs of the registers, the multiply time has been halved. It is also apparent from an examination of the two block diagrams in Figs. 12.12 and 12.14, that the stored shifted unit requires a considerably less complex controller. The asymmetric clock signal P for the stored-shifted controller should include a long setup time, t_s, and a short transfer time, t_t.

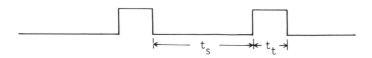

The setup time, t_s, takes the following propagation delay times into consideration:

a) the time required to pass data through the right-side zero-true element on the adder (2 units of gate delay).

b) the n-bit fast adder time (12 units for a 1st order carry-lookahead adder).

c) the time required to pass data through the input multiplexers on the data inputs to the AC and MR registers (2 units of gate delay).

d) the worst-case setup time for the AC and MR register (3 units of gate delay assuming edge-triggered D flip-flops are used).

e) wiring delay and safety factors as required to account for environmental changes (3 units of gate delay for wiring and 5 units of gate delay for a safety margin).

Since the control counter is initially in state zero, the decoding delay for the control counter can be neglected. The decoding delay for the STOP signal will overlap that required by the adder. Thus the setup time for the stored shifted unit will be less than that required by the preceding unit. An estimated reduction of eight units of gate delay is not unreasonable.

The transfer time, t_t, must be long enough to allow the new data to be properly transferred into the registers and control counters. (5 units of gate delay should be adequate.)

Depending on the logic configurations selected for the registers, adder, and counters, the *estimate* of 32 units of gate delay for this multiplier is easily attainable. Using this estimate, the multiply time for a stored-shifted multiplier would be *approximately* $32(N + 1)$ units of gate delay as compared with $(4N - 2)$ for an equivalent parallel multiplier. Thus, for reasonable values of N, we can conclude that a stored-shifted serial-parallel multiplier will be approximately 8 to 12 times slower than a parallel multiplier using equivalent logic gates.

12.4 SERIAL-PARALLEL DIVISION OF BINARY INTEGERS

Before we write out a binary division algorithm, let us establish a few simple conventions. The dividend A will be a $2n$-bit binary integer of the form

$$A = a_{2n-1}a_{2n-2} \cdots a_n a_{n-1} \cdots a_1 a_0.$$

The divisor B will be an n-bit binary integer of the form

$$B = b_{n-1} \cdots b_1 b_0.$$

The quotient Q and remainder R resulting from the division of A by B will be n-bit binary numbers of the form

$$Q = q_{n-1}q_{n-2} \cdots q_1 q_0$$

and

$$R = r_{n-1}r_{n-2} \cdots r_1 r_0.$$

Initially, the dividend A will be preloaded into a pair of n-bit registers. The left half or most significant n-bits of A, $A_{LH} = a_{2n-1} \cdots a_n$, will be placed in an n-bit register which we will refer to as the ACumulator register (AC). The right half or least significant n bits of A, $A_{RH} = a_{n-1} \cdots a_0$, will be placed in an n-bit register which we will refer to as the Quotient Register (QR). A_{LH} and A_{RH} will be used to denote the left half (LH) and right half (RH) of the dividend A. The accumulator register will be used during the division operations to hold temporary results and will contain the remainder R at the termination of the division algorithm. The divisor B will be loaded into an AuXiliary storage register (AX), where it will remain throughout the division operation. Two additional single-bit flip-flops which we will refer to as the link (L) and OVerflow (OV) flip-flops will also be utilized during the division process. The need for these registers will become apparent shortly.

Over-flow (OV)	Link (L)	Accumulator Register (AC)	Quotient Register (QR)	Auxiliary Register (AX)
1 bit	1 bit	n bits long	n bits long	n bits long

Before proceeding into a discussion of binary division algorithms, we must pause and consider a problem which is inherent in a division process requiring a fixed-bit length for the quotient. Let us introduce the problem by means of an example. When we divide $(250)_{10}$ by $(12)_{10}$, we obtain a quotient of $(20)_{10}$ and a remainder of $(\frac{10}{12})_{10}$. So far no problem; however, if we consider the same question, this time using binary numbers, the problem will become readily apparent.

$$\left(\frac{250}{12}\right)_{10} = (20)_{10} + \left(\frac{10}{12}\right)_{10}$$

$$\left(\frac{11110010}{1100}\right)_2 = (10100)_2 + \left(\frac{1010}{1100}\right)_2$$

The dividend $(250)_{10}$ is easily written as an 8-bit binary number $(11110010)_2$, and the divisor $(12)_{10}$ is also expressible as a 4-bit binary number $(1100)_2$. However, the quotient $Q = (20)_{10}$, is not expressible as a 4-bit binary number. In this case five bits are required with $(20)_{10} = (10100)_2$. The situation would be even worse if the divisor had been a one, since now up to eight bits would be required to express the quotient of all possible dividends.

Whenever the division of a $2n$-bit dividend by an n-bit divisor results in a quotient which requires more than n bits, a *division overflow* condition is said to exist. Division by zero should always yield such an overflow condition.

Fortunately, it is an easy matter to detect an overflow condition. *Anytime the left half of the dividend $A_{LH} = a_{2n-1} \cdots a_n$, treated as an n-bit binary integer, is greater than or equal to the n-bit binary integer divisor B, an overflow condition will occur.*

That this is true can best be indicated by means of an example. Letting $A = a_3a_2a_1a_0$ and $B = b_1b_0$, the division of A by B yields a quotient Q of the form $q_3q_2q_1q_0$

$$\frac{q_3q_2q_1q_0}{b_1b_0 \overline{\smash{\big)}\, a_3a_2a_1a_0}}$$

where q_3 and q_2 will be zero if and only if (b_1b_0) is strictly greater than (a_3a_2), i.e., $A_{LH} < B$. The extension of this argument to n-bit numbers is obvious.

Since A_{LH} will always be ≥ 0, it will not be necessary to provide a special test for division by zero.

In general, it will be necessary to determine the quotient bits by means of a trial-and-error procedure starting with the most significant position. The most common methods used when employing a pencil-and-paper technique are variations of the general base-R restoring division procedure outlined in Algorithm 12.4. This method tells us that the ith quotient bit can be obtained, beginning with the most significant bit, by means of repeated subtraction of B times R^i from the partial remainder. The subtractions continue until the partial remainder becomes negative.

The partial remainder is again made positive by adding back B times R^i and setting q_i equal to the number of subtractions prior to obtaining a negative result. A flow chart follows the formal description of the algorithm and is shown in Fig. 12.15. Two examples are also worked out in detail, the first using the more familiar base-10 integers and sign-plus-magnitude notation. In the second example an extra sign digit has been appended to the binary numbers so that two's complement arithmetic could be utilized for the subtraction operations.

ALGORITHM 12.4

A general base-R restoring division procedure for positive integers

$$\frac{A}{B} = Q + \frac{X}{B}$$

$$A = a_{2n-1}a_{2n-2} \cdots a_n a_{n-1} \cdots a_1 a_0.$$

$$B = b_{n-1}b_{n-2} \cdots b_1 b_0.$$

STEP   OPERATION

1.   Form $A - (B \cdot R^n)$.
        If the result is positive, i.e., $A \geq (B \cdot R^n)$, then set an overflow indicator and STOP, otherwise go to step 2.

2.   Set $i = n - 1$, $j = 1$ and $P_{n-1,0} = A$.

3.   Form the new partial remainder

$$P_{i,j} = P_{i,j-1} - (B \cdot R^i).$$

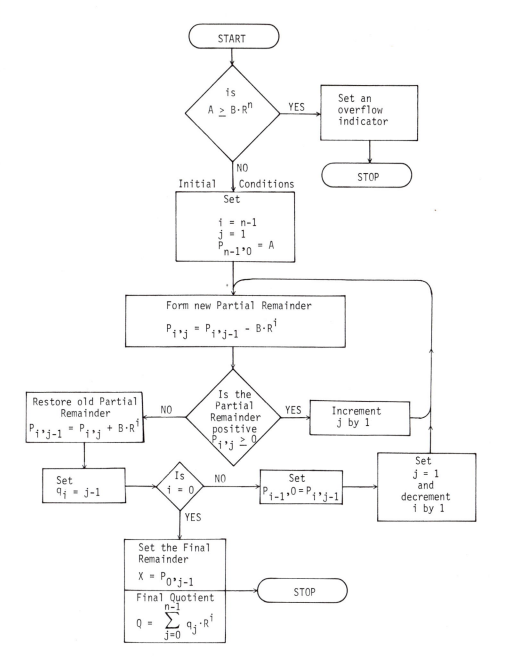

Fig. 12.15 Flow Chart for the General Base-*R* Restoring-Division Procedure Stated in Algorithm 12.4

4.   If $P_{i,j}$ is positive, then increment j by 1 and go back to step 3, otherwise go to step 5.

5.   Form $P_{i,j-1} = P_{i,j} + (B \cdot R^i)$ and set $q_i = j - 1$ (the ith quotient digit has now been determined).

6.   If i is equal to zero, go to step 8; otherwise set $P_{i-1,0} = P_{i,j-1}$ and go to step 7.

7.   Set $j = 1$, decrement i by 1 and then go back to step 3 to calculate the next quotient digit.

8.   Set the remainder X equal to $P_{0,j-1}$ and STOP. The n-bit quotient is given by

$$Q = \sum_{j=0}^{n-1} q_j \cdot R^j$$

$$Q = q_{n-1}q_{n-2} \cdots q_0.$$

Example 12.4 $A = (1624)_{10}, \quad B = (75)_{10}$

Since $A < 100 \cdot B$, overflow will not occur.

$$1624 < 7500$$

Initial conditions $i = 1, j = 1, P_{1,0} = A = 1624$

$P_{1,0}$	1624	$j = 1 \quad i = 1$
$-B \cdot 10^1$	-750	
$P_{1,1}$	874	$j = 2$
$-B \cdot 10^1$	-750	
$P_{1,2}$	124	$j = 3$
$-B \cdot 10^1$	-750	
$P_{1,3}$	-626	
$+B \cdot 10^1$	750	
$P_{1,2}$	124	$q_1 = j - 1 = 2 \quad P_{0,0} = P_{1,2} = 124$
$P_{0,0}$	124	$j = 1 \quad i = 0$
$-B \cdot 10^0$	-75	
$P_{0,1}$	49	$j = 2$
$-B \cdot 10^0$	-75	
$P_{0,2}$	-26	
$+B \cdot 10^0$	75	
$P_{0,1}$	49	$q_0 = j - 1 = 1$

Since $i = 0$, $Q = q_1 q_0 = 21$ and $X = P_{0,1} = 49$
therefore

$$\frac{1624}{75} = 21 + \frac{49}{75}.$$

Example 12.5 $A = (39)_{10} = (100111)_2$
$B = (6)_{10} = (110)_2$
$n = 3$

Since $A < 2^3 \cdot B$ no overflow will occur:

$$100111 < 110000$$

Initial conditions: $i = 2$, $j = 1$, $P_{2,0} = A = (0100111)_2$, $B = (0110)_2$, and $-B = (1010)_2$

(Note: an extra sign bit has been appended to A and B so that 2's complement arithmetic can be used.)

$P_{2,0}$	0100111	$j = 1$ $i = 2$
$-B \cdot 2^2$	1101000	
$P_{2,1}$	0001111	$j = 2$ (Note that we ignore the carry from
$-B \cdot 2^2$	1101000	the sign position in 2's
		complement arithmetic.)
$P_{2,2}$	1110111	
$+B \cdot 2^2$	0011000	
$P_{2,1}$	0001111	$q_2 = j - 1 = 1$ $P_{1,0} = P_{2,1} = 01111$
$P_{1,0}$	01111	$j = 1$ $i = 1$
$-B \cdot 2^1$	10100	
$P_{1,1}$	00011	$j = 2$
$-B \cdot 2^1$	10100	
$P_{1,2}$	10111	
$+B \cdot 2^1$	01100	
$P_{1,1}$	00111	$q_1 = j - 1 = 1$, $P_{0,0} = P_{1,1} = 0011$
$P_{0,0}$	0011	$j = 1$ $i = 0$
$-B \cdot 2^0$	1010	
$P_{0,1}$	1101	
$+B \cdot 2^0$	0110	
$P_{0,0}$	0011	$q_0 = j - 1 = 0.$

Since $i = 0$, $Q = (110)_2$, and $X = (011)_2$ (Note the sign bits have been dropped.)

therefore

$$\frac{A}{B} = Q + \frac{X}{B} = \left(\frac{100111}{110}\right)_2 = (110)_2 + \left(\frac{011}{10}\right)_2,$$

It is clear from Example 12.5 that this general algorithm is going to need some modification if it is to be suitable for binary-hardware implementation. The most obvious improvement would be the retention of the old partial remainder at each step, thereby eliminating the need for an extra addition step required to restore the old remainder.

We can also eliminate the double subscript notation for the partial remainders, since at each step in the case of binary numbers only one partial remainder is calculated. Let us designate the partial remainder in the binary case by using a script $\mathscr{R}$. Thus $\mathscr{R}_i$ will be used to denote the partial remainder following the determination of the ith quotient bit q_i, with $\mathscr{R}_n$ initially set equal to the dividend A. Thus we have in the binary case

$$\mathscr{R}_n = A,$$

$$q_i = 0, \quad \text{if} \quad \mathscr{R}_{i+1} < B \cdot 2^i$$

or

$$q_i = 1, \quad \text{if} \quad \mathscr{R}_{i+1} \geq B \cdot 2^i$$

and

$$\mathscr{R}_i = \mathscr{R}_{i+1} - (q_i \cdot B \cdot 2^i) \quad \text{for} \quad n - 1 \geq i \geq 0$$

Since there are only two choices for the ith quotient bit, the subtraction operation can be considered as a simple comparison operation.

It is also apparent from an examination of Example 12.5 that only the most significant four bits ($n + 1$ bits in general) of the partial remainder are involved in the comparison operation.

Thus, when two's complement arithmetic is employed, the comparison of $\mathscr{R}_{i+1}$ with $B \cdot 2^i$ is equivalent to comparing only the left $n + 1$ bits of $\mathscr{R}_{i+1}$, $r_{n+1}r_{n+1-1} \cdots r_i$, which will be denoted by $(\mathscr{R}_{i+1})_L$ against B. Thus q_i will be equal to a one whenever $(\mathscr{R}_{i+1})_L \geq B$.

It should also be noted that each time a new quotient bit is determined, the new partial remainder will also require one less bit. Thus the sum of the bits required to specify the ith partial remainder and the quotient bits $q_{n-1} \cdots q_{i+1}$ will be constant (n) throughout the division operation.

We can take advantage of this last observation in a hardware implementation by initially storing the dividend A in a double-length register. The left ($n + 1$) bits of this register can be compared with the divisor B by means of a conventional two's complement subtraction circuit. After each quotient bit is determined and the new partial remainder calculated,

$$\mathscr{R}_i = \mathscr{R}_{i+1} - (q_i \cdot B \cdot 2^i),$$

the partial remainder $\mathscr{R}_i$ can be shifted left one position with the quotient bit q_i

entered in the vacated rightmost position. Following the left shift operation, the top $(n + 1)$ bits of the new partial remainder will be aligned with those of B so that the next comparison can be readily performed. At the completion of n comparisons, the upper n bits of the register will contain the final remainder $\mathcal{R}_0$ and the lower n bits of the register will contain the final quotient Q.

Because of the need to check for an overflow condition prior to proceeding with the actual division operation, it is convenient, from a hardware viewpoint, to extend the length of the dividend register by one bit. We will refer to this extra bit as the link. This extra bit (link) will allow the overflow comparison test to use the same hardware configuration as that used in the division operation. In this case a dummy quotient bit $q_n = 0$ is shifted into the register when no overflow condition occurs. At the completion of the division cycle the final remainder will now reside in the extended bit (link) and the left $(n - 1)$ bits of the dividend storage register. The link will be initially set to zero.

An n-bit binary subtraction unit can also be utilized in the comparison operation by noting that whenever the most significant bit of the partial remainder (the link bit) is a one, then the next quotient bit will always be a one, since $(\mathcal{R}_{i+1})_L$ will always be greater than B. When the most significant bit of the partial remainder is a zero, then a carry out of an n-bit 2's complement binary subtracter can be used to determine whether $(\mathcal{R}_{i+1})_L \geq B$. A general logic circuit which will carry out this comparison operation and generate both the new partial remainder $(\mathcal{R}_i)_L$ and the quotient bit q_i is shown in Fig. 12.16.

The preceding comments are reflected in Algorithm 12.5, a binary-integer stored-shifted algorithm. This algorithm is written assuming the register configuration described earlier in this section. Other variations of this algorithm are possible including those which store the difference and then shift the partial remainder. The implementation of these variations will be left as an exercise for the reader. The stored-shifted algorithm used here was selected because it will not require a very complex control system.

ALGORITHM 12.5

A binary integer-comparison division procedure using a stored-shifted technique.

$$A = a_{2n-1} \cdots a_n a_{n-1} \cdots a_0$$

$$B = b_{n-1} \cdots b_0$$

$$\frac{A}{B} = Q + \frac{R}{B}$$

$$Q = q_{n-1} q_{n-2} \cdots q_0, \qquad R = R_0$$

where

$$R_n = A$$

and

$$R_i = R_{i+1} - (q_i \cdot B \cdot 2^i) \qquad n - 1 \geq i \geq 0$$

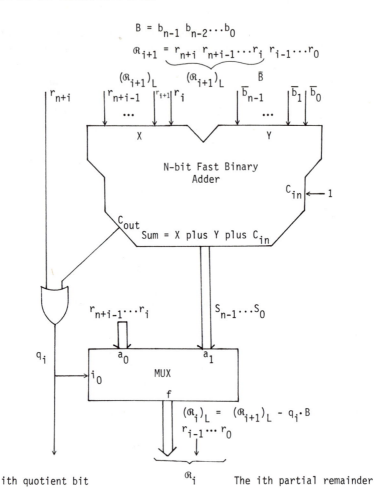

Fig. 12.16 General-Logic Circuit for Generating the *i*th Partial Remainder R_i

STEP   OPERATION

1.   Load Data

   a) Load A_{LH} into the accumulator register.

   $AC \leftarrow a_{2n-1} \cdots a_n$

   b) Load A_{RH} into the quotient register.

   $QR \leftarrow a_{n-1} \cdots a_0$

   c) Load B into the auxiliary register.

   $AX \leftarrow b_{n-1} \cdots b_0$

d) Clear the overflow (OV) and link (L) flip-flop to zero.

     link $\leftarrow 0$,   OV $\leftarrow 0$

(Note: R_n is now stored in link $AC_{n-1} \cdots AC_0$
$QR_{n-1} \cdots Q_0$ with $(R_n)_L$ stored in link $AC_{n-1} \cdots AC_0$.)

2.   Check for overflow. Set overflow flip-flop, OV, to a one, and STOP if $B = 0$ or $(R_n)_L \geq B$.
(Form AC minus AX, setting OV to 1 if $C_{out} = 1$.)

3.   Align the initial partial remainder R_n. Shift the link, accumulator register, and quotient register left one place shifting in a zero with

     link $\leftarrow AC_{n-1}$

     $AC \leftarrow AC_{n-2} \cdots AC_0 QR_{n-1}$

     $QR \leftarrow QR_{n-2} \cdots QR_0 0$.

4.   Divide loop. Repeat step 5 a total of N times then go to step 8. Initially set $i = N$, and decrement i by 1 just prior to each execution of step 5.

5.   The comparison operation: If $(R_{i+1})_L \geq B$, set q_i to a one and go to step 6. Otherwise, set q_i to a zero, and go to step 7. (Form $S = C_{out} S_{n-1} \cdots S_0 = AC$ minus AX with $q_i = 1$, if link OR $C_{out} = 1$.)

6.   Store the quotient, $q_i = 1$, and align the ith partial remainder R_i, then go back to step 4. Simultaneously store the sum and shift the result one place to the left, shifting in a quotient bit equal to one.

     link $\leftarrow S_{n-1}$

     $AC \leftarrow S_{n-2} S_{n-2} \cdots S_0 QR_{n-1}$

     $QR \leftarrow QR_{n-2} QR_{n-3} \cdots QR_0 1$.

7.   Store the quotient $q_i = 0$ and align the partial remainder R_i, then go back to step 4. Simultaneously shift the contents to the link, AC, and QR left one place, shifting in a zero quotient bit.

     link $\leftarrow AC_{n-1}$

     $AC \leftarrow AC_{n-2} \cdots AC_0 QR_{n-1}$

     $QR \leftarrow QR_{n-2} QR_{n-3} \cdots QR_0 0$.

8.   Realign the final remainder setting $R = R_0$ and then STOP.

     link $\leftarrow 0$

     $AC \leftarrow$ link $AC_{n-1} \cdots AC_1$

The quotient Q is now stored in the QR register with the remainder R stored in the AC register.

     A summary of Algorithm 12.5 is presented in flow-chart format in Fig. 12.17. A worked example using this flow chart is shown as Example 12.6. In this example the comments refer to the register information. It should be noted that, during the realignment of the final remainder in step 8, no information is lost,

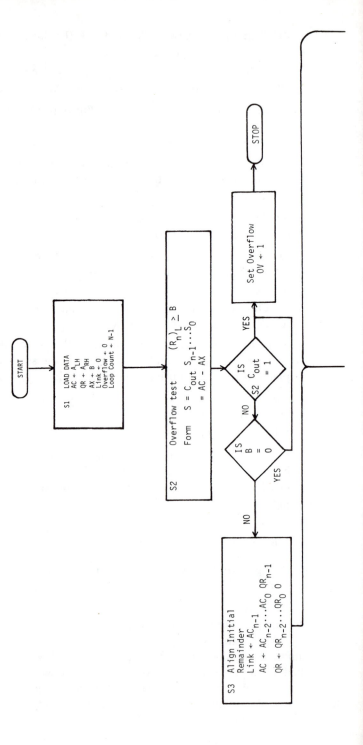

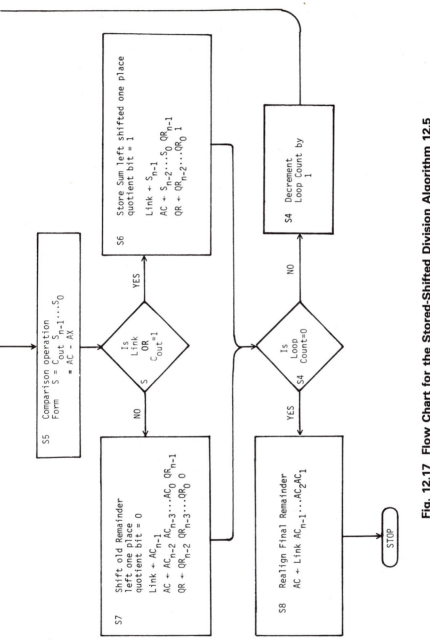

Fig. 12.17 Flow Chart for the Stored-Shifted Division Algorithm 12.5

since AC_0 will always be zero just prior to the right shift operation. This zero was inserted during the overflow check in step 2.

Example 12.6 Divide $A = 10101101$ by $B = 1100$ using Algorithm 12.5.

Step	OV	C_{out}	Sum	L	AC	QR	AX	LCNT	Comment
1	0	d	d	0	1010	1101	1100	3	Load data.
2	0	0	1110	0	1010	1101	1100	3	Check for overflow AC < AX.
3	0	d	dddd	1	0101	1010	1100	3	Shift left Link-AC-QR, QR ← 0.
5	0	0	1001	1					Form AC-AX. Link OR C_{out} = 1.
6				1	0011	0101			Store shifted left one place and set the quotient bit to a one.
4_1								2	Set LCNT to 2.
5		0	0111	1					Form AC-AX. Link OR C_{out} = 1.
6				0	1110	1011			Store the sum shifted left one place and set the quotient bit to a one.
4_2								1	Set LCNT to 1.
5		1	0010	0					Form AC-AX. Link OR C_{out} = 1.
6				0	0101	0111			Store the sum shifted left one place and set the quotient bit to a one.
4_3								0	Set LCNT to 0.
5		0	1001	0					Form AC-AX. Link OR C_{out} = 0.
7					1010	1110			Shift the Link, AC, and QR left one place and set the quotient bit to a zero.
4_4								0	Since LCNT = 0, go to step 8.
8					0101				Load Remainder into AC and STOP. Quotient is stored in QR.

$Q = 1110$ and Remainder is 0101

Let us implement Algorithm 12.5 using the same basic approach outlined in Section 12.3. Examination of Algorithm 12.5 reveals that the auxiliary register consists of a simple storage register needed to hold the divisor B with $AX \leftarrow b_{n-1} \cdots b_0$. The quotient register will require two separate data inputs, one to initially load the right half of the dividend $QR \leftarrow a_{n-1} \cdots a_0$ and the second to provide a left shift operation, with $QR \leftarrow QR_{n-2} \cdots QR_0 q$, where q is the value of the new quotient bit ($q = $ Link OR C_{out}). The accumulator register will require four data inputs. The first is required for the loading of the left half of the dividend A, with $AC \leftarrow a_{2-n} \cdots a_n$. A second input is required to store the new partial product shifted left when the quotient bit q is equal to a one, with $AC \leftarrow S_{n-2} \cdots S_0 QR_{n-1}$. A third data input is required to shift left the old partial remainder when the quotient bit q is equal to zero, with $AC \leftarrow AC_{n-2} \cdots AC_0 QR_{n-1}$. The fourth data input is required for the right shifting of the final remainder R_0 with $AC \leftarrow$ Link $AC_{n-1} \cdots AC_1$. The Link flip-flop will also require four data inputs, one to initially clear it to zero, Link $\leftarrow 0$. A second input is required to store the most significant bit of the new partial remainder when q is a one, Link $\leftarrow S_{n-1}$. A third input is required to store the most significant bit of the old partial remainder when $q = 0$, Link $\leftarrow AC_{n-1}$. Finally, the fourth input is used to clear the Link to zero during the realignment of the final remainder, Link $\leftarrow 0$. A simple delay flip-flop can be used for the overflow flip-flop.

The comparison circuit shown in Fig. 12.16 can be used to generate the partial remainders.

A basic hardware configuration for this stored-shifted division algorithm is shown in Fig. 12.18. Thirteen numbered control signals are indicated on this host configuration. These lines cause the following action to take place:

1. DSLX $\Big\}$ Data-select lines for the data-input multiplexer for the Link flip-
2. DSLY   flop, L.

Function	DSLX	DSLY	Input Selection
Load Data	0	0	S_{n-1}
Load Data	0	1	AC_{n-1}
Clear	1	0	0
Clear	1	1	0

3. ELINK   Clock input enable on the link flip-flop.
4. DSACX $\Big\}$ Data-select lines for the data-input multiplexer for the accumulator
5. DSACY   register.

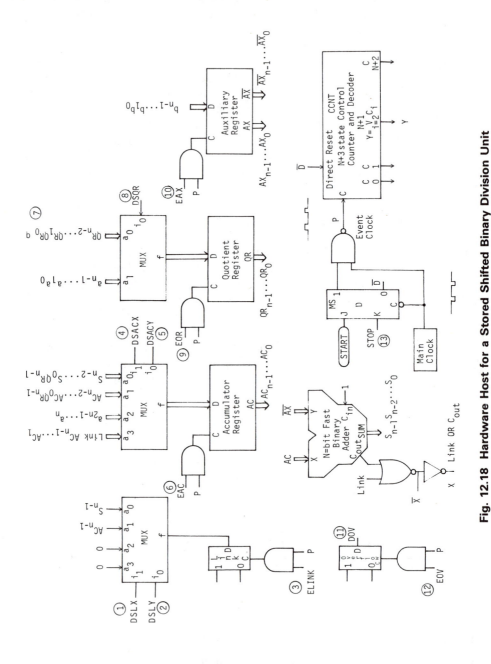

Fig. 12.18 Hardware Host for a Stored Shifted Binary Division Unit

Function	DSACX	DSACY	Input Selected
Store left shifted	0	0	$S_{n-2}S_{n-3} \cdots S_0 QR_{n-1}$
Shift left	0	1	$AC_{n-2} \cdots AC_0 QR_{n-1}$
Load data	1	0	$a_{2n-1} \cdots a_1 a_0$
Shift right	1	1	Link $AC_{n-1} \cdots AC_2 AC_1$

6. EAC   Clock input enable for the AC register.

7. q The new quotient digit.

8. DSQR   Data-select line for the quotient-register data-input multiplexer.

Function	DSQR	Input Selected
Shift left	0	$QR_{n-2}QR_{n-1} \cdots QR_0 \; q$
Load data	1	$a_{n-1} \, a_{n-2} \cdots a_1 a_0$

 9. EQR   Clock input enable for the quotient register.

10. EAX   Clock input enable on the overflow flip-flop.

11. DOV   Data input on the overflow flip-flop.

12. EOV   Clock input enable for the auxiliary register.

13. STOP   Stop enable (K input) on the division event flip-flop D.

    A control unit consisting of a divide event flip-flop D and a control counter with $(N+3)$ counting states is required to properly sequence these thirteen control signals. Once the divide flip-flop D is externally set by a START command, it will remain on (set to one) throughout the division operation. As long as D is a one, the event clock P will be enabled. The divide flip-flop, D, can be turned off (set to zero), at the completion of the algorithm by setting the STOP command to a one. Counter state C_0 will be used to load the various registers with data. Counter state C_1 will be used for the overflow check. Counter states C_2 to C_{N+1} will be used to control the n-step division loop. Finally, counter state C_{N+2} will be used to readjust the final remainder. Let us determine in detail the various control line inputs for this control unit.

When the control counter is in state C_0, the Link and overflow flip-flops should be cleared to zero, the dividend A loaded into the accumulator and quotient registers, and the divisor B loaded into the auxiliary register. These tasks can be simultaneously accomplished by setting ELINK, EOV, EAC, EQR, and EAX to ones, thereby enabling the clock inputs and properly selecting the various data inputs. By setting DSLX to a one and DSLY to a zero, the Link will be cleared to zero by the next clock pulse. The overflow flip-flop will be cleared to zero by placing a zero on DOV. The left half of the dividend will be placed in the AC by setting DSACX to a one and DSACY to a zero. The right half of the dividend will be placed in the QR by setting DSQR to a one. The divisor B is automatically loaded into the auxiliary register on the next clock pulse, since the AX register has only a single data input.

When the control counter is in state C_1, an overflow check is automatically carried out by the subtracter circuit, with X being set to one, if $A_{LH} \geq B$. When X is equal to one, the overflow flip-flop is set to a one, and a STOP command issued on the next clock pulse; thus DOV, EOV, and STOP are set to X. When X is equal to zero, the initial partial remainder must be shifted left one place with a zero entered from the right. This can be accomplished by setting the clock enables on the Link, AC, and QR to $\bar{X}$, thereby allowing these units to be clocked only when X is equal to zero. By setting DSLX, DSACX, and DSQR to zero and DSLY and DSACY to one, the left shift will occur on the next clock pulse. q will be set to X and thus will equal 0 whenever the shift takes place. The reason for setting q to X instead of zero will be apparent shortly. When the control counter is in states C_2 to C_{N+1}, indicated by Y being equal to a one,

$$Y = \bigvee_{i=2}^{N+1} C_i,$$

the divide loop is executed N times. The quotient q equal to $X = $ Link OR C_{out}, is formed automatically by the subtracter. When q is equal to one, the new partial remainder is formed by storing the difference left shifted one place with a one entered from the right (new quotient bit). When q is equal to zero, the old partial remainder is left shifted one place with a zero entered for the new quotient bit. Since the Link, AC, and QR change each time, their clock input enables; ELINK, EAC, and EQR are set to ones. The data select lines DSLX, DSACX, and DSQR are set to zero for these N clock cycles. DSLY and DSACY are set to X, thereby allowing either the old partial remainder ($X = 0$) or the new difference from the subtracter ($X = 1$) to be stored left shifted. The quotient bit q will be set equal to X (Link OR C_{out}).

Finally, when the counter reaches state C_{N+2}, we must turn off the event flip-flop D by setting the STOP enable to a one. Also, we should return the final remainder R_0 to the AC register and clear the Link. These latter operations can be accomplished by enabling the clock inputs on the Link and AC register and setting the data input on the Link and AC for a right shift. Hence, we should set DSLX, DSLY, DSACX, and DSACY to one.

The results of this discussion are summarized by the control matrix shown in Fig. 12.19. Note that if an overflow occurs during control state C_1, the dividend is not left shifted. It should also be noted, that the don't-care condition indicated on the control line 7 (q) can be set equal to X, thereby simplifying the logic.

The logic equations for the thirteen control lines required by this division unit are:

$$Y = \bigvee_{i=2}^{N+1} C_i$$

1. $\text{DSLX} = C_0 + C_{N+2}$ 2. $\text{DSLY} = C_1 + Y\bar{X} + C_{N+2}$

3. $\text{ELINK} = C_0 + C_1\bar{X} + Y + C_{N+2}$ 4. $\text{DSACX} = C_0 + C_{N+2}$

5. $\text{DSACY} = C_1 + Y\bar{X} + C_{N+2}$ 6. $\text{EAC} = C_0 + C_1\bar{X} + Y + C_{N+2}$

7. $q = X$ 8. $\text{DSQR} = C_0$

9. $\text{EQR} = C_0 + C_1\bar{X} + Y$ 10. $\text{EAX} = C_0$

11. $\text{DOV} = C_1 X$ 12. $\text{EOV} = C_0 + C_1 X$

13. $\text{STOP} = C_1 X + C_{N+2}$

The control-logic circuit for these equations is shown in Fig. 12.20. A detailed example using this hardware configuration is worked out as Example 12.7.

Example 12.7   Divide $A = (010010)_2$ by $B = (101)_2$.

Present Control State	D	CNT	OV	X	C_{out}	Sum AC minus AX	Link	AC	QR	AX	Comment
START	1	0	—	—	—	—	—	—	—	—	Turn on the Divide Flip-Flop
C_0	1	1	0	—	—	—	0	010	010	101	$OV \leftarrow 0$, Link $\leftarrow 0$, $AC \leftarrow A_{LH}$ $QR \leftarrow A_{RH}$, $AX \leftarrow B$.
C_1	1	2	0	0	0	101	0	100	100	101	$A_{LH} < B$ ∴ no overflow. Link $\leftarrow AC_2$, $AC \leftarrow AC_1AC_0QR_2$ $QR \leftarrow QR_1QR_0\,0$.
C_2	1	3	0	0	0	111	1	001	00<u>0</u>	101	$X = $ Link $+ C_{out} = 0$ ∴ $q = 0$. Link $\leftarrow AC_2$, $AC \leftarrow AC_1AC_0QR_2$ $QR \leftarrow QR_1QR_0\,0$.
C_3	1	4	0	1	0	100	1	000	00<u>1</u>	101	$X = $ Link $+ C_{out} = 1$ ∴ $q = 1$, Link $\leftarrow S_2$, $AC \leftarrow S_1S_0QR_2$ $QR \leftarrow QR_1QR_0\,1$.
C_4	1	5	0	1	0	011	0	110	<u>011</u>	101	$X = $ Link $+ C_{out}$ ∴ $q = 1$, Link $\leftarrow S_2$, $AC \leftarrow S_1S_0QR_2$ $QR \leftarrow QR_1QR_0\,1$.
C_5	0	0	0	—	—	—	0	011	011	101	Realign remainder Link $\leftarrow 0$. $AC \leftarrow$ Link AC_2AC_1. STOP.

QR
$Q = 011$ AC
             $R = 011$

Control Line → Control Counter State ↓	1 DSLX	2 DSLY	3 ELINK	4 DSACX	5 DSACY	6 EAC	7 q	8 DSQR	9 EQR	10 EAX	11 DOV	12 EOV	13 STOP	Comment
C_0	1	0	1	1	0	1	d(X)	1	1	1	0	1	0	Link $\leftarrow$ 0, AC $\leftarrow A_{LH}$, QR $\leftarrow A_{RH}$; AX $\leftarrow$ B, OV $\leftarrow$ 0
C_1	0	1	$\bar{X}$	0	1	$\bar{X}$	X	0	$\bar{X}$	0	X	X	X	Overflow check and alignment of initial Partial Remainder
C_2 to C_{N+1} $[Y = \overset{N+1}{\underset{i=2}{V}} c_i]$	0	$\bar{X}$	1	0	$\bar{X}$	1	X	0	1	0	0	0	0	Basic Divide loop
C_{N+2}	1	1	1	1	1	1	d(X)	0	0	0	0	0	1	Realign final remainder and stop

Note: X = Link OR C_{out}

d = don't care

Fig. 12.19 Control Matrix for the Division Unit Shown in Fig. 12.18

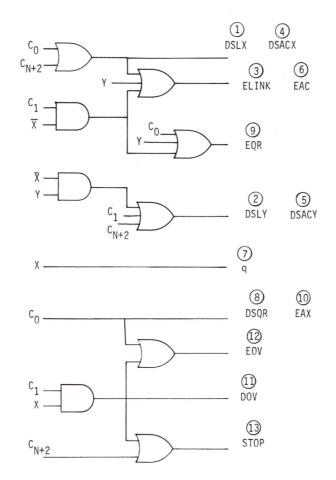

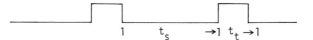

Fig. 12.20　Control Logic for the Stored-Shifted Division Circuit Shown in Fig. 12.18

In general, this hardware division circuit will require $(N + 3)$ clock pulses to complete the operations once the start signal has been issued. The clock pulse should be asymmetric with a relatively long setup time, t_s, and a relatively short

transfer time, t_t. The setup time, t_s, should include the following propagation delay times:

a) the time required for the n-bit fast adder to carry out the subtraction operation (12 units of gate delay assuming a first-order carry-lookahead-adder unit).

b) the time required for the new data input to be selected on the various input multiplexers (4 units of gate delay in the worst case).

c) the input setup time for the various registers, OV, LINK, AC, QR, and AX (3 units of gate delay assuming edge-triggered delay flip-flops are used).

d) wiring delay and a safety factor as required to account for environmental changes (3 units for wiring and 5 units for a safety margin).

(Typical delay estimates in terms of units of gate delay are shown in parenthesis.)

The transfer time, t_t, must be long enough to allow the new data to be properly transferred into all of the registers and counters. It must also be short enough so that changes in enable signals prior to the removal of the clock signal do not cause any false clock pulses to be issued. (Five units of gate delay should be adequate.)

Thus a conservative estimate of the clocking requirements yields a setup time of 27 units of gate delay and a transfer time of five units of gate delay, for a total clocking period of 32 units of gate delay. This estimate yields a division time of $32(N+3)$ units of gate delay. This compares favorably with the delay estimate of $32(N+1)$ units of delay for the unsigned stored-shifted multiply unit designed in Section 12.3. The division operation is slightly slower due to the extra cycles required for the overflow check and remainder realignment.

As was the case with the serial-parallel multipliers, this unit could easily be constructed as a large-scale integrated circuit. One such configuration is shown in Fig. 12.21.

The divide flip-flop D can be used to determine when the final quotient is done. As long as D is equal to one, the unit would be considered as busy. When D is returned to zero, the division operation would be complete with the result correct as long as OV (overflow) is equal to zero. Either an internal or an external

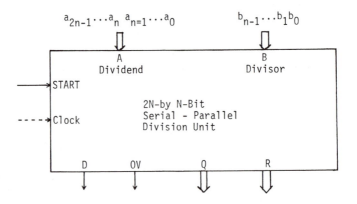

Fig. 12.21 Logic Symbol for a 2N by N Bit Serial Parallel Division Unit

clocking signal could be used to control the unit. The start signal should be removed prior to the time D returns to zero in order to prevent repeated division operation. Because most unsigned division algorithms are variations of this algorithm, the design of additional unsigned binary dividers will be left as exercises for the reader.

12.5 PROBLEMS

12.1 Modify the serial adder-subtracter discussed in Section 12.2 so that one's complement arithmetic can be used. Be careful about how and when you check for an overflow condition. Show the details for your control unit.

12.2 Modify the serial adder-subtracter discussed in Section 12.2 so that sign-and-magnitude arithmetic can be used. Include the details of your control unit.

12.3 Multiply the following numbers using Algorithm 12.2 and the format shown in Example 12.1.
   a) $A = 1111$ $B = 1111$
   b) $A = 1000$ $B = 0111$
   c) $A = 1101$ $B = 1100$
   d) $A = 0110$ $B = 0101$

12.4 Using the hardware circuit for Algorithm 12.2 and the format shown in Example 12.2, multiply the following set of binary numbers.
   a) $A = 0111$ $B = 1111$
   b) $A = 1000$ $B = 1000$
   c) $A = 1010$ $B = 1010$
   d) $A = 1011$ $B = 0110$

12.5 Using Algorithm 12.3 and the format given in Example 12.3, multiply the following sets of binary numbers.
   a) $A = 1101$ $B = 0111$
   b) $A = 1001$ $B = 1101$
   c) $A = 0110$ $B = 0011$
   d) $A = 1011$ $B = 1001$

12.6 Consider the following algorithm for signed binary multiplication of two n-bit numbers. By inserting the words *sign* and *magnitude, one's* or *two's* ahead of the word, *complement,* Algorithm 12.6 can be tailored to a particular arithmetic notation.

ALGORITHM 12.6

Signed binary-integer multiplication using a stored shifted approach

STEP   OPERATION

1.   a) Clear the accumulator register (AC).
         b) Load the multiplier A into the multiplier register (MR).
         c) Load the multiplicand B into the auxiliary register (AX).

      d) Store A_{n-1} in a flip-flop called A_{sign}.

      e) Store B_{n-1} in a flip-flop called B_{sign}.

2.   a) If A_{sign} indicates that A is negative, *complement* the contents of the MR register using the appropriate arithmetic.

      b) If B_{sign} indicates that B is negative, *complement* the contents of the AX register using the appropriate arithmetic.

3.   Repeat step 4 N times and then go to step 5.

4.   a) Form the sum

$$S = AC \text{ plus } (x \text{ times AX})$$

     where

$$x = 0 \quad \text{if} \quad MR_0 = 0$$

     or

$$x = 1 \quad \text{if} \quad MR_0 = 1.$$

      b) Replace the contents of the AC register with $C_{out}S_{n-1} \cdots S_1$ and simultaneously.

      c) Replace the contents of the MR register with $S_0 MR_{n-1} \cdots MR_1$.

5.   If the contents of A_{sign} and B_{sign} indicate that the product should be negative, then *complement* the product stored in the AC and MR registers using the appropriate arithmetic.

6.   STOP. The signed product is stored in the AC and MR registers with the most significant n bits in the AC register and the least significant n bits in the MR register.

Assuming sign-and-magnitude arithmetic, and n-bit numbers, implement Algorithm 12.6 using the procedure outlined in Sections 12.3.1 and 12.3.2. The hardware organization shown in Fig. 12.14 can be easily modified to handle signed numbers by using 4 input multiplexers on the data inputs to the AC and MR register and a 2-input multiplexer on the input to the auxiliary register.
Note: a two's complement adder should be used since all addition uses only positive numbers.

a) Draw a flow chart for your algorithm.

b) Draw a block diagram of your hardware organization showing the labeled control lines.

c) Construct a control matrix for your unit.

d) Write out the control line equations for your design, and show a logic circuit realization of these equations.

e) Estimate the multiply time for your unit, i.e., estimate the setup time and transfer time requirements for your clock line.

12.7 Using the format shown in Fig. 12.22, multiply the following sign-and-magnitude numbers. CCNT is the control counter or step count and should be incremented on each step.

a) $A = 1000 \quad B = 1000$

b) $A = 1011 \quad B = 0110$

c) $A = 0101 \quad B = 0111$

d) $A = 0110 \quad B = 1011$

$A = a_3a_2a_1a_0$					$B = b_3b_2b_1b_0$				
$A_{sign} = A_3$					$B_{sign} = b_3$				

Present Control State	Value of Register at the end of the clocking period							Comment
	M	C_{out}	SUM	AC	MR	AX	CCNT	
Start	1	-	-	-	-	-	0	Turn on multiply flip flop

Figure 12.22

12.8 Repeat Problem 12.6 using one's complement arithmetic. *Do not* replace the adder with a one's complement unit since all additions will involve only positive numbers.

12.9 Repeat Problem 12.7 assuming the numbers given are in one's complement format.

12.10 Repeat Problem 12.6 using two's complement arithmetic. Note: you will have to design special registers for the AC, MR, and AX which can load data from two

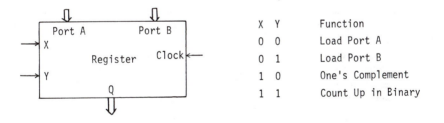

X	Y	Function
0	0	Load Port A
0	1	Load Port B
1	0	One's Complement
1	1	Count Up in Binary

separate input ports, take the one's complement of the register contents, and also count up in binary. Also be careful how you accomplish the post-complementing operation.

12.11 Repeat Problem 12.7 assuming the numbers given are in two's complement format.

12.12 Using Algorithm 12.5 and the format shown in Example 12.6, divide the following sets of binary numbers.
   a) $A = 01100110$ $B = 0111$
   b) $A = 11101100$ $B = 1111$
   c) $A = 00110011$ $B = 0101$
   d) $A = 01011011$ $B = 0100$

12.13 Following is an algorithm for signed binary division. The words *sign* and *magnitude*, *one's* or *two's* should be placed ahead of the word *complement* in order to tailor this algorithm to a particular arithmetic convention.

ALGORITHM 12.7

Signed binary integer division using a stored shifted approach

$$A = a_{n-1}a_{2n-2} \cdots a_n a_{n-1} \cdots a_0$$

$$B = b_{n-1}b_{n-2} \cdots b_0$$

$$\frac{A}{B} = Q + \frac{R}{B}$$

$$Q = q_{n-1} \cdots q_0 \qquad R = R_0$$

where

$$R_n = A$$

$$R_i = R_{i+1} - (q_i \cdot B \cdot 2^i) \qquad n - 1 \geq i \geq 0$$

STEP   OPERATION

1.   Load data
       a) Load A_{LH} into the accumulator register: $AC \leftarrow a_{2n-1} \cdots a_n$.
       b) Load A_{RH} into the quotient register: $QR \leftarrow a_{n-1} \cdots a_0$.
       c) Load B into the auxiliary register: $AX \leftarrow b_{n-1} \cdots b_0$.
       d) Clear the overflow flip-flop to zero: $OV \leftarrow 0$.
       e) Store a_{2n-1}, the sign of A, in a flip-flop called A_{sign}.
       f) Store b_{n-1}, the sign of B, in a flip-flop called B_{sign}.

2.   Precomplement operation
       a) If A_{sign} indicates that A is negative, *complement* the dividend stored in the
          accumulator and quotient registers using the appropriate arithmetic conven-
          tion.
       b) If B_{sign} indicates that B is negative, *complement* the divisor stored in the
          auxiliary register using the appropriate arithmetic conventions.
       (Note: at the end of step 2, the AX will contain $|B|$ and the AC and QR will
       contain $R_n = |A|$, with $(R_n)_L$ stored in the AC register.)

3.   Initial overflow check
       Set the overflow flip-flop to a one and STOP if $B = 0$ or if $(R_n)_L \geq B$. (Form AC
       minus AX, setting OV to a 1 if $C_{out} = 1$.)

4.   Align the initial partial remainder R_n. Shift the accumulator register and quotient
       register left one place shifting in a zero.

$$AC \leftarrow AC_{n-2} \cdots AC_0 QR_{n-1}$$

$$QR \leftarrow QR_{n-2} \cdots QR_0\, 0$$

5.   Divide loop
       Repeat step 6 a total of N times and then proceed to step 9. Initially set $i = n$
       and decrement i by 1 just prior to each execution of step 6.

6.   The comparison operation
       If $(R_{i+1})_L \geq B$, set q_i to a one and go to step 7; otherwise set q_i to a zero and go to
       step 8.
       (Form $S = C_{out}S_{n-1} \cdots S_0 = AC$ minus AX and set $q_i = C_{out}$.)

7. Store the quotient $q_i = 1$ and align the ith partial remainder R_i, then go back to step 5. Simultaneously store the sum shifting the result one place to the left, shifting in a quotient bit equal to one

$$AC \leftarrow S_{n-2}S_{n-3} \cdots S_0 \, QR_{n-1}$$
$$QR \leftarrow QR_{n-2} \cdots QR_0 \, 1$$

8. Store the quotient $q_i = 0$ and align the ith partial remainder R_i, then go back to step 5. Simultaneously shift the contents of the AC and QR left one place shifting in a quotient bit equal to 0.

$$AC \leftarrow AC_{n-2} \, AC_{n-3} \cdots AC_0 QR_{n-1}$$
$$QR \leftarrow QR_{n-2} \, QR_{n-3} \cdots QR_0 0$$

9. Realign the final remainder setting

$$R = R_0$$

Right shift the AC register one place

$$AC \leftarrow 0 \, AC_{n-1} \cdots AC_2 AC_1$$

10. Postcomplement operation
    a) If $A_{sign} \oplus B_{sign}$ equals one, then *complement* the quotient Q stored in the Quotient Register.
    b) If A_{sign} equals one, then complement the remainder R stored in the accumulator register using the appropriate arithmetic conventions. The remainder thus has the same sign as the original dividend.

11. Final overflow check
    *[If the quotient equals 0, go to step 12, otherwise] if $A_{sign} \oplus B_{sign} \oplus Q_{n-1} = 1$, then set the overflow flip-flop OV to a one.
    * For two's complement arithmetic, add the step between the brackets.

12. STOP
    As long as the overflow flip-flop OV is zero, the correctly signed quotient, Q, will be in the quotient register and the correctly signed remainder R will be in the accumulator register.

    Using Algorithm 12.7 and assuming sign-and-magnitude numbers, divide the following sets of binary numbers.
    a) 01000000   1000
    b) 00011001   0101
    c) 10011001   0101
    d) 01101101   0111
    e) 10011101   1110
    f) 01100000   0111

12.14 Repeat Problem 12.13 assuming the numbers given are expressed in one's complement notation.

12.15 Repeat Problem 12.13 assuming the numbers given are expressed in two's complement notation.

12.16   Implement Algorithm 12.7 given in Problem 12.13 assuming sign-and-magnitude notation.

12.17   Implement Algorithm 12.7 given in Problem 12.13 assuming one's complement notation.

12.18   Implement Algorithm 12.7 given in Problem 12.13 assuming two's complement notation.

12.19   Comment on the problems which can occur if it is decided to round the quotient obtained in an unsigned integer binary division operation, using the rule that if the $|R| \geq |B|$ then add one to the quotient.

12.20   Comment on the problems which can occur if it is decided to round the quotient obtained using a signed binary division algorithm. Consider the three arithmetic notations in your discussion.

12.6 BIBLIOGRAPHY

Braun, E. L., *Digital Computer Design*. New York: Academic Press, 1963.

Cardenas, A. F., L. Presser, and M. A. Marin, *Computer Science*. New York: Wiley, 1972.

Chu, Y., *Digital Computer Design Fundamentals*. New York: McGraw-Hill, 1962.

Flores, I., *Computer Design*. Englewood Cliffs, N.J.: Prentice-Hall, 1963.

Flores, I., *The Logic of Computer Arithmetic*. Englewood Cliffs, N.J.: Prentice-Hall, 1963.

Knuth, D. E., *The Art of Computer Programming: Vol. 2, Seminumerical Algorithms*. Reading, Mass.: Addison-Wesley, 1969.

MacSorley, O. L., "High-Speed Arithmetic in Binary Computers," *Proc. IRE*. vol. 49, 67–91, January, 1961.

Peatman, J. B., *The Design of Digital Systems*. New York: McGraw-Hill, 1972.

Stein, M. L., and W. D. Munro, *Introduction to Machine Arithmetic*. Reading, Mass.: Addison-Wesley, 1971.

Index to
Algorithms and Rules

Algorithm	Title	Page
1.1	Conversion of a k-digit base-R integer to a base-10 integer	4
1.2	Conversion of a base-10 integer to a base-R integer	6
1.3	Conversion of a j-digit base-R fraction to a base-10 fraction holding an accuracy of $\pm(.1)_R^j$	10
1.4	Conversion of a k-digit decimal fraction to a base-R fraction, holding an accuracy of $\pm(.1)_{10}^k$	12
1.5	Conversion of a k-digit base-B integer to a base-A integer	15
1.6	Conversion of a base-A integer to a base-B integer	15
1.7	Conversion of a j-digit base-B fraction to a k-digit base-A fraction holding an accuracy of $\pm(.1)_B^j$	16
1.8	Conversion of a k-digit base-A fraction to a j-digit base-B fraction, holding an accuracy of $\pm(.1)_A^k$	16
1.9	Sign-and-magnitude complement (negation) of a base-R number	20
1.10	For taking the diminished-radix complement (negation) of a base-R number.	21
1.11	For taking the radix-complement of a base-R number.	23
2.1	Crutch addition of two unsigned base-R numbers	31

Rule		
2.1	Addition of two base-R numbers using radix notation	37
2.2	Subtraction of two base-R numbers using radix notation	37
2.3	Addition of two base-R numbers using diminished-radix notation.	42

Rule	Title	Page
2.4	Subtraction of two base-R numbers using diminished-radix notation.	43
2.5	Addition of two base-R numbers using sign-and-magnitude notation.	47
2.6	Subtraction of two base-R numbers using sign-and-magnitude notation	47

Algorithm

	Title	Page
3.1	For generating the fundamental sum-of-product form (FSP)	99
3.2	For generating the fundamental product-of-sum form (FPS)	100
3.3	*Rule* for writing fundamental sum-of-product form from a truth table.	102
3.4	*Rule* for writing fundamental product-of-sum form from a truth table.	102
3.5	*Rule* for conversion from product-of-sum form to sum-of-product form	104
3.6	*Rule* for conversion from sum-of-product to product-of-sum form	104
5.1	General procedure for minimizing multivariable single-output Boolean functions.	163
5.2	Quine-McCluskey method for finding prime implicants	173
8.1	SR analysis procedure	308
8.2	Sequential circuit synthesis procedure using clocked SR flip-flops	315
9.1	JK flip-flop analysis procedure	367
9.2	Procedure for designing clocked synthesis sequential circuits using JK flip-flops from known next-state equations using excitation tables	370
9.3	Synthesis procedure for designing clocked sequential circuits using JK flip-flops from known next-state equations using expansions	375
11.1	Synthesis procedure for the generation of an N-bit periodic sequence using feedback shift registers	468
11.2	Synthesis procedure for the generation of an N-bit periodic sequence using resettable counters.	470
12.1	An algorithm for a 4-bit serial addition and subtraction unit.	481
12.2	Add and then shift binary multiplication algorithm	489
12.3	A stored-shifted binary integer multiplication algorithm	502
12.4	A general base-R restoring division procedure for positive integers	508
12.5	A binary integer comparison division procedure using a stored-shifted technique	513
12.6	Signed binary integer multiplication using a stored-shifted approach	527
12.7	Signed binary integer division using a stored-shifted approach	530

Index

Absorption, 66, 67, 160
Accumulator register
  for binary division, 507
  for binary multiplication, 489
Adder
  binary full adder, 201
  binary half adder, 197
  binary serial, 479–486
  decimal full adder, 261–267
Addition
  base-R, rules, 31
  diminished radix, 38–43
  radix, 33–38
  sign-and-magnitude, 43–47
Algorithm, *see* Algorithm index
Analog to digital converter, 418
Analysis
  of JK flip-flops, 367
  of SR flip-flops, 308–311
  of toggle flip-flops, 365
AND, 70
AND-OR-INVERT logic
  for multiplexers, 237
    as part of multifunction register, 447
Arbitrary modulo-N counter, 399–407
Arithmetic
  diminished-radix, 38–43

radix, 33, 38
sign-and-magnitude, 43–47
Arithmetic units
  binary carry-lookahead, 208–222
  binary ripple adders, 207–208
  comparison of types, 232
  decimal, 261–266
  first-adder carry-lookahead adder,
    225–232
  ripple-lookahead adders, 222–225
Associative law, 68
Autonomous counters, 389
Auxiliary register, for binary division, 507
Axioms of duality, 91

B_n, 78
Base conversion, 3
Base-R full adder, 48
Basic counting cell, 386
Basic logic blocks, 194–197
BCD full adder, 261–267
Binary arithmetic circuits, 197–207
Binary counter
  asynchronous, 386–389
  direct resetting, 408–414
  down, 414–418
  Gray code counters, 419

535

modulo-N, 391
modulo-5, 407
modulo-10, 400
modulo-10 excess-3, 419
modulo-16, 395
ring counters, 419
self-starting, 398
self-starting ring, 421
self-starting switch-tail, 424
switch tail, 424
synchronous, 389–399
up-down, 415
Binary counting registers, 449–453
Binary full adder, 201
Binary half adder, 197
Binary storage elements
delay flip-flop, 336–341
edge triggered flip-flops, 356–360
JK flip-flop, 336–346
master-slave flip-flop, 346–356
NOR-gate latch, 297
set-reset flip-flop (SR), 298–303
toggle flip-flop, 341–342
Binary variable, 70
Binary-coded decimal, 242
biquinary code, 250
code conversion, 254–257
ripple adder, 267–270
self-complementing code, 243
Biquinary code, 249
Boolean algebra
basic properties, 61–69
canonic product term, 73
canonic sum term, 76
definition, 61
fundamental product term, 73, 98
fundamental sum term, 76, 100
maxterm, 76, 89
minterm, 73, 84
operators, 70
product term, 73, 98
sum term, 76, 100
Boolean function
complementation, 95
conversion between standard forms, 104
definition, 73
enumeration, 73

general procedure for minimization, 163
incompletely specified, 186–189
K-map method for minimization,
162–172
literal of, 95
map specification, 82
minimization, 155–189
minimization of multi-output, 170–172
multiplexer realization, 136–147
nonredundant expression, 162
prime implicant, 161
product-of-sum form, 76, 100
Quine-McCluskey minimization method,
162–172
self dual, 107, 202
standard forms, 98–104
sum-of-product form, 75, 98
test for implication, 161
truth-table specifications, 81, 82
weight of, 172
Bounce-free switch, NAND implemented,
329

Canonic product term, 73
Canonic sum term, 76
Carry group size, 212
Carry-lookahead adder, 208–222
carry group size, 212
fan out, 221
general equations, 215
generate function, 210
parts count, 218–220
propagate function, 210
Characteristic number, 77–81, 98
dual of, 94
CLA(n), 208–222
Clocked flip-flops
delay, 336–341, 361–363
edge-triggered, 356–367
JK, 359–361
master-slave, 346–356
self-reset (SR), 305–308
toggle, 363–367
Combinational logic circuit, 193
Commutative law, 61, 62
Complement, 20, 62, 66
Boolean function, 95

diminished-radix, 21
 logic circuit, 97
 one's, 23, 24
 radix, 23
 sign-and-magnitude, 20
 truth-table, 96
 two's, 23, 24
Complementing register, 453–456
Control matrix
 for store then shift multiplication, 497
 for stored-shifted division, 524
Control unit
 for serial adder, 485
 synchronous counter for serial adder,
  486
Controlled
 diminished-radix complement unit, 50
 one's complement circuit, 197
 radix complement unit, 50
Conversion
 between Boolean forms, 104
 characteristic number to product-of-sum
  form, 78
 characteristic number to sum-of-product
  form, 78
Core, 163
Cost criteria for minimization, 162
Counter, *see also* Binary counters
 arbitrary modulo-N binary, 399–407
 autonomous, 389
 binary asynchronous, 386–389
 binary synchornous, 389–399
 direct resetting, 408–414
 excess-3, 419
 Gray code, 419
 modulo-N, 391
 resettable binary, 401–403
 ring, 419
 self-starting binary, 398
 self-starting ring, 421
 self-starting switch tail, 424
 switch tail, 424
Critical race condition, 301
Crutch addition, 31

D_n, 78
Data set-up time, master-slave JK, 356

Decimal adding circuits, 261–267
Decoders, 147–148
Delay flip-flop, 336–341
 edge-triggered, 361–363
 gated latch, 337
 as part of multifunction register, 447
 synthesis, 340
DeMorgan's law, 69
Demultiplexers, 147–148
Digital clocks, 324–328
Digital one shots, 319–324
Digital stop watch, design of, 427–434
Diminished-radix
 arithmetic, 38–43
 complement unit, 49
 extended precision, 27
 hardware complexity, 56
 notation, 21
 rules for addition, 42
 rules for overflow, 42
 rules for subtraction, 43
Direct-resetting binary counters, 408–414
 with clock filter, 412–413
 with decoding spikes, 411
 with inverter chain filter, 411
Distributive law, 61
Division
 algorithm for signed binary numbers,
  530
 general algorithm for, 508
 overflow, 511, 513, 518, 522
 serial parallel, 506–527
 stored-shifted
  algorithm, 513
  control matrix, 524
  control requirement, 521–523
  flow chart for, 517
  hardware requirement, 419–521
  timing requirements, 507
Down counters, 414–418
Dual
 characteristic number, 94
 logic tables, 90
 K-maps, 91
 symbols, 90
Duality, 62
 axioms of, 91

Edge-triggered
  delay flip-flop, 361–363
  JK flip-flop, 359–361
  SR flip-flop, 356–367
  toggle flip-flop, 363–367
End-around-carry, 40, 208, 269
Equivalent circuits, 132–136
Error-correcting codes, 244,
    Hamming, 250–254
Error-detecting codes, parity, 249
Essential prime implicant, 163
  on a K-map, 164
Excess-3 code, 242
Excess-6 code, 242
Excess-3 counter, 419
Excitation table
  for SR flip-flops, 313–315
  for JK flip-flops, 370
Exclusive NOR, 119–124
Exclusive OR, 119–124
  basic properties, 120
  complement of, 122–123
  dual of, 122–123
  tree circuits, 122
Expressions, Boolean, 62

Fan in, 156
Fan out, carry-lookahead adder, 221
Feedback shift register, as timing
      generator, 464–469
First-order carry-lookahead adder
  equations for, 227
  fan out, 231
  first-order generate functions, 225
  first-order propagate functions, 225
  parts count, 229-231
Fized-point arithmetic, 30
Fixed-point number, 1
Flip-flop
  delay, 336–341
  edge-triggered delay, 361–363
  edge-triggered JK, 359–361
  edge-triggered SR, 356–360
  edge-triggered toggle, 363-367
  gated latch, 337
  JK from SR, 342–346

  master-slave JK, 348
  master-slave SR, 347
  toggle, 342
  with preclear inputs, 377–380
  with preset inputs, 377–380
Floating-point number, 1
Fractional number conversion, accuracy,
    7
Full adder,
  binary, 201
  for serial adder, 842
  using multiplexers, 205
Full subtracter, binary, 205
Fundamental product-of-sum, 100
  algorithm for generating, 100
  from truth table, 102
Fundamental product term, 73, 98
Fundamental sum-of-product, 98, 99
  algorithm for generating, 99
  from truth table, 102
Fundamental sum term, 76

G_i, 210
Gated latch, 337
Gated oscillater, 288
Generate function G_i, 210
Gray code, 243
Gray code counter, 419
Guaranteed noise margins, 281

H_n, 78
Half adder, binary, 197
Hamming codes, 250–254
Hardware complexity
  base-R arithmetic, 48–57
  diminished-radix base R, 56
  radix-arithmetic base R, 56
  sign-and-magnitude base R, 56
Hazard-free circuits, 284
Hazards,
  dynamic, 281–283
  static, 281–282
Huntington's Postulates, 61

Idempotency, 160
Idempotency law, 65
Identity, 66

Implication, 161
  test for, 161
Incompletely specified functions, 186–189
Input hold time, master-slave JK, 356
Integer, number conversion
  base-R to base-10, 4
  base-10 to base-R, 6
Integer division, serial parallel, 513
Integer multiplication
  parallel, 233
  serial parallel, 489, 502
Involution, 66
  use in NAND-NOR logic design, 114

JK flip-flops
  analysis of, 367
  excitation table, 370
  master-slave, 348
  as part of a multifunction register, 447
  present-state/next-state table, 371
  with preset and preclear inputs, 378
  specification, 343
  from SR, 342–346
  synthesis of sequential circuit using, 370

Karnaugh Map, 83–90
K-map, 83–90
  minimization of incompletely specified
    functions, 186–189

Law of complements, 66
Law of identity, 66
Law of involution, 66
Law of zeroes and ones, 63
Laws of absorption, 65, 67
Leading-edge detector, 286
Logic equivalent circuits, 132–136
Logic symbols, 119–131
Logical completeness, 124
Logical superposition, 457–462
Logic-delay units, 319–324
Link, 489
  for binary division, 507
Literal, 95
Lumped logic model, 294

Mahoney Map, 88–89

Master-slave flip-flop
  data-setup time, 356
  input-hold time, 356
  nonisolated JK, 349
  with preset and preclear inputs, 378,
    409–413
  timing requirements, 348–356
  timing malfunctions, 348–356
Master-slave principle, 346–356
Matching table, 173
Maxterm, 76, 89
Minimization, 155–189
  contraints, 156
  core, 163
  cost criteria, 162
  essential prime implicant, 163
  general procedure, 163
  of incompletely specified functions,
    186–189
  Karnaugh map method, 163–172
  matching table, 173
  multi-output Boolean function, 170–172
  nonessential prime implicant, 163
  nonredundant expression, 162
  objective, 156
  prime implicant table, 174
  product-of-sum, 182
  Quine-McCluskey method, 172–186
  reasons for, 156–160
  Table of Choices, 179–181
  two-level logic circuits, 162–170
Minimum product-of-sum expression from
    K-map, 165
Minimum sum-of-product expression from
    K-map, 164
Minterm, 73, 84
Modulo-N, 391
Modulo-N counter
  arbitrary, 399–407
  arbitrary synchronous, 396
  resettable binary, 401–403
  using coincidence gates, 404
  using a decoding network, 405
  using NAND-gate input switching, 406
Modulo-4 ring counter, 426
Modulo-4 self-starting ring counter, 426
Modulo-8 switch-tail counter, 416, 417

Modulo-10 binary counter, 400
Modulo-16 binary counter, 395
Monostable multivibrators, 319–324. *See
    also* One shot
Multifunction register
  design of, 443-449
    using delay flip-flops, 444
    using JK flip-flops, 447
    using multiplexers, 451–453
    using NAND-wired-AND gates, 447
Multiplexers, 136–147
  full-adder circuit, 205
  half-adder circuit, 200
  as logic elements, 138–141
  as part of multifunction registers,
      451–453
  as read-only memories, 138, 139
  two-level circuits, 143–144
Multiplication
  algorithm for signed binary numbers,
      527
  binary parallel, 233–234
  decimal parallel, 273
  serial-parallel techniques, 487–506
  store then shift, 489–500
    algorithm, 489
    control matrix, 497
    control requirements, 494–498
    flow chart, 491
    hardware requirements, 492–494
    timing requirements, 497
  stored shifted
    algorithm, 502
    control requirements, 505
    flow chart, 503
    hardware requirements, 504–505
    timing requirements, 505
Multiplier register, 489

NAND logic, 111
  basic properties, 112
  design with, 113–119
  multilevel circuits, 118
NAND-wired-AND, full-adder circuit,
    206
Negate, 20
Negative logic, 125

Negative numbers, 19
  conversion, 3
  diminished-radix notation, 21
  one's complement notation, 23, 24
  radix notation, 22
  radix-minus-one notation, 21
  sign convention, 20
  sign-and-magnitude, 20
  two's complement notation, 23, 24
Nine's complement
  arithmetic, 267
  BCD logic unit, 244
Noise immunity margins, 280
Nonessential prime implicant, 163
Nonredundant expression, 162
NOR-gate latch, 297
NOR logic, 111
  basic properties, 112
  design with, 113–119
  multilevel circuits, 119
NOT, 70
Number conversion
  extended precision, 27
  fractional
    base-A to base-B, 16
    base-B to base-A, 16
    base-R to base-10, 10
    base-10 to base-R, 10
  hexadecimal-to-octal, 14
  integer
    base-A to base-B, 15
    base-B to base-A, 15
    base-R to base-10, 4
    base-10 to base-R, 6
  short-cut techniques, 13
  special problems, 26
Number systems, comparison between, 24

0_n, 78
One shot, 319-324
  NAND-implemented, 334
  NOR-implemented, 320-322
OR, 70
Oscillating state, 290
Overflow,
  diminished-radix arithmetic, 42
  division, 507

sign-and-magnitude arithmetic, 47
    radix-arithmetic, 37

P_i, 210
Parity codes, 248
    even-, 248
    odd-, 248
Partial remainder, for division, 508
Perfect induction, 72
Pierce arrow, 111
Positive logic, 125
Preclear inputs, 377–380
Present-state/next-state table, 293
Preset inputs, 377–380
Prime implicants
    hazard-free circuits, 284
    on a K-map, 163
    Quine-McCluskey method, 173
    table, 179
    theorem, 162
Principle of duality, 90–94
    definition, 90
Product term, 73, 98
Product-of-sum form, 76, 100
    from K-map, 103
Propagate function, 210
Propagation delay, 275
Pure delay model, 278

Quine-McCluskey method, 164, 177–186
    for incompletely specified functions,
        186–189
    matching table, 173
    prime implicant table, 179
    table of choices, 179–181
Quotient register, for binary division, 507

$R(n)$, 207
Radix arithmetic, 33–38
    base-R, 52
    hardware complexity base-R, 56
    rule for addition, 37
    rule for overflow, 37
    rule for subtraction, 37
Radix notation, 22
    rules for extended precision, 27
Radix-minus-one notation, 21

Register
    binary counting, 449–453
    clear before setting, 440
    complementing, 453–456
    data storage, 434–449
    definition, 439
    design of multifunction, 443–449
    design for serial adder, 482
    design using logical superposition,
        457–462
    one-step-read, 442
    set before clearing, 442
    shift, 443
    two-step-read, 440
Relation
    equivalence, 61
    reflexive, 61
    symmetric, 61
    transitive, 61
Resettable binary counters, 401–403
    as timing-pattern generators, 469–472
Ring counter, 419
    self-starting, 419
    for serial adder control unit, 486
Ripple adders, 207
    binary coded decimal, 267–270
Ripple lookahead adders, 222–225
    fan out, 222
    part count, 222
RLA(n:a,b), 222–225

Self-complementing code, 243
Self-dual, 107, 202
Self-starting
    counters, 398
    ring counter, 421
    switch-tail counter, 424
Sequential circuits
    critical race condition, 301
    definition, 194
    dynamic hazards, 281–287
    guaranteed noise margins, 281
    introduction, 287–296
    JK flip-flop model, 368
    lumped-logic model, 294
    noise margins, 280

oscillating state, 290
present-state/next-state table, 293
propagation delay, 275
pure delay model, 278
SR flip-flop model, 311–313
stable state, 290–294
state, 289
state diagram, 289
state transition diagram, 290
static hazards, 281–287
time-dependent equations, 287
transition time, 275
transitional state, 290
Serial adder
binary, 479–486
control unit, 485
full-adder/subtracter, 482
register design, 482
Serial parallel
division, 506–527
multiplication, 487–506
Set-Reset (SR) flip-flop, 298–303
Seven-segment indicators, 257–261
Shaffer stroke, 111
Shannon's generalized DeMorgan's law,
99
Shift register, 443
Short cut method, for JK flip-flops,
375–377
Sign convention, 20
Sign-and-magnitude arithmetic
base-R, 54
hardware complexity base-R, 56
notation, 20
rules for addition, 47
rules for extended precision, 27
rules for overflow, 47
rules for subtraction, 47
SR flip-flop
analysis of, 308–311
as a delay flip-flop, 336–341
edge-triggered, 356–360
logic equations, 303
master-slave, 347
present-state/next-state table, 301
with preset and preclear inputs, 378
synthesis of sequential circuits, 311–319

timing chart, 302
using NAND gates, 303–305
using NOR gates, 298–303
Stable state, 290, 294
State, 289
State diagram, 289
State transition diagram, 291
Stop watch, design of digital, 427–434
Store then shift, *see* Multiplication
Store shifted, *see* Multiplication; Division
Subtracter, binary, 205
Subtraction
diminished-radix arithmetic, 43
radix arithmetic, 37
sign-and-magnitude arithmetic, 47
Sum-of-product form, 75, 98
from K-map, 103
Sum term, 76
Superposition, logical, 457–462
Switch-tail counter, 424
Symbols, logic, 129–131
Synchronous binary counter, arbitrary
modulo-N, 396
Synchronous counter, control unit for
serial adder, 486
Synthesis of sequential circuits
using delay flip-flops, 340
using JK flip-flops, 370, 375
short-cut method, 375–377
using SR flip-flops, 311–319
using toggle flip-flops, 365

Table of choices, 164, 179–181
Table of correspondence, 3
Ten's complement arithmetic, 267
Time-dependent logic equation, 287
Timing-pattern generators, 462–472
using feedback shift registers, 464–469
using resettable counters, 469–472
Toggle flip-flop, 341–342
analysis, 365
edge-triggered, 363–367
synthesis, 365
trailing-edge, 387–389
Trailing edge
detector, 286
toggle flip-flops, 387–389

Transition time, 275
Transitional state, 290
Truth table, 81

Up-down counters, 415
   for analog-to-digital converter, 418

Weight, 172
Wired gates, 126
Wired-AND, 126
Wired-OR, 126

XOR, *see* Exclusive OR